Species Tree Inference

Species Tree Inference

A Guide to Methods and Applications

EDITED BY
LAURA S. KUBATKO AND
L. LACEY KNOWLES

PRINCETON UNIVERSITY PRESS
Princeton and Oxford

Published by Princeton University Press
41 William Street, Princeton, New Jersey 08540
99 Banbury Road, Oxford OX2 6JX

press.princeton.edu

Library of Congress Cataloging-in-Publication Data

Names: Kubatko, Laura S. (Laura Salter), editor. | Knowles, L. Lacey, editor.
Title: Species tree inference : a guide to methods and applications / edited by Laura S. Kubatko and L. Lacey Knowles.
Description: Princeton : Princeton University Press, [2023] | Includes bibliographical references and index.
Identifiers: LCCN 2022026581 (print) | LCCN 2022026582 (ebook) | ISBN 9780691207599 (hardback) | ISBN 9780691207605 (paperback) | ISBN 9780691245157 (ebook)
Subjects: LCSH: Phylogeny. | Biology—Classification.
Classification: LCC QH367.5 S64 2023 (print) | LCC QH367.5 (ebook) | DDC 576.88—dc23/eng/20220808
LC record available at https://lccn.loc.gov/2022026581
LC ebook record available at https://lccn.loc.gov/2022026582

British Library Cataloging-in-Publication Data is available

Editorial: Alison Kalett and Hallie Schaeffer
Production Editorial: Natalie Baan
Cover Design: Heather Hansen
Production: Danielle Amatucci
Publicity: Charlotte Coyne and Matthew Taylor
Copyeditor: Eva Silverfine

Jacket image: Universal Images Group North America LLC / Alamy Stock Photo.

To all the students and researchers who revel in the messiness of genomic data and all that it can teach us about evolution

Short Contents

Contents

Preface

Estimating evolutionary relationships among a collection of organisms remains a central focus of much of evolutionary and ecological study within the field of biology as these relationships provide the background for subsequent hypotheses in these fields. For example, support for different hypotheses about early animal evolution is contingent upon the phylogenetic relationships among the earliest diverging animal lineages. Such hypotheses include questions about the evolution of sophisticated cell types, such as nerve and muscle cells, and specifically whether the complex cell types of Ctenophora and bilaterians represents a shared ancestry or evolved repeatedly, and independently. Likewise, accurate time and rate estimation of species divergence form the basis for a variety of questions in ecology and evolution about why species diversity differs across space, time, and among groups of taxa. Potential tests for such differences in species diversity include whether there have been shifts in diversification rates and/or the mechanisms that might drive diversification. Clearly, accurate estimation of phylogenetic relationships that can leverage all available data within a firm inferential framework are crucial to addressing such questions.

Within the last 20 years, the field of phylogenetics has grown rapidly, both in the quantity of data available for inference and in the number of methods available for phylogenetic estimation. Our first book, *Estimating Species Trees: Practical and Theoretical Aspects*, published in 2010, gave an overview of the state of phylogenetic practice for analyzing multilocus sequence data at the time, but much has changed since then. Indeed, the rapid pace at which the field has advanced in the intervening time has led to the need for an updated reference. We intend this book both to serve as an update on current practices and challenges within the field and to provide a timely look toward the future.

The book is organized into three parts. The first part is devoted to chapters describing recent analytical and methodological developments. Chapters in this section provide both general descriptions of the challenges inherent in making species-level phylogenetic inference from large-scale genomic data as well as specific methods for inference. The second part focuses on providing empirical examples that highlight the challenges and potential for the application of methods for species tree inference to answer compelling questions in empirical systems. The final part of the book consists of a collection of chapters that go beyond species tree inference to address questions that require an evolutionary framework more broadly. The parts are prefaced with an introductory chapter that is designed to orient the novice to the history of the field, to provide some preliminary definitions and concepts, and to set the stage for the topics to be discussed in the remainder of the book.

While the chapters are focused broadly around species tree estimation and often reference one another in order to highlight connections among topics, each chapter can generally be read independently of the others. Some readers may find it useful to work through the book in a different order, perhaps by starting with part II or part III to get a feel for the problems that can be addressed with methods for inferring species trees before returning to part I to dive into the methodological details. Others may prefer to get a firm grasp on methods before considering applications. Our separation of topics into parts aims to guide readers to approach the book in whatever way is most comfortable for them given their background and goals.

While the pace of analytical and genomic development provides a diverse range of opportunities for scientific discovery, it also poses notable challenges to staying current in the field. This book can ease the reader's path, whether for empirical inference or for applications of phylogenetic data, while enabling and encouraging readers to tackle questions in statistically sophisticated ways that maximize biological insight.

Laura S. Kubatko and L. Lacey Knowles
December 2021

Acknowledgments

We thank our editor and assistant editor at Princeton University Press, Alison Kalett and Hallie Schaeffer, for all of their assistance in the preparation of this manuscript.

We are grateful for the thoughtful contributions of our chapter authors, without whom this book would not exist.

Contributors

Paul D. Blischak, Data Scientist, Bayer Crop Science

Jeremy M. Brown, Associate Professor, Department of Biological Sciences, Louisiana State University

Zhen Cao, Graduate Student, Department of Computer Science, Rice University

Alison Cloutier, Department of Organismic and Evolutionary Biology and Museum of Comparative Zoology, Harvard University

Kerry Cobb, Graduate Student, Department of Biological Sciences, Auburn University

Alexandria A. DiGiacomo, Graduate Student, Department of Organismic and Evolutionary Biology, Harvard University

Deren A. R. Eaton, Assistant Professor, Department of Ecology, Evolution, and Environmental Biology, Columbia University

Scott V. Edwards, Professor, Department of Organismic and Evolutionary Biology, Harvard University

Kyle A. Gallivan, Professor, Department of Mathematics, Florida State University

Daniel J. Gates, Checkerspot, Inc., Alameda, California

Phil Grayson, Banting Postdoctoral Fellow, Department of Biological Sciences, University of Manitoba

L. Lacey Knowles, Robert B. Payne Collegiate Professor, Department of Ecology and Evolutionary Biology, and Curator of Insects, Museum of Zoology, University of Michigan

Laura S. Kubatko, Professor, Department of Statistics and Department of Evolution, Ecology, and Organismal Biology, Ohio State University

Xinhao Liu, Graduate Student, Department of Computer Science, Princeton University

Patrick F. McKenzie, Graduate Student, Department of Evolution, Ecology, and Environmental Biology, Columbia University

Siavash Mirarab, Assistant Professor, Department of Electrical and Computer Engineering, University of California–San Diego

Erin Molloy, Assistant Professor, Department of Computer Science, University of Maryland–College Park

Genevieve G. Mount, NSF Postdoctoral Researcher, Department of Biology, Utah State University, Museum of Vertebrate Zoology and Department of Integrative Biology, University of California Berkeley

Luay Nakhleh, Professor, Department of Computer Science and William and Stephanie Sick Dean of the George R. Brown School of Engineering at Rice University

Jamie R. Oaks, Assistant Professor and Curator, Department of Biological Sciences and Museum of Natural History, Auburn University

Huw A. Ogilvie, Assistant Research Professor of Computer Science, Rice University

James B. Pease, Assistant Professor, Department of Biology, Wake Forest University

Diana Pilson, Associate Professor, School of Biological Sciences, University of Nebraska

Timothy B. Sackton, Director of Bioinformatics, FAS Informatics Group at Harvard University

Stacey D. Smith, Associate Professor, Department of Ecology and Evolutionary Biology, University of Colorado–Boulder

Stephen A. Smith, Associate Professor, Department of Ecology and Evolutionary Biology, University of Michigan

Claudia Solís-Lemus, Assistant Professor, Wisconsin Institute for Discovery, Department of Plant Pathology, University of Wisconsin–Madison

David L. Swofford, Visiting Scientist, Florida Museum of Natural History, University of Florida

Coleen E. Thompson, Research Assistant, Department of Molecular Genetics, University of Cincinnati

Emiko M. Waight, Research Technologist, University of Nebraska Medical Center

Joseph F. Walker, Assistant Professor, Department of Biological Sciences, University of Illinois at Chicago

Tandy Warnow, Co-Chief Scientist, C3.ai Digital Transformation Institute, Grainger Distinguished Chair in Engineering, and Associate Head, Department of Computer Science, University of Illinois at Urbana–Champaign

Ellen I. Weinheimer, Graduate Student, Department of Biology, Wake Forest University

James C. Wilgenbusch, Director of Research Computing, Minnesota Supercomputing Institute

Andrea D. Wolfe, Professor, Department of Ecology and Evolution, Ohio State University

Zhi Yan, Graduate Student, Department of Computer Science, Rice University

Species Tree Inference

CHAPTER 1

Introduction to Species Tree Inference

L. Lacey Knowles and Laura S. Kubatko

1.1 Introduction

Estimation of the evolutionary relationships among a collection of organisms remains a central focus of much of evolutionary and ecological study within the field of biology as these relationships provide the background for testing hypotheses in these fields. For example, support for different hypotheses about early animal evolution, and in particular the evolution of sophisticated cell types such as nerve and muscle cells, was contingent upon the phylogenetic relationships among the earliest diverging animal lineages. Especially important in addressing these questions was the placement of Ctenophora because of their shared complex cell types with bilaterians [642]. As another example, accurate time and rate estimation forms the basis for questions in ecology and evolution [468], with shifts in rates being central to tests about the drivers of diversification (e.g., [143, 596]). Clearly, accurate estimation of phylogenetic relationships that can leverage all available data within a firm inferential framework are crucial to addressing questions such as these.

Within the last 20 years, the field of phylogenetics has grown rapidly, both in the quantity of data available for inference and in the number of methods available for phylogenetic estimation. Our first book, *Estimating Species Trees: Practical and Theoretical Aspects*, published in 2010, gave an overview of the state of phylogenetic practice for analyzing multilocus sequence data at the time, but much has changed since then. Indeed, the rapid pace at which the field has advanced in the intervening time has led to the need for an updated reference. We intend this book both to serve as an update on current practice within the field and to provide a timely look toward the future.

We begin this chapter with a brief recap of the history of species tree estimation, including definitions and basic terminology. We next discuss both opportunities and challenges in the field. This discussion includes a critical look at the limitations currently imposed by data availability and computational power and how these might be expected to change in the future, but it also addresses uncertainty surrounding sampling and data analysis in the wake of the big data wave sweeping phylogenetics. We then consider inference beyond the species tree, highlighting the important problems that a genome-scale phylogeny and underlying data allow us to address in a rigorous inferential framework. We conclude with an overview of the book and its organization.

1.2 Background and Terminology

Prior to the routine collection of DNA sequence data, the fields of population genetics and phylogenetics were largely viewed as distinct as they addressed questions at different evolutionary time scales. Much of the mathematical and statistical development of models at the within-population scale was undertaken in the 1980s, through contributions by Kingman [364, 365, 363] and others (e.g., [746, 745]) that resulted in what is now known as *Kingman's coalescent model*, a continuous-time approximation of the Wright–Fisher (and other) population-level models. Kingman's coalescent today forms the theoretical basis for many of the methods used for species tree inference.

Following these developments, several authors noted that when Kingman's coalescent model was applied across species, inferred evolutionary relationships might vary from gene to gene. Important contributions to the development of these ideas, including mathematical details, were provided by [743], [784], [744], and [559], among others. However, much of this work went unnoticed by the phylogenetics community until the mid-1990s, when a seminal paper by Maddison [455] provided clear descriptions of the possible causes of differences in gene-level and species-level phylogenies. This coincided with a decrease in the cost of DNA sequencing, and the subsequent availability of multi-locus sequence data prompted several authors to highlight the need for new inferential frameworks to accommodate these data properly [813, 538, 633, 634].

Importantly, the potential for differences between gene trees and species trees were also recognized to result not only from the coalescent process but also from other evolutionary processes, such as horizontal transfer and gene duplication and loss. By the early 2000s, several papers highlighted the possibility of variation in the evolutionary history across the genome in carefully annotated empirical data sets (e.g., [134, 630, 213]), and the need for methodology that specifically aimed to estimate species-level phylogenetic trees became well accepted by many in the community.

1.2.1 DEFINITIONS AND TERMINOLOGY

A species tree or species phylogeny can be defined as a rooted bifurcating phylogenetic tree for which the tips of the tree represent species and the internal nodes represent speciation events. The times associated with internal nodes of the tree represent the times of speciation events, and branch lengths along the species phylogeny represent the amount of time between speciation events. Speciation times are often given in coalescent units, which can be defined as the number of $2N_e$ generations, where N_e is the effective population size. The advantage of using coalescent units to describe speciation times is that a standardized unit can be discussed in such a way that characteristics associated with this unit can be "translated" to any species of interest once the generation time in years and the effective population size are specified. When N_e varies across the tree, it may be more difficult to define an appropriate unit (number of generations is a reasonable choice, see [446]). Mutation units, the unit commonly used for gene tree inference that is given by the number of substitutions per site per unit time, are also sometimes used. Figure 1.1 shows an example species phylogeny for three taxa, labeled A, B, and C (shaded, thicker tree in each panel).

A gene tree represents the evolutionary history for an individual gene, where a gene is defined as a stretch of contiguous sequence of any length. The tips of a gene tree represent sequences collected from individuals sampled from a particular species, while the internal nodes represent gene divergence times (looking forward in time) or common

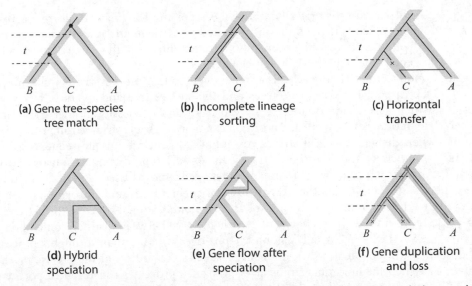

(a) Gene tree-species tree match

(b) Incomplete lineage sorting

(c) Horizontal transfer

(d) Hybrid speciation

(e) Gene flow after speciation

(f) Gene duplication and loss

Figure 1.1. Relationships between gene trees and species trees. In each panel, the species tree is represented by the shaded, thicker tree. Speciation events are indicated with horizontal dotted lines, and the length of time between speciation events is denoted by t. Gene divergence, or coalescent, events are indicated in panel (a) by black circles. Each panel shows a possible relationship between the gene tree and the species tree resulting from a specific evolutionary process: (a) The gene tree and species tree share the same topology. (b) The topologies of the gene and species trees are discordant due to incomplete lineage sorting. Tracing the lineages sampled from species B and species C back in time, we see that they fail to coalesce in the immediately ancestral population, and instead the lineage sampled from species C coalesces with that sampled from A in the common ancestral population. (c) Genetic information is transferred horizontally across the phylogeny from species A to species C, leading to a gene tree that is discordant with the species tree. (d) A species network in which species C is a hybrid of species A and B is shown. For the particular gene sampled, species C inherited its genetic material from species A. Owing to the hybrid speciation event, it is possible for C to inherit genetic information directly from either B or A, even in the absence of incomplete lineage sorting. (e) Gene tree discordance due to gene flow from A to C following speciation. (f) A gene duplication event, marked by a star, occurs after the separation of the lineage leading to A from the ancestor of B and C; the duplicated lineage is sampled in A and C, while the original lineage is sampled in B, leading to discordance between the gene tree and species tree. See also figure 7.1.

ancestor events for the sampled sequences (looking backward in time). These are sometimes also called coalescent events. A gene tree may have many more tips than a species tree because multiple individuals may be sampled within each species included in the species phylogeny. A gene tree may differ from the species tree that gives rise to it both in terms of its topology (branching pattern) and in terms of the times associated with its nodes. Differences in topology between gene trees and the species tree can result from many different evolutionary processes. For example, incomplete lineage sorting (i.e., the failure of lineages to coalesce in their immediately ancestral population) can lead to gene trees with topologies that differ from the species tree (see figure 1.1b). This form of gene tree discordance is typically modeled by applying Kingman's coalescent across the phylogeny (which is then commonly referred to as the *multispecies coalescent*) and is well studied; in particular, the probability distributions of both gene tree topologies [179] and gene genealogies [601] have been derived.

Horizontal transfer (figure 1.1c) is another evolutionary process that is well-known to generate discord between gene trees and the species tree and refers to any process by which genetic information is moved from one species to another by means other than modification with descent. For example, in bacteria, horizontal transfer occurs when distinct bacterial strains recombine to generate unique sequences that include genetic material from both strains. In sexually reproducing organisms, horizontal transfer can occur when a virus or other vector moves a segment of DNA from one species' genome to another. Hybridization (figure 1.1d) and introgression/gene flow (figure 1.1e) can also be thought of as forms of horizontal transfer, in that these processes both involve the exchange of genetic material between distinct, contemporaneous species (i.e., "horizontally" along the phylogeny) rather than through a process of descent with modification within a single species. Regardless of the precise mechanism by which the horizontal transfer occurs, such processes can result in portions of the genome that are inherited differently than others. For example, introgressed loci will show a pattern of inheritance from a species different than that of the majority of the genome if the introgression occurs between non-sister taxa (e.g., figure 1.1c). In the absence of other processes, the extent of discordance due to horizontal inheritance will depend on the extent to which genetic material has been transferred from one species to another throughout the evolutionary history of the set of species under consideration.

The process of gene duplication and loss (figure 1.1f) provides another evolutionary mechanism that results in differences between gene trees and species trees. When a gene is duplicated in a genome, the two versions of the gene subsequently evolve independently of one another, and in descendent species one or both versions of the gene may be present in the genome being sampled. Depending on which copy is sampled, the gene tree for the locus under consideration may differ from the true species-level relationship. Loss of one copy of a duplicated gene may also lead to incongruence between the gene tree and the species tree, or may result in missing data for the locus under consideration, depending on the time that has passed since the duplication and loss events. Gene duplication and loss is prevalent in many species and provides an important mechanism for the generation of new gene function (e.g., a duplicated copy of the gene is under less evolutionary constraint and may evolve to provide a new function in the organism). Thus, consideration of this evolutionary process at the stage of species tree inference is crucial, and many methods have been and continue to be proposed for inference in the presence of duplication and loss.

Closely related to the concept of a species tree is that of a species network, in which relationships between species are depicted by a sequence of speciation events, as in a species tree, but in which species may arise from more than one immediately ancestral species. This may result from evolutionary processes such as hybrid speciation (figure 1.1d), extensive gene flow between distinct species (figure 1.1e), or other forms of horizontal transfer. Much recent work has focused on carefully defining species networks and developing methods of inferring such networks from phylogenomic data, often within a coalescent framework (see, e.g., [845, 841, 843, 713, 861], as well as chapters 5 and 6 of this volume).

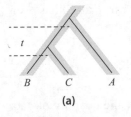
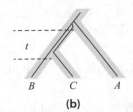

(a)

(b)

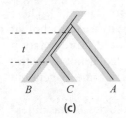
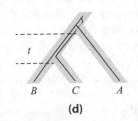

(c)

(d)

Figure 1.2. Four coalescent histories compatible with a three-taxon species tree. Note that the histories in (a) and (b) share the same topology as the species tree, while those in (c) and (d) do not.

1.2.2 AN INTRODUCTION TO THE MULTISPECIES COALESCENT

As mentioned in the previous section, the multispecies coalescent model underlies many of the methods for species tree inference that are commonly applied to multilocus data. Rather than provide a complete mathematical description of this model, we provide here an introduction to the main ideas for three-taxon trees. Readers wishing to see a more full description can consult [383, 770, 289].

Figure 1.2 shows the same three-taxon species trees as shown in figure 1.1. Embedded within the species tree are the four possible *coalescent histories* consistent with this species tree, where coalescent histories refer both to the gene tree topology and the species tree branch lengths along which coalescent events occur. Note that the history in figure 1.2a is the only one in which the first coalescent event occurs within the species branch of length t. Under Kingman's coalescent, times to coalescent events follow an exponential distribution with rate given by $\binom{n}{2}$ when n lineages are available to coalesce. Since $n = 2$ lineages are available to coalesce in the interval of length t in figure 1.2a, the probability of observing this history is the probability that an exponential random variable with rate 1 is less than t, which is $1 - e^{-t}$.

Since the probability associated with all four histories must sum to 1, this leaves e^{-t} of the probability to be distributed over the other three histories, shown in figure 1.2b–d. Note that these three histories all involve the first coalescent event occurring above the root of the species tree, and all three lineages are available to coalesce within this ancestral population. Under Kingman's coalescent, each pair of lineages is equally likely to be the first to coalesce, and thus each of these histories has probability $\frac{1}{3}e^{-t}$.

Finally, we note that the first two histories (figure 1.2a and b) have the same gene tree topology. Thus to derive the probability distribution of gene tree topologies, we can add these two probabilities. The coalescent model then specifies that for three species, the gene tree topology that matches the species tree occurs with probability $1 - \frac{2}{3}e^{-t}$, while the two nonmatching gene trees each have probability $\frac{1}{3}e^{-t}$. Noting that $1 - \frac{2}{3}e^{-t} \geq \frac{1}{3}e^{-t}$ with equality only when $t = 0$, we can identify a common pattern for which the coalescent model is a good fit: a dominant gene tree topology that occurs with highest frequency (the one matching the species tree) with the two alternative topologies occurring in lower and approximately equal frequencies. Such a pattern has been observed for empirical data [565, 145], and deviation from this pattern has been used as evidence for introgression [652].

1.2.3 DATA TYPES AND TECHNOLOGIES FOR GENERATING PHYLOGENOMIC DATA

New data collection techniques have driven shifts in not only the quantity of data but also in the types of data available for phylogenetic inference, with a variety of high-throughput phylogenetic data collection technologies to choose from (table 1.1). These range from different types of targeted sequencing technologies (e.g., hybrid enrichment strategies; [422, 611]) to random genomic sequencing (e.g., reduced representation restriction site-associated DNA sequencing [RADseq]) or targeted genotyping-by-sequencing (GBS) (e.g., RAPTURE; see [8, 60]) and whole transcriptome or genome sequencing.

One important factor in deciding among the different technologies is the differences in their costs, both in terms of the initial time investments and expense but also associated costs when expanding to large numbers of taxa (or individuals). For example, amplifying targeted amplicons involves substantial costs for setup, but it is relatively inexpensive to capture sequences, whereas random genomic sequences from RADseq technologies are economical and provide a universal approach for collecting comparative genomic data. As sequencing costs drop, whole transcriptome and genome sequencing are becoming more widely applied [853, 425]. Alternatively, RADseq can generate very large numbers of loci (i.e., in the thousands to millions of loci) while being scalable to large sample sizes [414], including hundreds of thousands of individuals with targeted genotyping-by-sequencing, and, because of the short sequence reads, they are amenable for applications to museum specimens for which DNA degradation can preclude large amplicons [837].

Another primary consideration for choosing a technology (besides the cost and ease of setup) is differences in their utility. For example, the very large numbers of loci generated by technologies like RADseq become highly desirable for estimation of phylogenetic relationships at recent time scales (e.g., [475, 465]). However, their utility drops as the evolutionary distances between taxa increase (but see [779]) because of allele dropout (but see [210]), which will result in missing data among more distantly related taxa (i.e., homologs will not be sequenced in some taxa because of mutations in the enzyme cutter sites, although new technologies guard against allelic dropout; see [84]). Decisions about what threshold of missing data to use for analysis of RADseq data is complicated. Eliminating loci with a lot of missing data can result in a biased data set with an overrepresentation of loci with low mutation rates [318], which means the data set may not contain the actual loci that are phylogenetically informative for resolving relationships among taxa that diversify rapidly—that is, loci with the highest rate of evolution. On the other hand, discordant relationships have been shown to be disproportionately represented among loci with missing data [413], suggesting that they may be less reliable for phylogenetic inference. Whole-genome or transcriptome sequencing has the appeal of providing not just a lot of data for phylogenetic inference but also information to address questions provided by the phylogenetic framework, including questions about genome evolution [428]. However, in addition to assembly challenges, such data also pose new challenges because of the potential heterogeneity of processes contributing to genomic differences among taxa, making model misspecification a more pressing problem compared with the relatively small data sets (e.g., hundreds to a few thousand loci). In contrast, targeted amplicon approaches such as

Table 1.1. Summary of sequencing technologies.

Method	Description	Data	Reference
Targeted enrichment	Also known as "hybrid enrichment," "anchored enrichment," and "enrichment"; may also be referred to as "capture." Probes or baits complementary to the sequence of interest are used to hybridize with the target sequence, which is then enriched via PCR and sequenced on a HTS platform.	Typical enrichment/capture data sets for phylogenomics contain tens to hundreds of taxa and hundreds to thousands of loci.	Mertes et al. 2011 (https://doi.org/10.1093/bfgp/elr033); Cronn et al. 2012 (https://doi.org/10.3732/ajb.1100356); Folk et al. 2015 (https://doi.org/10.3732/apps.1500039)
Transcriptome sequencing	This method involves sequencing RNA reads (typically via RNA-Seq on a short read HTS platform) and assembling the reads into the transcriptome. When transcriptomes are used for phylogenomic inference, a necessary step is orthology inference to ensure that markers have sequence homology.	This method can generate hundreds to thousands of loci. The number of taxa that can be included is more variable among studies but is typically fewer than 100 due to costs.	Wen et al. 2013 (https://doi.org/10.1371/journal.pone.0074394); Shen et al. 2017 (https://doi.org/10.1093/gigascience/gix116)
Whole-genome sequencing	Whole-genome sequencing for phylogenomics has the advantage of obtaining an unbiased representation of genomic data relative to reduced representation methods. As costs for genome sequencing fall, and computational efficiency increases, this method is becoming increasingly accessible to many researchers. Orthology assessment is also a necessary step when using loci from whole-genome sequencing.	Currently, whole-genome uses similar amounts of data as transcriptome sequencing—thousands of loci for tens to hundreds of individuals.	Allen et al. 2017 (https://doi.org/10.1093/sysbio/syw105)

(continued)

Table 1.1. (*continued*)

Method	Description	Data	Reference
RAD-Seq / GBS	This is a reduced representation library approach in which 1–3 restriction enzymes are used to fragment the genome. Adapters with built-in barcodes are ligated to the fragments, then the fragments are pooled and size selected. The pool is typically sequenced on a short-read HTS platform (e.g., Illumina).	The number of loci sequenced at a given depth depends on the genome size and number of individuals multiplexed in a sequencing run. For example, a user could reasonably expect to get 5,000–20,000 loci with 10X coverage when multiplexing 1000 individuals from a species with a 1 Gb genome in a lane of Illumina HiSeq.	Andrews et al. 2016 (https://doi.org/10.1038/nrg.2015.28)
Parallel microfluidic PCR + HTS	This approach uses primer pairs designed for PCR and utilizes microfluidic PCR to amplify loci in parallel using HTS. The primers can be designed for organelles or the nuclear genome, but they must have similar annealing temperatures since the PCR will be processed simultaneously in parallel. This method bypasses traditional library preparation.	Typical studies use 48–96 loci for approximately 50–100 taxa.	Uribe-Convers et al. 2016 (https://doi.org/10.1371/journal.pone.0148203); Kates et al. 2017 (https://doi.org/10.1016/j.ympev.2017.03.002)

Note: HTS = high-throughput sequencing, RADseq = restriction site-associated DNA sequencing, and GBS = genotyping by sequencing.

hybrid enrichment approaches avoid the problems of missing data by relying on conserved sets of priming sites to amplify sequences. They also present less of a challenge for assembly, modeling, and analysis compared with technologies like RADseq and whole-genome/transcriptome sequencing. However, they also result in substantially fewer loci, and because they rely on specific priming sites, they are nonrandom samples of the genome, which may make them less desirable for some questions.

These different data set properties (e.g., SNP-based information content, or inherent heterogeneity in underlying evolutionary model with genomic-scale sampling, and/or differing amounts and distributions of missing loci in data sets) are likewise driving different analytical and theoretical areas in phylogenetic inference. These new areas range from exciting new approaches for phylogenetic estimation and the evaluation of the confidence of such relationships (e.g., assessing phylogenetic signal; [423, 771]) to determination of the different processes contributing to locus-specific patterns of ancestry (e.g., [88, 371, 771]) and identification of subsets of data for phylogenetic inference from genome-scale data sets [675, 192, 319]. The analytical methods that might be applied will also differ depending on the technology used to generate the data. For example, the short sequence reads of RADseq means that they are not generally amenable to gene tree estimation but instead are analyzed as SNP data, whereas standard gene tree estimation methods are applied to sequences generated from technologies like hybrid enrichment because those technologies target specific genomic regions of longer read lengths. Likewise, with genome-scale data sets, computational challenges restrict the types of analyses that might be done [503].

The new technologies and unprecedented abundance of data they generate is changing phylogenetic inference and no doubt providing better resolved and more reliable phylogenetic inference in some cases. However, recalcitrant nodes persist (e.g., [798, 590]). Moreover, with phylogenetic estimates differing as a function of analysis, data set design, or inclusion/exclusion of loci, genome-scale data sets are raising many questions with no clear answers. For example, how might genome-scale data be analyzed to provide reliable phylogenetic estimates? If subsets of the data are to be analyzed, how should such data be identified (both in terms of loci and taxa)? These are some of the questions that are explored in this book, as researchers contend with the uncertainty surrounding sampling and data analysis in the big data era. Despite these unknowns, it is clear that along with these complicated questions come some amazing opportunities that extend beyond a focus on the species tree itself. As we look to the future, and in the following chapters, we emphasize this expanded role of genome-scale data—that is, next-generation inference, which will no doubt become the new focus of researchers as next-generation sequencing becomes routine (table 1.1).

1.3 Overview of Current Methods for Species Tree Inference

Given the processes described above, the precise mechanism by which data arise must be taken into consideration in the development of methods for inferring species-level phylogenies. Regardless of the process(es) responsible for gene tree–species tree discordance, it is usually assumed that gene trees arise from evolutionary processes occurring along the species tree, and DNA sequence data are subsequently generated from the gene trees associated with individual loci. Thus, DNA sequences observed from loci that are freely recombining can be viewed as conditionally independent of one another, where the conditioning is based on their underlying gene trees arising from a shared species phylogeny. Inference then proceeds in the "reverse" direction—that is, given a set of observed DNA sequence data from multiple loci, it is desired to obtain an estimate of the species tree. Although gene trees are not directly observed, it is clear that they play an important role in the data-generation mechanism. For this reason, methods for estimating species trees are commonly categorized according to how they account for uncertainty in the gene trees in carrying out inference.

One class of methods for species tree inference is referred to as *summary statistics methods* or *summary methods* because these methods carry out species tree inference in two distinct steps, the first of which represents a summarization of the data. In this first step, a gene tree is estimated for each locus in the data set using one of the standard methods for phylogenetic tree estimation (e.g., maximum likelihood). The gene trees estimated in this first step are then used as input to the second step of the procedure, and a species tree estimate is obtained using only the information contained in these input gene trees. Such methods have the advantage of being computationally efficient. In the first step, the gene trees for the individual loci can be estimated in parallel, as each depends only on the sequence alignment for that gene under the conditional independence assumption mentioned in the previous paragraph. The second step is typically carried out by assuming some model for the relationship between the gene trees and the species tree. The features of the models that are commonly used for inference typically lead to computationally tractable algorithms for inference. The drawback of summary methods is that uncertainty in the gene tree estimates is typically unaccounted for in the second step of the procedure, making estimation of the uncertainty in the species-level phylogenetic estimate difficult to quantify. Though some suggestions have been made to remedy this (e.g., using as input bootstrap samples of gene trees rather than only a single point estimate of the gene trees), these strategies have not definitively been shown to improve the overall inference.

The second class of methods for species tree inference is referred to as *coestimation methods*, because they jointly estimate gene trees and the species tree under probabilistic models. The current methods within this class employ a Bayesian framework and use Markov chain Monte Carlo (MCMC) to carry out inference. Such methods have the advantages that fairly complex models can be fit and that estimates of all model parameters and associated measures of uncertainty are naturally obtained as part of the MCMC inference procedure. However, these methods can be computationally intensive, particularly as the size of the data and the complexity of the model increase. Most current methods cannot be feasibly run on genome-scale data for more than 20 or so species, but as computational power continues to grow, the ranges of data set sizes and models that can be successfully analyzed in this framework will continue to expand.

A third class of methods are those based on site pattern frequencies, which generally involve the use of genome-scale data to compare data features expected under various evolutionary models to those found in observed data sets. Importantly, site-pattern-based methods are distinct from summary methods as they are applied directly to the sequence data with the need to estimate gene trees first. Such methods have been shown to be computationally feasible for large data sets and thus show promise for carrying out efficient inference for large data sets for which analysis in the Bayesian coestimation framework is computationally prohibitive.

Across all of these classes of methods, the most common model assumed for the relationship between gene trees and the species tree is the coalescent process; however, other processes have been considered. For example, in the class of summary statistics methods, a model of duplication and loss can be used to carry out inference under the parsimony criterion by selecting the species tree that minimizes the number of gene duplication or loss events required to explain the set of estimated gene trees.

Of course, important issues arise, such as whether duplications and losses should be weighted equally and how the space of possible species trees should be searched to find the tree that minimizes the number of duplications and losses. A challenge for the future will be the development of methods that can simultaneously account for many of the evolutionary processes known to generate gene tree–species tree discord in a computationally feasible manner.

1.3.1 CONTROVERSIES IN THE ESTIMATION OF SPECIES TREES

Careful testing that employs both simulated and empirical data and that includes comparisons between methods are important components of the development of methods for the problem of species tree inference, and work in this area has predictably led to disagreements among researchers about the appropriateness of various methods. One source of controversy has involved the utility of the multispecies coalescent model in improving the accuracy of estimated species-level phylogenies. Two observations have contributed to this viewpoint. First, when the possibility of discord in the gene trees is ignored and the data from all of the sampled loci are concatenated and analyzed with a method for gene tree inference, such as maximum likelihood, the resulting tree is often an accurate estimate of the species tree topology, particularly when the length of time between speciation events is large. The second, and related, observation is that incomplete lineage sorting is most commonly observed in empirical data for recent speciation events that have occurred in quick succession (e.g., species are often not monophyletic when multiple individuals are sequenced), which has led to the speculation that incomplete lineage sorting, and therefore the need for methods that explicitly model the coalescent process, is not relevant for nodes "deeper" in a tree. However, the coalescent model predicts that the amount of incomplete lineage sorting depends upon the length of time between speciation events, regardless of whether speciation was recent or occurred deep in the past (i.e., regardless of the depth in the species tree). Thus, it is not reasonable to conclude that the possibility of incomplete lineage sorting can be disregarded in the inference stage for deeper histories without knowledge about the time separating speciation events.

In regard to the observation that concatenated methods often infer the species tree topology with high accuracy, a more nuanced view is required to relate this observation to the performance of such methods. First, concatenating multilocus data and carrying out, for example, a maximum likelihood analysis on the resulting concatenated alignment makes the assumption that all loci share a single underlying phylogenetic tree from which data evolve according to the nucleotide substitution model specified during the analysis. Yet several empirical examinations (e.g., [134, 630, 213]) have established that the phylogenetic history varies across the phylogeny in a way that is often consistent with the multispecies coalescent. While one might argue that phylogenetic inference commonly requires simplifying assumptions (for example, methods for estimating gene trees from single-locus alignments typically assume that sites evolve independently), these assumptions are generally necessary because computationally efficient methods that implement more appropriate models are lacking. The variety of computationally tractable methods described within this book demonstrates that this is not a limiting factor in the case of modeling gene tree discordance.

Second, and more importantly, while the topology may be estimated fairly accurately, other important quantities may not be. One example is the quantification of uncertainty in the estimated tree, which is typically carried out by bootstrapping when a method like maximum likelihood is used to analyze the concatenated alignment. Two issues arise when using the bootstrap on the concatenated alignment. First, the bootstrap assesses uncertainty in the repeated application of a particular method to data from the population of interest and thus cannot address uncertainty that results from inaccurate assumptions in the model used [222]. Given that concatenation represents an incorrect modeling assumption (i.e., that all loci share a common gene tree), the bootstrap support values are difficult to interpret. Second, bootstrap support from large concatenated alignments will tend to overestimate the actual support for a node. For example, consider a node supported by 55% of the sites in a data set while 45% of the sites favor some other arrangement, and suppose that the data contain one million sites (thus, 550,000 bp favor the node of interest, while 450,000 bp favor the other relationship). Most bootstrap samples will contain a majority of sites favoring the node of interest, and thus the bootstrap support for the node under consideration will be near 100%. Yet, the data show much more even support for the two alternative relationships. For example, significant underlying conflict was masked by high support values of concatenated analyses when conflicting phylogenetic relationships were actually strongly supported (e.g., [675, 774, 426]).

Finally, species divergence times (or speciation times) will be inaccurately estimated using a concatenation method as such a method assumes that the common ancestor for all loci in the concatenated alignment was identical. However, under a model such as the coalescent, it is clear that the gene divergence events (i.e., common ancestor times for the individual loci) must all predate the time of the speciation event in the absence of gene flow or some other form of horizontal transfer (see figure 1.1). Use of a coalescent model allows estimation of the speciation time after accounting for this, while concatenation methods do not. An example of the potential consequences of this is given in [390] (their table 5) for an empirical data set of *Sistrurus* rattlesnakes, for which the speciation times estimated for a concatenated versus coalescent-based analysis differ by as much as 70%. In addition, tree-associated parameters other than the speciation times, such as overall evolutionary rates, effective population sizes, and rates of gene flow, could potentially be affected by model misspecification when loci are concatenated.

1.4 A Look to the Future

1.4.1 CURRENT LIMITATIONS AND FUTURE PROSPECTS

The primary challenge facing current methods of species tree inference arises from the conflict between the desire to fit increasingly realistic, and therefore complex, models to the data and the computational resources required to fit such models as the size of data sets grows both in terms of taxa and genomic coverage. Overcoming this challenge will require new approaches to these problems, and these approaches will need to be designed specifically for the problem at hand. Here, we highlight several of the important issues to be addressed when the goal is to infer a species tree using phylogenomic data.

First, we note that new methods must be designed by carefully considering the data characteristics specific to phylogenomic data. In particular, properties of methods developed for inference of gene trees from single loci may need to be re-evaluated when applied to genome-scale data. As a first example, we note that when carrying out analyses of single loci, it is common to use a model selection procedure (e.g., ModelTest [584]) to choose a model of nucleotide substitution prior to inference of the gene tree. However, for a phylogenomic data set of several hundred or several thousand genes, selection of specific models for individual genes to be specified in the downstream species tree inference procedure might yield little increase in the statistical power for tree inference at the expense of computational time in the model-fitting stage.

As a second example, consider the case of *phylogenetic invariants*, which were proposed in the late 1980s by [398], [124], and [125] as a possible method for inferring gene trees for samples of three or four taxa. Although promising from a theoretical standpoint, later work by [333] showed that invariants-based methods lacked power for gene tree inference and were outperformed by other methods in common use at the time. One explanation for the result of Huelsenbeck and Hillis is that phylogenetic invariants are formed from polynomials in the site pattern frequencies, where the site pattern frequencies are estimated from the data. For single loci that may be only a few hundred to a few thousand base pairs in length, accurate estimation of site pattern frequencies may be difficult, and polynomials formed from these estimates may have high variance, making them ineffective at differentiating among trees. For genome-scale data, however, a wealth of data are available, and estimates of site pattern frequencies are expected to be much more accurate, making invariants a reasonable tool for examining species-level phylogenetic relationships. The success of invariants-based methods, such as the ABBA-BABA test [208], in addressing complex problems (i.e., hybridization) indicates that the performance of such methods warrants new examination in light of the very different data structure provided by multiple loci.

A second challenge in developing methodology for species tree inference is that the availability of large quantities of sequence data provides the opportunity to use subsets of the data selectively to address specific questions of interest. Using only portions of the data can clearly result in increased precision in the resulting inferences (e.g., excluding loci with errors in alignment, assembly, or orthology detection; see [95]), but it also risks the introduction of bias when the data are not appropriately sampled. As described above, sequencing technologies in current use often result in large quantities of data, but the quality of the data for phylogenetic inference may differ. For example, some loci might be characterized by large amounts of missing data, or some loci may be involved either in a disproportionate amount of discord throughout the tree or discord in parts of the tree may arise by processes other than those captured by the phylogenetic model used for inference (e.g., [700, 95]). Depending on the method used for species tree inference, the investigator may be required to make a decision about whether to exclude taxa and/or loci because of the pattern of missing data (see, e.g., [638]). However, few methods currently incorporate a model of missing data explicitly in the inference procedure. Similarly, although data filtering is also becoming more common during the preprocessing of data in preparation for species tree inference, methods that explicitly incorporate the filtering step into formal inference model are lacking. The filtering may involve inclusion/exclusion of genes due to level of variation, the possibility of horizontal transfer or gene duplication, or numerous other reasons. However, when the process of sampling data is not correctly incorporated into the underlying models used for inference, the inferred phylogeny may be biased. It is unknown how much bias may be introduced by the nonrandom inclusion of data, and there is as yet no consensus on how filtering decisions should be made [371].

1.4.2 BEYOND THE SPECIES TREE

Even though a major challenge to species tree resolution is the gap between the data we collect for phylogenetic analyses (i.e., large-scale transcriptomic and genomic data) and the methods that accommodate the inherent complexity of big data (i.e., processes in addition to incomplete lineage sorting that contribute to discord among loci), such data is also an unprecedented opportunity to better illuminate the processes that shape the tree of life. That is, big data and all its complexities when studied in a phylogenetic framework open new opportunities to address questions beyond the primary goal of resolving the species tree.

The characterization of patterns of discordance and the contribution of different processes to this discordance is itself of interest for generating hypotheses about the role of lateral gene transfer, gene duplication, and incomplete lineage sorting during the divergence of different taxa [627, 697]. Such hypotheses include those about the distribution of duplication across the tree of life and its potential association with the shifts in diversification rates or its concentration at the origin of major clades (as opposed to being dispersed across taxa). Likewise, by focusing on the processes that lead to gene tree discord, we can test whether lateral gene transfer is commonly associated with hypothesized ecological transitions versus evidence of convergent molecular evolution (e.g., [249]).

In the future, and as comparative genomics expands [425], the species tree and all its ancillary applications will only grow (e.g., dissecting evolution and disease; [491]). Such novel biological questions depend upon seeking evolutionary explanations for the distribution of discord in the gene trees, which means that the heterogeneity of comparative phylogenetic data sets should be embraced (as it is arguably the key to accurate resolution of recalcitrant phylogenetic relationships).

1.5 Organization of This Book

In the book we highlight, by example, not only how species tree estimation differs from a phylogeny estimated from concatenated multilocus data but also the issues that arise more generally from a mismatch between phylogenomic data, with all its complexities, and the models used for inference. This includes both conceptual and practical issues related to improving species tree estimates, as well as inferences that are nonbifurcating (i.e., networks). The book devotes five chapters to methodological developments, whereas the latter portion of the book, a total of eight chapters, focuses on empirical applications, including those that consider questions beyond the species tree. Some of the contributors were participants in a workshop offered at the 2018 Society of Systematic Biology Standalone meeting, whereas others were invited to cover topics identified by workshop attendees from questionaires they completed to assure that the book reflects the experiences, interests, and concerns of the diverse community that is engaged in species tree inference.

Through the set of chapters (authors representing their own perspective on aspects of species tree estimation relevant to their individual research programs), a diversity of perspectives and backgrounds are represented. This diversity means that the book speaks to people with varying levels of familiarity with the topic of species tree estimation, but it does not (nor is it intended to) provide a comprehensive overview of the subject. The combination of theoretical and empirical work is meant to provide readers with a level of knowledge of both the advances and limitations of species tree inference that can guide researchers in applying the methods while also inspiring future advances among those researchers with an interest in methodological development. Such cross talk (between empiricists and theoreticians/mathematicians) is vital to the growth of phylogenomics as it refocuses attention on the biological history of diversification (i.e., the timing and pattern of species divergence), the processes generating the observed patterns of genetic variation (e.g., sorting of ancestral polymorphism and gene flow, in addition to mutation models of nucleotide evolution), and the vast opportunities of study that includes and goes beyond a focus on the species tree itself.

PART I

Analytical and Methodological Developments

The more than 10 years since publishing our first edited volume on the topic of species tree inference have seen a rapid proliferation in the methodology developed to infer the history of species divergence from genetic data. Much of this development has been driven by continued technological advances for collecting large-scale genetic data (e.g., whole-genome sequencing) coupled with increases in the capacity and availability of high-performance computation. These developments have spurred innovation in both the models and algorithms utilized for the complex problem of inferring the evolutionary relationships among collections of species. In the first part of this volume, we highlight several advances in the areas of model and algorithm development since the publication of our last volume.

The first chapter in this section, contributed by Malloy and Warnow, provides a comprehensive overview of the progress to date when the goal is to infer very large species trees, in which the authors not only describe different approaches but also discuss the strengths and weaknesses of different classes of methods that are used to estimate very large species trees. The chapter begins with a breakdown of the various types of approaches applied for estimating species trees that differ depending upon the type of molecular data. These approaches include the practice of concatenating sequence data from multiple genes, methods that either rely on summaries of the sequence data or coestimate gene trees and the species tree thereby avoiding the information loss that accompanies summaries of the sequence data, and methods that are site based. Among these different model classes, the authors briefly describe particular methods implemented in commonly used software packages (e.g., ASTRAL and ASTRID within the class of summary methods and SVDQuartets and SVDQuest within the class of site-based methods). The chapter also includes a description of several divide-and-conquer approaches that can be used when either the number of loci or the number of species is too large to be analyzed with other approaches. Last, by making some comparisons among the methods, again with a focus on the large data setting and the use of parallel computing to make inference feasible, the chapter provides valuable advice on choosing among methods for inference. These general guidelines are accompanied by an appendix that includes links for available tutorials and for downloading the software, as well as specific information about approaches.

The next two chapters focus on two widely used approaches for species tree inference that are popular due to their computational efficiency and provide in-depth descriptions with an eye toward the practical issues that empiricists are likely to encounter in analyzing their data. In chapter 3, Mirarab provides an overview of ASTRAL, the most widely used summary method for species tree inference, carefully explaining the underlying motivations for the methods and algorithms employed. With this summary of ASTRAL's theoretical properties, a demonstration based on computer simulations of ASTRAL's performance in various scenarios that might be common in practice is provided. Reflecting on the sensitivities of the method, the chapter concludes with a valuable discussion of issues commonly encountered when preparing data for analysis with ASTRAL, as well as a detailed description of how ASTRAL's results should be interpreted. As such, this chapter is essential reading for those planning to use ASTRAL to estimate species-level phylogenies from multilocus data.

A complement to this efficient summary method is described in chapter 4 by Swofford and Kubatko, who present two distinct site-based methods for analyzing genome-scale data in a coalescent framework. The first is SVDQuartets, a method for estimating the species tree topology from either multilocus or SNP data. In addition to a comprehensive description of the method and its implementation in PAUP*, the authors provide practical advice for empiricists, including when the goal is to estimate a species tree in which the terminals are species versus a tree with individuals as the terminals. The second method they discuss is one that estimates the timing of species divergences (i.e., species tree branch lengths). This approach can be paired in a complementary way with the species tree inferred with SVDQuartets to estimate speciation times, but as the authors note, the method provides a more general framework for estimating speciation times, and as such it can be applied to any species-level phylogeny, irrespective of the method used to estimate the tree. The separation of the estimation of species trees from the estimates of branch lengths, as opposed to their joint estimation, makes this approach computationally efficient and amenable to analyses of large-scale data. However, as the authors caution the user, these approaches should not be applied to small data sets; fortunately, MCMC-based methods such as StarBEAST2 [298, 549] or SNAPP [98] are often computationally feasible in these cases.

Rather than focusing on a bifurcating history of divergence, chapters 5 and 6 focus on estimation of species networks, a timely topic as the extent and importance of nonvertical evolutionary processes (e.g., introgression and lateral gene transfer) is now widely recognized. In chapter 5, Cao, Liu, Ogilvie, Yan, and Nakhleh provide a practical introduction to the use of the PhyloNet software for inferring species-level phylogenetic networks. The chapter details various estimation frameworks, including parsimony, likelihood, and Bayesian frameworks, and addresses important practical considerations, such as running time, the size of data sets that might be analyzed, summarization and interpretation of the results, as well as the analysis of polyploids. In chapter 6, Solís-Lemus provides guidance about estimating phylogenetic networks versus phylogenetic trees as well as presents issues for consideration when deciding the types of network frameworks that might be applied based on explicit comparisons across frameworks, with particular attention (and a detailed description) of a pseudolikelihood approach for inferring species-level networks using the PhyloNetworks package. Together, these chapters cover state-of-the-art approaches for estimating species-level phylogenetic networks, which are increasingly being applied in empirical studies to infer relationships that deviate from those that can be represented by a typical phylogenetic tree.

Those following the current literature in this field will note that none of the chapters in this part cover methodological and algorithmic developments in the area of Bayesian species tree inference, an area that has also seen substantial improvements in the computational requirements for effective species tree inference over the last 10 years. Bayesian approaches have some advantages over many of the methods presented here in that they employ a fully model-based framework for which Markov chain Monte Carlo (MCMC) algorithms can be used to estimate posterior probability distributions for any parameters of interest. These parameters might include the species phylogeny as well as parameters such as the effective population sizes along branches of the phylogeny, the divergence times represented by internal nodes in the species tree, and even parameters in the underlying substitution models. While MCMC-based Bayesian approaches generally require substantial computing effort for large data sets, the advantage of estimating posterior distributions is that this automatically provides a natural measure of uncertainty in the estimated values, in contrast to the methods of chapters 3 and 4, for which additional analyses are required to quantify uncertainty.

One of the most widely used Bayesian methods for multilocus phylogenetic analyses is StarBEAST2 [298, 549]. StarBEAST2 is included in the BEAST2 [85] family of methods for general phylogenetic inference and has undergone a series of algorithmic and computational innovations that have allowed it to scale up to reasonably sized problems. For example, Ogilvie et al. [549] report that it is $\sim$ 13–30 times faster than the original version (*BEAST; [298]), with successful application of StarBEAST2 to data sets consisting of tens of species and tens to hundreds of loci. In addition, the models in StarBEAST2 allow for inclusion of several biologically realistic features, such as species tree relaxed clock models that enable estimation of species-specific substitution rates [549]. Another popular Bayesian method is Bayesian Phylogenetics and Phylogeography (BPP) [602, 243], which has recently been expanded to include inference of the species-level phylogeny as an option [602, 243]; BPP was originally designed for delimiting species boundaries. Although it incorporates only simple models of nucleotide substitution (e.g., JC69), it allows inference of phylogenetic networks under the coalescent [244]. An important aspect of inference in a Bayesian framework is the need to specify prior distributions for parameters of interest. Both StarBEAST2 and BPP include a range of choices for these prior distributions, with sensible default settings that are designed to be relatively uninformative. Still, users will want to consider choices carefully for these distributions to ensure robustness of the resulting inference. Many excellent tutorials for both StarBEAST2 and BPP are available, for example https://github.com/genomescale/starbeast2 (StarBEAST2) and [244] (BPP).

The area of Bayesian phylogenetic inference more generally has recently seen several innovations in methodology for inference, including the use of sequential Monte Carlo methods [184] and variational methods [849]. Although such approaches have not yet been applied in the setting of species tree inference, they show enormous potential for improving the computational requirements associated with phylogenetic inference in a Bayesian context in a general sense, and we look forward to seeing whether they can provide scalable approaches for species-level phylogenetic inference for genome-scale data.

CHAPTER 2

Large-Scale Species Tree Estimation

Erin Molloy and Tandy Warnow

2.1 Introduction

Phylogenetic trees, whether for genes or for species, are graphical models of evolution that can be used to address many biological questions of interest, including how traits evolve and how species adapt to their environments [151]. The conclusions of biological research can therefore be affected by the accuracy of the *estimated* phylogenetic tree used in the study.

In practice, phylogenies (whether species trees or gene trees) are nearly always estimated using statistical methods, such as maximum likelihood, under models that assume all the sites in the input multiple sequence alignment evolve down some common (but unknown) tree under some sequence evolution model, such as the generalized time reversible (GTR) model [747]. If the true evolutionary model fits the assumed evolutionary model and if the input alignment is correct, then such approaches can be *statistically consistent*. A specific method is said to be statistically consistent under a specific model, if, as the amount of data generated under the model goes to infinity, the probability that the method (given the data as input) returns the true tree goes to one. For example, maximum likelihood, if solved exactly, is statistically consistent under the GTR model, but maximum parsimony is not [239]. In contrast, the true tree is not identifiable under the *no common mechanism* model of Tuffley and Steel [757], and so even maximum likelihood is not statistically consistent under this model. These assumptions—that the alignment is correct, that all the sites in the input alignment evolve down the same model tree, and that the statistical model governing the evolution of sites is sufficiently well behaved—are significant and may not hold for any given data set. For example, multiple sequence alignment can be very difficult, especially on large data sets [433, 506, 521, 537], and biological evolution is much more complicated than standard sequence evolution models (e.g., these models do not include heterotachy, selection, dependencies between sites, etc. [386, 449, 482]).

Another basic challenge is that biological processes, such as gene duplication and loss [224], incomplete lineage sorting [455], horizontal gene transfer [229], and hybrid speciation [77], can cause different parts of the genome to have different histories, with the consequence that the true evolutionary tree for any single genomic region (referred to as a gene tree) may not match the true species tree! As a result, species trees are usually estimated from multiple genomic regions in the hope that phylogenetic information from across the genome will help identify the species tree [464].

Although there are multiple sources for discord between gene trees and species trees, much of the focus in terms of method development has been on incomplete lineage sorting (ILS), largely because of the well-developed mathematical theory regarding species tree estimation in the presence of ILS, which is modeled by the multispecies coalescent (MSC) [365]. Under this model, the true species tree is a rooted binary tree with leaves representing species, internal nodes representing speciation events, and branch lengths given in coalescent units. Each model species tree defines a distribution on gene tree topologies, and every possible gene tree topology has strictly positive probability of being generated by the species tree. Furthermore, the species tree topology and branch lengths are identifiable from the distribution on gene tree topologies [12]. Of course, in practice, we observe molecular sequences rather than gene trees. Thus, species evolution is typically modeled hierarchically: the model species tree generates gene trees under some model of gene evolution (e.g., the MSC model), and then each gene tree generates molecular sequences under some model of sequence evolution (e.g., the GTR model). Many methods have been developed to estimate species trees in the presence of ILS, and some of these methods have been proven to be statistically consistent under the MSC model or under some hierarchical models under which gene trees evolve within the species tree under the MSC [177, 217, 368, 441, 511].

Yet correspondingly less is known about the statistical estimation of species trees under gene duplication and loss (GDL). Some methods have been developed for species tree estimation from multicopy genes (e.g., [88, 133, 786]), and a formal model for GDL was presented in [35]. However, until recently it was not even known if the species tree topology was identifiable from the distribution it defines on gene family trees (i.e., trees with multiple leaves labeled by the same species because of gene duplication events). In late 2019, Legried et al. [418] proved that the species tree topology is indeed identifiable under the GDL model in [35], and proofs of statistical consistency under some models of GDL were provided for some methods [418, 514].

Despite the development of phylogeny estimation methods that take gene tree heterogeneity into account, concatenation analysis using maximum likelihood remains the dominant approach to estimating species trees. In a concatenation analysis, multiple sequence alignments are estimated for each genomic region separately and then concatenated together into a single supermatrix, which can be given as input to a maximum likelihood heuristic. However, there are two drawbacks to such analyses. First, concatenation analyses do not account for gene tree heterogeneity and thus are not statistically consistent in the presence of ILS, GDL, et cetera. Second, the better heuristics for maximum likelihood are computationally expensive—prohibitively so when the number of loci and number of species are both large. Indeed, the computational cost of running a concatenation analysis on some multilocus data sets with fewer than 100 species, such as the avian phylogenomics data set [343], may require access to a supercomputer. Hence, part of the motivation for this chapter is to explore computationally efficient alternatives to concatenation analyses.

The rest of the chapter focuses on methods for computing point estimates of the species tree from large data sets. In section 2.2 and section 2.3, we provide an overview of selected methods for species tree estimation in the presence of ILS and GDL, respectively, focusing on methods that have been proven statistically consistent under models that account for gene tree heterogeneity. The impact of parallel computing on scalability is discussed in section 2.4, and divide-and-conquer approaches to scale species tree estimation methods to large data sets are discussed in section 2.5. In section 2.6, we discuss how to choose between methods with respect to the conditions under which methods are proven statistically consistent and their relative performance on large data sets. We conclude in section 2.7 with a summary of our observations as well as a discussion of opportunities for future research. Finally, an appendix includes a discussion of "Big-O" notation and how it is used in running time analyses.

2.2 Species Tree Estimation Methods Addressing ILS

2.2.1 OVERVIEW

Broadly speaking, there are three different types of approaches to species tree estimation that address gene tree heterogeneity resulting from ILS. The first type operates by computing gene trees (one tree for every locus) and then uses the resulting gene trees to estimate a species tree. Since these methods operate by computing summary statistics from the input gene trees, they are called "summary methods." ASTRAL [506, 507, 851] is one of the most frequently used summary methods, but there are many others, including NJst [439], MP-EST [440], and ASTRID [761]. The second type coestimates the gene trees and species tree; the first of these coestimation methods was probably BEST [435], with *BEAST [298] and StarBEAST2 [549] now the most well-known examples. The third type infers the species tree directly from site patterns (e.g., in the concatenated alignment), and so bypasses the challenge of estimating gene trees entirely; SVDQuartets [140] is the most well-known example of these "site-based methods" (although SNAPP [98] and METAL [160] are other examples). Although species tree estimation methods use different techniques, many have been proven to be statistically consistent under models that account for gene tree heterogeneity due to ILS.

2.2.2 SUMMARY METHODS

Summary methods infer a species tree from a collection of input gene trees, and some of these methods can be very fast, once the gene trees are computed. Some summary methods can be applied to unrooted gene trees, but others require that the input trees be rooted. Because rooting gene trees can be difficult, methods that operate on unrooted gene trees are the most flexible. Furthermore, many summary methods are statistically consistent under the MSC model, making them an appealing approach to species tree estimation in the presence of ILS.

While summary methods all use estimated gene trees as input, they vary in their techniques. Distance-based methods (e.g., NJst [439], ASTRID [761], GLASS [522], iGLASS [345], STAR [442], and STEAC [442]) operate by computing distances between pairs of species based on the input gene trees and then inferring a species tree from the resulting distance matrix. In contrast, other summary methods construct the species tree by characterizing each gene tree in terms of its induced triplet (i.e., rooted three-leaf) trees or induced quartet (i.e., unrooted four-leaf) trees and then searching for a tree that shares as many of these induced subset trees as possible. However, finding the species trees that maximizes quartet support in the input gene trees is an NP-hard optimization problem [396], and thus methods, such as ASTRAL [506] and BUCKy [405], implement heuristics for this optimization problem. Importantly, the heuristic used to assemble the subset trees can have a profound effect on the computational requirements and on the criterion score of the returned tree for this type of summary method.

Finally, some summary methods define likelihood or pseudolikelihood functions based on the MSC model and then seek the model species tree that maximizes this function. Such methods are very appealing due to their statistical interpretability, although, in practice, they have been shown to be more computationally intensive than the two types of methods previously discussed. For example, MP-EST, which stands for "maximum pseudolikelihood estimation," uses a heuristic to search for the model species tree that maximizes the probability of generating the set of triplet trees defined by the input (rooted) gene trees. Other likelihood-based methods have been developed, including STEM [388], STELLS [816], and STELLS2 [571], but these methods have been shown to be even more computationally intensive than MP-EST in practice.

In the remainder of this section, we discuss two summary methods: ASTRAL and ASTRID. Both have been proven statistically consistent under the MSC, have been shown to have high accuracy in simulation studies (although this depends on the model condition), and have been used to analyze very large data sets reasonably quickly. The main thing to realize, with respect to computational requirements, is that when using one of these fast summary methods the vast majority of the time is typically spent computing gene trees; as discussed in section 2.4, the computational requirements for computing gene trees using a maximum likelihood heuristic are typically less than the computational requirements for concatenation analyses using the same maximum likelihood heuristic. Thus, for phylogenomic data sets with thousands of loci, summary methods can have a clear computational advantage over concatenation analyses. Furthermore, in many biological analyses the gene trees are of independent interest, and so they are computed anyway. Thus, summary methods are generally a small fraction of the computational cost of species tree estimation from phylogenomic data sets.

2.2.2.1 ASTRAL

ASTRAL, which was introduced in [506] and subsequently improved in [507, 851], is now one of the most commonly used species tree methods that addresses gene tree heterogeneity due to ILS. For the purposes of this chapter, we will focus on the aspects of the ASTRAL algorithm that impact running time and scalability to large data sets.

The input to ASTRAL is a set $\mathcal{T} = \{T_1, T_2, \ldots, T_k\}$ of gene trees. Letting $\mathcal{L}(t)$ denote the leaf set of tree t and $S = \bigcup_i \mathcal{L}(T_i)$, the input to ASTRAL has size $O(nk)$, since each of the k gene trees has at most $|S| = n$ leaves. ASTRAL returns a species tree T on leaf set S as follows. To begin, ASTRAL uses the set $\mathcal{T}$ of input gene trees to compute a set X of "allowed bipartitions" on leaf set S. Note that each edge e in a tree t defines a bipartition $\pi(e) = A|B$ on its leaf set $\mathcal{L}(t)$ (i.e., deleting e from t produces two disjoint sets of leaves A and B); we denote the set of bipartitions defined by the edges in t as $Bip(t)$. Later versions of ASTRAL have varied in how the set X is computed, but the first version defines X to contain exactly the bipartitions for any gene tree T_i that is complete, meaning that $\mathcal{L}(T_i) = S$ (otherwise, T_i must be missing one or more species in S, and so we say that T_i is incomplete).

After computing the set X, ASTRAL solves the following optimization problem exactly: Given $\mathcal{T} = \{T_1, T_2, \ldots, T_k\}$ and X, find a tree T on leaf set S such that

$$T = \arg\max \sum_{i=1}^{k} |Q(T) \cap Q(T_i)|,$$

where $Q(t)$ denotes the set of quartet trees induced by t and where T is subject to the constraint $Bip(T) \subseteq X$. In other words, ASTRAL finds a species tree T that agrees with as many of the quartet trees induced by the gene trees in $\mathcal{T}$ as possible, but the search space for T is constrained by the set X. As previously mentioned, the unconstrained version of this optimization problem, where X is the set of *all possible* bipartitions on S, is NP-hard [396].

ASTRAL solves its optimization problem using an algorithmic technique called dynamic programming. The running time for ASTRAL-I [506] was shown to be $O(|X|^2 n^2 k)$, and modifications presented in ASTRAL-II [507] improved the running time to $O(|X|^2 nk)$. This upper bound on the asymptotic running time was subsequently tightened to $O(|X|^{1.726} nk)$ [354, 851]. Hence, the running time for the current version of ASTRAL (ASTRAL-III) is linear in the number of gene trees, linear in the number of species, but close to quadratic in $|X|$, the number of the allowed bipartitions.

The size of the constraint space X, and thus the running time of ASTRAL, increases with heterogeneity among the input gene trees. If all the trees in $\mathcal{T}$ are identical (i.e., there is some tree T^* so that $T_i = T^*$ for all $i = 1, 2, \ldots, k$), then $X = Bip(T_1) = Bip(T_2) = \ldots = Bip(T_k)$. Because ASTRAL returns a tree satisfying $Bip(T) \subseteq X$, ASTRAL will return T^*, and its running time will be very fast, as $|X| = n - 3$. On the other hand, if all the trees in $\mathcal{T}$ are very different, then $|X|$ will be much larger; in particular, $|X| = O(kn)$, as it is possible for each of the k gene trees to differ on each of their internal edges.

It is worth considering the causes for heterogeneity in the input set $\mathcal{T}$ that result in $|X|$ being large. While gene tree heterogeneity may be due to ILS (or other biological processes that result in true gene trees that differ from the true species tree and from each other), heterogeneity may also be due to *gene tree estimation error*. Gene trees are typically estimated under standard sequence evolution models (e.g., the GTR+GAMMA model) using maximum likelihood heuristics, and thus estimation error can result from many causes, including errors in the input gene sequence alignment, inadequate phylogenetic signal (perhaps due to insufficient sequence length), insufficient running time for the maximum likelihood heuristic to find a good tree (perhaps due to terminating the tree search before convergence to a local optimum), et cetera. Another cause for gene tree estimation error that is becoming increasingly apparent is model misspecification, as standard models, including the GTR+GAMMA model, make assumptions about sequence evolution that can be violated by biological data sets [375, 531]. The use of parameter-rich models for gene tree estimation, such as general heterogeneous evolution on a single topology (GHOST) [156], may reduce error due to model misspecification, but inference under these more complex models is more computationally intensive and has not yet become commonplace in phylogenomic studies.

To summarize, heterogeneity in $\mathcal{T}$ can result from several causes, and substantial heterogeneity is generally to be expected *except* when ILS is low and gene tree estimation error is *also* low. These two conditions can occur in some data sets, but genome-scale data sets are unlikely to exhibit both properties, at least in part because some loci will be slowly evolving (and so will have inadequate phylogenetic signal) and some loci will evolve so quickly that both alignment and tree estimation will become difficult. In other words, gene tree heterogeneity will be high in many (and perhaps even most) multilocus data sets, and thus the set X will typically be large.

Later versions of ASTRAL expanded the set X to ensure that a larger part of the solution space was explored. This improved accuracy but also increased running time. Subsequently, various algorithmic techniques were developed to reduce the running time, including techniques that did not expand the set X quite so much. The current version of ASTRAL has been well optimized and can run on many large data sets.

However, even this current version can be impacted by gene tree heterogeneity, since it still requires that X contain all the bipartitions found in all complete gene trees. For example, there are conditions with very high ILS, explored in [439], in which ASTRAL v.5.6.1 (i.e., ASTRAL-III) did not complete within 48 hours on some data sets with 1000 species and 1000 loci; however, on data sets with the same size but with lower ILS, ASTRAL completed within just a few hours [439]. While ASTRAL's running time can vary significantly depending on gene tree heterogeneity, there are many data sets for which it is quite fast, and so it is a good method to try when analyzing large multilocus data sets.

2.2.2.2 ASTRID

Like ASTRAL, the input to ASTRID [761] is a set $\mathcal{T}$ of unrooted gene trees, and the output is a species tree that contains one leaf for every species in S. ASTRID operates by first computing the average gene tree internode distance (AGID) matrix: an $n \times n$ matrix D, where $D[i, j]$ is the average number of internodes on the path between leaves i and j in the gene trees in $\mathcal{T}$ that contain both i and j. Whenever none of the gene trees in $\mathcal{T}$ contain both leaves i and j, then entry $D[i, j]$ is undefined or missing. If there are no missing entries in D, then ASTRID seeks a tree under the balanced minimum evolution (BME) criterion using FastME [417]; otherwise ASTRID runs BIONJ* [153]. Given the input gene trees, the running time is $O(kn^2)$ to compute the AGID matrix and then $O(n^2 \log n)$ to run BME, as implemented within FastME [417]; thus, the total running time of ASTRID is $O(n^2(k + \log n))$. In practice, ASTRID is typically faster than ASTRAL-II, especially for data sets with high gene tree heterogeneity.

An earlier gene tree summary method, NJst is identical to ASTRID except that it runs neighbor joining [645] on the AGID matrix. Because neighbor joining does not allow entries of the distance matrix to be missing, this restricts NJst to only those inputs with no missing entries. Thus, the main difference between ASTRID and NJst is that ASTRID can run on all data sets (because BIONJ* is designed for data sets with missing entries). On those data sets for which NJst and ASTRID can both run, the difference in accuracy depends on whether neighbor joining or BME (as implemented with FastME) is more accurate. Although it is not completely clear when one method will be more accurate than the other, prior studies suggest that BME typically has a slight advantage over neighbor joining in terms of topological accuracy [183, 761, 777].

2.2.3 COESTIMATION METHODS

Bayesian species tree methods coestimate the gene trees and the species tree under a hierarchical model of evolution that addresses ILS (e.g., gene trees evolve within a species tree under the MSC model and then sequences evolve down each gene tree under a standard sequence evolution model). *BEAST, perhaps the most well-known of these Bayesian coestimation methods, uses a Markov Chain Monte Carlo (MCMC) approach to sample from the distribution of gene trees and species trees. Hence, the input to *BEAST is a set of multiple sequence alignments (one for each locus), and the output is a set of gene tree distributions (one distribution for each locus) and a distribution of species trees. One advantage that Bayesian coestimation methods have over other methods is that they return distributions rather than a single point estimate, and these distributions enable the user to evaluate the statistical support on branches (or, equivalently, uncertainty) in a natural way. These distributions can then be used to produce point estimates of the gene trees and the species tree using several standard techniques.

The time needed for *BEAST to reach good ESS values (a statistic that is used to assess whether the method has converged to the stationary distribution) is nondeterministic, and so a Big-O analysis cannot be provided. However, the amount of time for coestimation methods to converge has been shown to increase substantially with number of species and/or the number of loci. In general, *BEAST is too computationally intensive to use except on moderately sized data sets, and data sets with 100 or more genes and 50 or more species may require several months of CPU time [56, 484, 415]. A recently developed version of *BEAST, called StarBEAST2 [549], is reported to be 33 times faster than *BEAST on simulated data sets. While this reduction in running time is important progress toward making coestimation more computationally feasible, improving the scalability of Bayesian coestimation methods continues to be an open research challenge.

2.2.3.1 BBCA: Scaling *BEAST to Large Numbers of Loci

A simple approach for addressing scalability to large numbers of loci is BBCA [864]. As found by [56, 864], point estimates of gene trees produced by *BEAST can be more accurate than maximum likelihood gene trees, and species tree estimated using summary methods on these *BEAST gene trees can be as accurate as point estimates of species trees produced using *BEAST. These observations suggested an approach for multilocus species tree estimation that makes it possible to use *BEAST (and other coestimation methods) without the same computational effort.

- Step 1: Randomly partition the set of genes into "bins" of the desired size (e.g., at most 25 genes per bin).
- Step 2: On each bin, use the desired coestimation method (e.g., *BEAST) to produce a gene tree (i.e., a point estimate) for each gene in the bin.
- Step 3: Run a summary method (e.g., ASTRAL) on the set of estimated gene trees.

Of these steps, the first is obviously very fast. The second step, in most cases, will be the most expensive, since coestimation methods such as *BEAST tend to be computationally intensive, even on data sets with only 25 genes. The third step will be fast, provided that a fast summary method is used. Therefore, the running time for a BBCA analysis is largely dominated by the second step.

In [864], BBCA was studied with simulated data sets with 11 species, 100 genes, and heterogeneity between true gene trees and species trees due to ILS. True gene sequence alignments, with varying lengths, were randomly divided into four bins, each with 25 genes. *BEAST was then run for 24 hours on each bin after which the resulting distribution for each gene tree was transformed into a point estimate. Then, the summary method MP-EST [440] was used to combine these estimated gene trees into a species tree. This BBCA analysis was compared with running *BEAST on the full set of 100 genes for 96 hours (i.e., the same total amount of time); the result was that BBCA produced species trees that were at least as accurate and often more accurate than the point estimate of the species trees produced by *BEAST. The explanation offered for the improved accuracy of BBCA over *BEAST is that by running *BEAST on 25 genes at a time, it was able to converge more quickly than on the full set of 100 genes; this explanation is supported by the improvement in ESS values obtained for each of the bins compared with ESS values for the full data set. In a second experiment, *BEAST was allowed to run for 168 hours (i.e., longer than the total BBCA time) on the full set of 100 genes. Even with the longer running time, *BEAST did not return more accurate species trees than BBCA.

The BBCA algorithm has several parameters that can be adjusted by the user. For example, the algorithm is based on random partitioning into bins, but the choice of the number of bins (or equivalently, the number of genes in each bin) is up to the user. Based on previous results, increasing the number of genes in each bin will improve accuracy, provided that the user can wait a sufficiently long time for *BEAST to converge. Thus, the choice of the number of bins is a trade-off between running time and accuracy. The first step of the BBCA algorithm is based on random partitioning of the loci into subsets because the theoretical guarantee for *BEAST assumes that the loci are drawn randomly from the genome. On the other hand, it is certainly possible that empirical performance could be improved through nonrandom partitioning. In the second step of BBCA, the user selects a coestimation method to produce a distribution on gene trees for each locus in each bin; the BBCA algorithm was tested with *BEAST, but this step could be performed using StarBEAST2 or any other coestimation method. Finally, the user must select the technique for producing the point estimate of each gene tree given the gene tree distribution and the summary method for combining gene trees. BBCA was studied with MP-EST, but other summary methods could be used as well, including ASTRAL or ASTRID.

To summarize, BBCA is a general framework for enabling computationally intensive coestimation methods to run on data sets with large numbers of loci. Its algorithmic design, which is based on random partitioning of the loci into small bins, means that it can be trivially parallelized, and since the second step dominates the others in terms of running time, parallelization (by assigning different bins to different processors) should result in a near-linear speedup. An important limitation of BBCA is that it produces only a point estimate of the species tree, rather than a distribution; hence, the use of BBCA instead of *BEAST does not provide the full power of a Bayesian method. Nevertheless, when the phylogenomic data set has too many loci for running *BEAST to be feasible, this technique can make it possible to have some of the advantages of coestimation without the computational cost. Finally, BBCA addresses scalability with respect to large numbers of loci but does not address the case when the number of species is also large.

2.2.4 SITE-BASED METHODS

Site-based methods use statistical properties of the MSC model to estimate the species tree directly from site patterns (e.g., in the concatenated alignment). Unlike summary methods, site-based methods do not require estimated gene trees, and unlike standard concatenation analyses under maximum likelihood, they specifically address heterogeneity due to ILS; thus, they are in a distinctly different class of method. One of the very appealing properties of site-based methods is that by avoiding the need to estimate gene trees, they are robust to conditions, such as low phylogenetic signal in each locus, that lead to gene tree estimation error. This is an important property, as gene trees estimated for several recent biological studies have been shown to have low bootstrap support values on average; see table 1 in [511] for the results from four biological studies [80, 343, 314, 731]. In the remainder of this section, we focus on two site-based methods: SVDQuartets and its recent extension, SVDquest.

2.2.4.1 SVDQuartets

The site-based method SVDQuartets [140, 141] uses the Singular Value Decomposition, a linear algebra technique, to compute four-taxon species trees directly from site patterns. SVDQuartets was recently proven to be statistically consistent under some hierarchical models of evolution, whereby gene trees evolve within a species tree under the MSC model and then a *single site* evolves down each gene tree under some standard sequence evolution model (e.g., the GTR model); see [783] for details. Thus, SVDQuartets has the distinctive advantage over summary methods in that it can be applied to SNP data.

In order to compute a species trees on a set S of more than four species, SVDQuartets can be run as follows.

- Step 1: Restrict the alignment to four species and then run SVDQuartets to compute a species tree; repeat for every four species in S, thus producing a set Q of quartet trees.
- Step 2: Apply a quartet amalgamation method to construct a species tree on leaf set S from Q.

This two-step pipeline has been implemented within PAUP* [741] with several different options for the quartet amalgamation method.

The input to a quartet amalgamation method is a set of quartet trees, and the objective is to find a tree on the full set of leaves that displays as many of the input quartet trees as possible. This is an NP-hard optimization problem [347], and so all quartet amalgamation methods are heuristics. Quartets MaxCut [706] and QFM [608] are probably the most well-known but also see [62, 99, 309, 453, 577, 603, 822]. Note that SVDQuartets can also output the three possible quartets each with an associated weight so that a weighted quartet amalgamation method, such as Weighted Quartets MaxCut [38], can be used.

The running time of the SVDQuartets pipeline is the time to compute a set of quartet trees plus the time to assemble the quartet trees into a species tree. SVDQuartets can compute a single quartet tree in an amount of time that is linear in the size of the input (i.e., the running time of SVDQuartets is $O(M)$, where M is the number of sites). Although each quartet tree can be computed in linear time and in parallel, the computation of all $O(n^4)$ quartet trees is still burdensome when the number n of species is large. Currently, the SVDQuartets pipeline, as implemented in PAUP*, can be run on data sets with large n by randomly sampling a subset of the possible quartets. This approach has definite advantages in terms of running time, but the accuracy of trees computed using sparsely sampled quartets may be reduced compared with using the full set of quartets [739]. An important direction for future research is the development of quartet sampling approaches that can improve the scalability of the SVDQuartets pipeline while maintaining accuracy.

2.2.4.2 SVDquest: Improving the Search Strategy for SVDQuartets

A recent development that enables the SVDQuartets pipeline to be run using a different quartet amalgamation method than those provided within PAUP* is SVDquest [762]. Specifically, SVDquest uses the following pipeline to utilize the same constrained optimization approach as ASTRAL, which guarantees a provably optimal solution with the constrained search space. The input to SVDquest is a set of multiple sequence alignments (one for each locus), and the output is species tree on leaf set S, computed as follows.

- Step 1: Compute a gene tree on each locus (e.g., using maximum likelihood).
- Step 2: Build a set X of allowed bipartitions (e.g., using the same approach as ASTRAL).
- Step 3: Run SVDQuartets on the concatenated alignment restricted to every possible set of four species; this creates a set Q of quartet trees.
- Step 4: Optionally, run PAUP* on Q to obtain an estimate of the species tree on leaf set S, and add its bipartitions to the set X.
- Step 5: Use dynamic programming to find a tree T on S, within the constrained search space (i.e., $Bip(T) \subseteq X$), that agrees with the largest number of quartet trees in Q.

Note that SVDquest can technically run on SNP data by treating each SNP as a genomic locus with one site, but it has not yet been tested in this context; furthermore, SVDquest relies on the ability to constrain the search space, and achieving this without the ability to estimate gene trees is likely to present challenges for SVDquest.

When the optional step is included, then by design SVDquest is more computationally intensive than PAUP* (since it calls PAUP*). In fact, it is substantially more expensive, as unlike the SVDQuartets pipeline, SVDquest requires gene tree estimation. However, SVDquest is guaranteed to find an optimal tree within its constraint space, which is the set of all binary trees whose bipartitions are found in X. Therefore, if SVDquest includes the optional step (i.e., run PAUP* and add the bipartitions from the resulting tree into X), SVDquest is guaranteed to produce a tree that agrees with *at least as many* quartet trees as the tree found by PAUP*. As shown in [762], SVDquest often returns trees with better criterion scores than does PAUP*, and so may be valuable for some inputs.

2.2.5 EVALUATION OF BRANCH SUPPORT IN SPECIES TREES

The focus so far has been on obtaining a highly accurate point estimate of the species tree on large data sets. Yet, phylogeny estimation is characterized by uncertainty, and assessing the uncertainty in a point estimate of the species tree is an important aspect to consider. Bayesian coestimation methods enable this assessment directly, but (as discussed in section 2.2.3) the current versions of these methods are too computationally expensive to use on large numbers of species or loci. One generic technique for assessing branch support that can be used with all the methods discussed here is *bootstrapping*, but this increases the running time, as methods must be run on many bootstrapped versions of the input data set. Furthermore, traditional bootstrap approaches can inflate branch support values under some conditions and can reduce them under others (e.g., when the data set has rogue taxa) [423]. If gene trees have been estimated, then a fast technique for assessing branch support based on quartet frequencies in the gene trees (called the "local posterior probability") can be used to evaluate branch support for a given species tree [657]. This approach, which has been implemented within ASTRAL, has been shown to perform better than multilocus bootstrapping, which can reduce the accuracy of the estimated species tree [504]. Even with the recent advances for quantifying uncertainty in species trees estimated from large phylogenomic data sets, this remains an important area for future research.

2.3 Species Tree Estimation under GDL

The previous section addressed species tree estimation when gene trees can differ from the true species tree because of ILS; however, other biological processes also result in gene tree heterogeneity and so affect species tree estimation. In particular, the methods described in the previous section assume that genes evolve without duplication or loss events, and thus exactly one copy of each gene appears in the genome for any given species. This assumption does not hold for many biological data sets; for example, the preliminary version of the 1000 plant transcriptome data set included over 9000 genes but less than 500 of them contained at most one copy for each species [798]. Thus, the vast majority of genes were excluded prior to species tree estimation with ASTRAL, illustrating the need for scalable methods that can infer species trees in the presence of GDL.

Bayesian coestimation methods offer an appealing and statistically rigorous approach for species tree estimation when there is gene tree heterogeneity (section 2.2.3). PHYLDOG [88] is a well-known Bayesian method that can be used to coestimate gene trees and the species tree in the presence of GDL and has been shown to have high accuracy. However, PHYLDOG is computationally intensive; for example, PHYLDOG was limited to data sets with only 10 species owing to computational issues in a study evaluating existing species tree estimation methods addressing GDL [131]. A more scalable alternative is first to estimate gene trees and then combine them into a species tree. We refer to these gene trees as gene family trees or mul-trees because gene trees can have more than one leaf labeled by the same species. Several different methods have been developed that can infer species trees from gene family trees, but until recently, little was known about whether any of these methods were statistically consistent under models of GDL.

In Fall 2019, Legried et al. [418] proved that the model species tree is identifiable under a standard stochastic model of GDL [35] and that ASTRAL-multi [595] (a recent extension of ASTRAL that allows it to handle multi-allele data sets) is a statistically consistent method for estimating the species trees under this model. Furthermore, ASTRAL-multi runs in polynomial time and, as shown in [418], typically achieves better accuracy than DupTree [786], a leading method for gene tree parsimony, especially when gene tree estimation error was high.

Another approach to species tree estimation from gene family trees is based on extending the Robinson-Foulds supertree (RFS) problem [46] to mul-trees. This optimization problem, proposed by Chaudhary et al. [133], is called the RFS-multree problem. Recently, Molloy and Warnow [514] proved that the solution to the RFS-multree problem is a statistically consistent estimate of the species tree under a generic duplication-only or loss-only model of gene evolution. Molloy and Warnow [514] also presented FastMulRFS, which like ASTRAL and SVDquest uses dynamic programming to find an optimal solution to its optimization problem within a constrained search space in polynomial time. This algorithmic approach enables FastMulRFS to find good solutions quickly—and FastMulRFS has been shown to be much faster than MulRF [133], a hill-climbing heuristic for the RFS-multree problem. FastMulRFS has been shown to have very good accuracy (as compared with ASTRAL-multi and Dup-Tree) under a range of model conditions, including moderate levels of both gene duplication *and* loss, low/moderate levels of ILS, and high levels of gene tree estimation error.

In 2020, a new quartet-based method for GDL-based species tree estimation, called ASTRAL-Pro [852], was demonstrated to be at least as accurate as the previously mentioned methods (often more accurate) and to have good scalability. ASTRAL-Pro has two phases: preprocessing and optimization. In the preprocessing phase, ASTRAL-Pro operates by rooting and "tagging" each gene family tree (i.e., it labels each node in each gene family tree as either a duplication or a speciation node). These labels are used to modify how quartet support is calculated during the optimization phase of the method. ASTRAL-Pro has been proven statistically consistent for GDL-based species tree estimation, provided that all gene family trees are correctly rooted and tagged. ASTRAL-Pro was followed by DISCO [803], which uses ASTRAL-Pro to root and tag the gene family trees and then decomposes the multi-copy gene family trees into single-copy gene trees. These single-copy gene trees can then be given to a selected species tree estimation method that takes single-copy gene trees as input; thus, following the decomposition with ASTRAL produces ASTRAL-DISCO, and following the decomposition with ASTRID produces ASTRID-DISCO. Like ASTRAL-Pro, ASTRAL-DISCO is statistically consistent for GDL-based species tree estimation whenever ASTRAL-Pro correctly roots and tags all the gene trees. Using DISCO generally reduces runtime, and ASTRID-DISCO is particularly fast and accurate, even on very large data sets.

In summary, the last few years have shown dramatic advances in GDL-based species tree estimation without requiring knowledge of orthology: some methods (e.g., ASTRAL-multi) have been proven statistically consistent and others have not yet been proven consistent but produce even better accuracy and scalability than ASTRAL-multi.

2.4 Parallel Implementations for Species Tree Estimation

Here, we discuss parallel implementations of two of the common approaches to multilocus species tree estimation: ASTRAL and concatenation analyses using maximum likelihood. In particular, we consider the impact of using additional processors: that is, can a species tree estimation method run in less time if given access to a greater number of processors? Ideally, running a method using p processors (instead of one processor) would decrease the running time by a factor of p, corresponding to 100% parallel efficiency. However, in practice, some operations depend on the previous operations having been performed (i.e., the work is serial), and in this case, parallel efficiency will be less than 100%. Amdahl's law [19] states that parallel efficiency is governed by the fraction of serial work; in particular, when the fraction of serial work is greater than zero, parallel efficiency goes to zero as the number of processors goes to infinity. In other words, as the number of processors increases, the execution time becomes dominated by the serial work! Although disappointing, this observation is useful for evaluating the scalability of parallel algorithms; see review in [295]. Here, we discuss recent advances in parallel codes for large-scale phylogenomic analyses with an emphasis on the serial work performed by these methods to illustrate open challenges.

2.4.1 ASTRAL-MP

ASTRAL-MP is a recent development within the ASTRAL suite of methods [838]. While previous versions of ASTRAL used only a single thread, ASTRAL-MP implements parallelism using vectorization (AVX2), CPU multithreading, and GPU multithreading (OpenCL). As noted in section 2.2.2.1, ASTRAL uses dynamic programming to solve its optimization problem exactly within a constrained search space. At the high level, ASTRAL operates by computing a set X of allowed bipartitions (which it uses to construct a set Y of allowed tripartitions), weights each tripartition in Y based on the input gene trees, and then uses the weighted tripartitions to construct the tree with bipartitions drawn from the set X. Importantly, the majority of time is typically spent weighting tripartitions, but as each tripartition can be weighted independently, this step lends itself to parallelism. Indeed, the amount of time required for tripartition weighting is greatly reduced by adding more processors, and on some data sets, the computation time can become dominated by other parts of the ASTRAL algorithm that are less parallelized (figure 1 in [838]). To achieve good scaling with large numbers of cores, ASTRL-MP changes some parts of the ASTRAL algorithm but does so without sacrificing statistical consistency. In summary, the AVX2 vectorization and CPU multithreading implemented in ASTRAL-MP enables speedups (compared with ASTRAL-III) on most modern machines, and users with access to GPU machines will benefit from even greater speedups. For example, the GPU version of ASTRAL-MP achieved speedups of 158x (compared with ASTRAL-III) and enabled the analysis of a data set with 147,800 gene trees and 144 species in less than two days [838]. Data sets with larger numbers of species (10,000 species and 1000 genes) were also able to be analyzed in less than two days. Thus, ASTRAL-MP is an exciting development in large-scale species tree estimation.

2.4.2 MULTILOCUS SPECIES TREE ESTIMATION USING MAXIMUM LIKELIHOOD

Maximum likelihood is a very common approach to multilocus species tree estimation; indeed, it is the dominant approach in the sense that publications performing multilocus species tree estimation typically present the tree computed using some concatenation analysis (either a maximum likelihood method or in some cases a Bayesian method) as their main species tree. Here we discuss the issues involved in running maximum likelihood on large multilocus data sets, and in particular, the use of parallelism in speeding up these methods.

Maximum likelihood methods seek a model tree that has the highest probability of generating the input multiple sequence alignment. This optimization problem is NP-hard [621], and thus maximum likelihood heuristics typically use local search strategies based on a combination of hill climbing and randomization. The tree search continues until a stopping criterion is met, indicating convergence to a local optimum, and because the running time of this approach is nondeterministic, a Big-O running time analysis cannot be provided.

For multilocus data sets, maximum likelihood tree estimation can be performed in various ways. In an unpartitioned analysis, all the sites are assumed to evolve down a single model tree (described by a tree topology with branch lengths, a substitution rate matrix, and a distribution of rates across sites). In a fully partitioned analysis, different numeric parameters can be estimated for each locus, and only the tree topology is assumed to be constant across all the loci. Unpartitioned analysis is faster than a parameter-rich partitioned analysis and hence is more commonly applied. Regardless of the approach, maximum likelihood on the concatenated alignment can be statistically inconsistent and even positively misleading (i.e., return the wrong tree with probability converging to 1) when there is gene tree discord due to ILS, as established in [624] (for unpartitioned maximum likelihood) and [622] (for fully partitioned maximum likelihood). See [780] for a discussion about the differences between partitioned and unpartitioned analyses in multilocus analyses.

2.4.2.1 RAxML

One of the most widely used maximum likelihood methods for phylogenomic analyses is RAxML; parallel versions of RAxML have existed for many years. Recall that maximum likelihood heuristics take as input a multiple sequence alignment and perform a heuristic search of tree space, computing the likelihood for candidate trees. Because the probability of observing a site pattern given a candidate tree is *independent* of the other sites in the alignment, the probability for each site can be computed in parallel. Parallelism across sites can be implemented at many different levels, and RAxML version 8 [720] uses pthreads, vector extensions (SSE3, AVX and AVX2), and other techniques to reduce the amount of time required to compute tree log-likelihood, which can be computationally intensive when there are many unique site patterns in an alignment. Such optimizations are critical, especially when inferring trees under more complex models; however, these optimizations do not impact the number of candidate trees evaluated during the tree search. RAxML version 8 can search from multiple starting trees in parallel (using MPI); however, this parallelism does not reduce the number of candidate trees that need to be evaluated for any one of these searches to converge to a local optimum; thus, searching tree space (using hill climbing and randomization) is effectively serial work. Because number of possible tree topologies increases exponentially with the number of species, the tree search phase may become quite long for data sets with large numbers of species; thus, despite significant optimizations, running RAxML can be computationally intensive on some data sets.

2.4.2.2 ExaML

For some very large multilocus data sets, the concatenated alignment may not fit into the memory of a single compute node. In this case, researchers with access to a distributed-memory system can run a different MPI version of RAxML, called ExaML [379]. ExaML operates by partitioning the alignment across sites and distributing these partitions across multiple processors. Thus, computing the tree log-likelihood given the *entire* alignment and coordinating the tree search requires some amount of communication (i.e., the sending/receiving of messages between different processes). For example, if the alignment was distributed across p processors, then $\log_2(p)$ communication steps would be required to compute the log-likelihood given the entire alignment using a standard global reduction; these $\log_2(p)$ steps are effectively serial work. Furthermore, the amount of time required for communication is significant compared with other operations, which is why avoiding communication [182], overlapping communication and computation [308], and modeling communication [158, 7, 279] are topics of interest to the high-performance computing community. Although significant optimizations may be made to improve communication performance, there is still the issue of navigating tree space effectively. Thus, ExaML is a significant advance in large-scale maximum likelihood tree estimation, but there are still open challenges to large-scale maximum likelihood tree inference.

2.4.2.3 Other Maximum Likelihood Methods and Recent Developments

Many other maximum likelihood heuristics have been developed including PHYML [285], GARLI [866], FastTree-2 [586], and IQTree [536]. Of these methods, FastTree-2 and IQTree have been able to analyze very large data sets, and in particular, FastTree has been used to construct a tree on the million-taxon RNASim data set [505, 537]. FastTree-2 operates by computing a minimum evolution tree and then performing a maximum likelihood tree search, limiting the search to $O(\log n)$ rounds of nearest-neighbor interchange (NNI) moves, where n is the number of species. IQTree is a newer method that has been shown to be similar to RAxML in terms of accuracy and speed. The fast version of IQTree (option: -fast) is similar to FastTree, in that it computes a distance-based starting tree and then uses a polynomial number of NNI moves during the tree search phase. Thus, these fast maximum likelihood heuristics improve running time by limiting the tree search rather than by utilizing additional processors. Finally, we note that the most recent version of RAxML, called RAxML-NG [380], re-implements the tree search algorithm inside of RAxML in order to improve speed and functionality.

2.4.2.4 Concatenation versus Summary Methods

Because many summary methods are quite fast (e.g., ASTRAL and ASTRID), from a computational viewpoint, the difference between concatenation and summary methods comes down to whether it is faster or slower to compute k gene trees rather than a single tree on the concatenated multiple sequence alignment for the k loci. For simplicity, we assume that the tree search is the same regardless of the approach; that is, the same candidate trees are evaluated in the same order for the concatenated alignment as well as each of the multiple sequence alignments for the k different loci. We also suppose that the concatenated alignment is large enough that ExaML must be used to perform the analysis, but the alignment of each locus is small enough that RAxML can be used to perform each gene tree analysis, which was the case in the Avian Phylogenomics Project [343]. Thus, each gene tree analysis (using RAxML) can be performed in an embarrassingly parallel fashion, but the concatenation analysis (using ExaML) will require some amount of communication to coordinate the tree search across multiple processes. Because the cost of communication is significantly greater than the cost of other operations (e.g., floating-point arithmetic), the concatenation analysis should take longer than analyzing each of the gene trees independently, as gene tree estimation can be performed in parallel without any communication. It is also possible that analyzing each locus individually could result in better memory locality (e.g., cache-oblivious algorithms [255]), further improving performance. Finally, as different parts of the genome can support different model trees, we conjecture that the tree search may converge to a good local optimum more slowly for the concatenated alignment compared with the alignments for individual loci. Finally, improvements in maximum likelihood methods for gene tree estimation, including the recently introduced method ParGenes [381], are likely to tip the balance further in favor of gene tree estimation followed by summary methods. While concatenation analyses often provide excellent accuracy and are still the norm for biological systematics, for those data sets with thousands (or tens of thousands) of loci, gene tree estimation followed by summary methods are likely to be more computationally efficient. The issue of whether summary methods or concatenation will provide better accuracy is discussed in section 2.6.

2.5 Divide-and-Conquer Species Tree Estimation

While many of the methods previously discussed are well optimized and relatively efficient in terms of their asymptotic running time, they can still be computationally intensive on large multilocus data sets. Divide-and-conquer has been proposed as a potentially important tool for large-scale species tree estimation in [71], and here we present a few divide-and-conquer approaches that have been studied in the context of multilocus species tree estimation.

2.5.1 DIVIDE-AND-CONQUER USING SUPERTREE METHODS

In the traditional divide-and-conquer approach, the species set is divided into overlapping subsets, trees are computed on the subsets, and then the subset trees are combined together using a supertree method (see [70, 180, 651] for an entry into the literature on supertree methods). One method that operates in this fashion, DACTAL [532] uses novel algorithmic techniques to perform key steps in this pipeline and combines divide-and-conquer with iteration (i.e., the tree from the previous iteration is used to define subsets for the subsequent iteration). While DACTAL was originally studied in the context of computing trees from unaligned sequences, more recently it was one of the first divide-and-conquer methods tested in the context of multilocus species tree estimation. Specifically, Bayzid et al. [54] implemented DACTAL (using MP-EST [440] to estimate subset trees and using SuperFine [738] to combine subset trees into a supertree) and tested it on simulated data sets. They tested this version on simulated data sets with gene tree discord due to ILS, finding that not only did DACTAL reduce the running time of MP-EST but the species trees estimated using DACTAL were more accurate than species trees estimated by MP-EST on the full data set [54]. While DACTAL is a promising approach to divide-and-conquer species tree estimation, unfortunately, software for DACTAL is not available at this time.

2.5.2 DIVIDE-AND-CONQUER USING DISJOINT TREE MERGER METHODS

Recently, a new divide-and-conquer strategy was proposed that decomposes the species set into *pairwise disjoint* subsets, estimates trees on subsets, and then merges the subset trees into a tree on the full data set. Because the trees are pairwise disjoint leaf sets, standard supertree methods cannot be used to combine subset trees. Instead, this step is accomplished using disjoint tree merger (DTM) methods, which require some additional information (e.g., evolutionary distances between pairs of species) as input in order to merge trees.

Divide-and-conquer pipelines using DTM methods have several advantages. First, they avoid solving the supertree problem, which is important as the most accurate supertree methods are typically heuristics for NP-hard optimization problems. To put this in perspective, ASTRAL could be used as a supertree estimation, but even ASTRAL can be computationally intensive on large data sets; in fact, later we will discuss results that show DTM methods can be used to speed up species tree estimation with ASTRAL. Second, DTM methods enforce the subset trees as topological constraints on the output tree, and so we typically refer to the subset trees as constraint trees. It is always possible to construct a tree on the full leaf set that satisfies the constraints because the constraint trees can simply be connected through edge additions. If a DTM method *requires* constraint trees to be connected through a sequence of edge additions, then we say that it does *not* allow "blending"; however, unless otherwise noted, the DTM methods discussed here allow the constraint trees to blend. Because the most accurate (but most computationally intensive) methods are used to estimate subset trees, respecting the subset trees as topological constraints is advantageous. Third, using DTM methods enables users to design divide-and-conquer species tree estimation pipelines that are provably statistically consistent under the MSC (although the specific requirements for pipelines to be provably statistically consistent depends on the DTM method [513, 512, 688]). In particular, divide-and-conquer pipelines that maintain statistical consistency have been developed for each of the DTM methods described below. Hence, DTM methods may be of great value for large-scale species tree estimation.

The first implemented DTM method, NJMerge [512], is a polynomial time modification of the neighbor joining method [645]. Specifically, NJMerge uses an input distance matrix to build a tree agglomeratively by attempting to accept only siblinghood proposals that are compatible with the input constraint trees. Because testing the compatibility of *more than two* unrooted trees is NP-hard [725], NJMerge uses a heuristic to decide whether or not to accept a siblinghood proposal. Thus, it is possible for NJMerge to accept siblinghood proposals that are not compatible with the input constraint trees, in which case, NJMerge will fail to a return a tree. However, when NJMerge successfully returns a tree, it tends to be highly accurate [512].

An improvement on NJMerge, TreeMerge [513] is guaranteed to return a tree that satisfies the input constraint trees. TreeMerge achieves this by running NJMerge on *pairs* of constraint trees, thus producing a new set of trees (on overlapping leaf sets) that satisfy the input constraint trees. TreeMerge then uses a fast technique that requires estimated branch lengths in order to combine the trees produced by NJMerge. This approach has two major advantages over NJMerge: it always outputs a tree that satisfies the input constraint trees, and it is faster than NJMerge on large data sets [513]. On the other hand, when NJMerge completes, it typically produces slightly more accurate trees than TreeMerge.

Constrained-INC [856, 411] is a DTM that operates by computing quartet trees from the input and then uses the resulting quartet trees to merge the constraint trees together. Constrained-INC has mainly been studied in the context of gene tree estimation under the GTR model, which presents different challenges.

The final DTM is guide tree merger (GTM) [688], which combines constraint trees through the addition of edges (i.e., blending is not allowed). Thus, the pipeline using GTM operates in a somewhat different manner than the previously discussed methods: a guide tree is computed from the multilocus data set, and then constraint trees are combined together by adding edges so as to minimize the Robinson-Foulds [620] distance to the guide tree (i.e., GTM does not allow the subset trees to be blended). As shown in [688], an optimal solution to the unblended supertree problem can be computed in polynomial time, while the general problem (in which blended supertrees are allowed) is NP-hard.

In a simulation study, NJMerge, TreeMerge, and GTM were evaluated with gene trees differing from the true species tree because of ILS. All three methods reduced the running time for several different species tree methods (including ASTRAL and RAxML) on data sets with 1000 species and 1000 loci [512, 513, 688]. For example, on high ILS data sets with 1000 species and 1000 genes, the entire divide-and-conquer pipeline using ASTRAL-III (to compute subset trees) and using TreeMerge (to combine subset trees) completed in about 4 hours. In contrast, using ASTRAL-III on the full data set typically did not complete within 48 hours, and when ASTRAL-III did complete, it used close to the 48 hours, the maximum allowed time on the Blue Waters supercomputer [513]. Most importantly, running ASTRAL-III within this divide-and-conquer approach resulted in species trees that had similar accuracy to those produced by running ASTRAL-III on the full data set [513].

It is worth noting that GTM was much faster than TreeMerge and NJMerge (although this speedup was negligible in the context of the divide-and-conquer pipeline, as the vast majority of the time was spent estimating subset trees). Furthermore, pipelines using GTM had similar accuracy to those using NJMerge and TreeMerge on the data sets evaluated in [688]; this is interesting given the restriction of GTM to combining subset trees only through edge additions. Whether GTM will have comparable accuracy to the other DTM methods under other conditions remains to be seen, and future research is needed to explore this issue.

2.6 Choice of Method

Given a particular multilocus data set, the choice of which method(s) to use to estimate a species tree can be challenging. For example, knowing the conditions under which a method is guaranteed to be statistically consistent (and so will converge to the true species tree as the amount of data increases) can be important, but the empirical performance of the method is also important. Furthermore, the specific properties of the data set also inform the choice of method. Here we discuss both types of issues for the methods we have discussed in this chapter.

2.6.1 STATISTICAL CONSISTENCY

Statistical consistency for multilocus species tree estimation is an important consideration when choosing between methods but, as we will argue, is also surprisingly complex. Recall that statistical consistency addresses whether a method is provably guaranteed to converge to the true tree as the amount of data generated under a particular model increases. For multilocus data sets, increases in available data can occur in two ways: the number of loci may increase or the number of sites per locus may increase. Distinguishing between these two ways in which data can increase is important, as it affects whether a method is statistically consistent or not. Many methods have been established to be statistically consistent when the number of sites per locus *and* the number of loci both increase, but fewer have been established to be statistically consistent when only the number of loci increase and all the loci are bounded in length by some arbitrarily chosen L. The first type of consistency, which assumes that the number of sites per locus and number of loci both increase, is referred to as "weak" statistical consistency, and the second type is referred to as "strong" statistical consistency [780].

Interestingly, although summary methods are commonly said to be statistically consistent, the proofs for consistency have assumed that the input are true gene trees, and hence these proofs establish statistical consistency in the "weak" sense. Furthermore, as shown in [622], standard summary methods (and all summary methods that satisfy some rather mild assumptions) can be positively misleading when the number of sites per locus is bounded; hence, standard summary methods are not statistically consistent in the strong sense.

The question of statistical consistency for concatenation analyses using maximum likelihood is also interesting to consider. It has long been known that this approach can be inconsistent (and even positively misleading) in the presence of gene tree heterogeneity due to ILS [624], but this proof is for unpartitioned analyses, which assume all sites evolve down the same model tree. Since maximum likelihood analyses can also be performed using partitioned analyses, which allow the numeric parameters for each locus to be estimated independently, we also need to consider whether partitioned analyses can avoid this problem. Perhaps surprisingly, if all the sites evolve down the same model tree (i.e., there is no gene tree heterogeneity) but the number of sites per locus is bounded, then unpartitioned maximum likelihood is statistically consistent but fully partitioned maximum likelihood *can be positively misleading* [622]. This disturbing result comes from the fact that maximum likelihood can also be affected by long branch attraction when the number of sites per locus is bounded. Overall, if there is no gene tree heterogeneity, then unpartitioned concatenation (but not partitioned concatenation) is statistically consistent—but if there is gene tree heterogeneity, then no concatenation analysis using maximum likelihood is provably statistically consistent.

This discussion reveals that when the number of sites per locus is bounded, then both concatenation analyses and summary methods can be inconsistent and even positively misleading. Here, we see a clear advantage of site-based methods. For example, SVDQuartets is statistically consistent in the strong sense as the proof of statistical consistency for SVDQuartets assumes the input is a *single* site per locus.

The discussion above has been based on the literature for species tree estimation when gene tree heterogeneity is due to ILS, but similar results are likely to be true when gene tree heterogeneity is due to GDL. In practice, gene tree heterogeneity may be due to multiple biological processes (e.g., ILS, GDL, horizontal gene transfer [HGT], gene flow, etc.), and no method has yet been established to be statistically consistent under such conditions; indeed, it is possible that many methods may be provably statistically inconsistent or even positively misleading (e.g., [713]). Overall, we conclude that while statistical consistency is a desirable property, few methods have this property in the strong sense (i.e., being provably statistically consistent where the number of loci increases but the number of sites per locus does not increase). Thus, statistical consistency for multilocus species tree estimation is a complex issue.

2.6.2 EMPIRICAL PERFORMANCE

Statistical consistency is a complicated issue, and methods that are statistically consistent may not have similar accuracy on finite amounts of data; thus, it is important to consider how each method has performed (in terms of accuracy and computational requirements) on simulated and biological data sets. One of the key questions in selecting a method is whether there is likely to be gene tree heterogeneity in the data set, and if so, what the biological cause of that heterogeneity is. Therefore, we discuss these issues first considering gene tree heterogeneity due to ILS and then considering multiple biological causes of gene tree heterogeneity.

Addressing ILS. When the number n of species is sufficiently small, then nearly all the methods we discussed (i.e., ASTRAL, ASTRID, SVDQuartets, SVDquest, and concatenation using maximum likelihood) can run, although coestimation methods such as *BEAST may have problems if the number k of loci is large. BBCA is a divide-and-conquer technique that enables *BEAST (and other methods for coestimating gene trees and species trees) to scale to data sets when k is large but does not address the computational challenge when n is large. Moreover, even if n is relatively small (e.g., under 100) but the number of loci is sufficiently large, then maximum likelihood methods, such as RAxML or ExaML, can be computationally intensive, requiring terabytes of shared memory, years of CPU time, and the use of supercomputers (as our experience with the Avian Phylogenomics Project revealed). In contrast, summary methods (ASTRAL and ASTRID) and site-based methods (SVDQuartets within PAUP* and SVDquest) remain practical techniques for analyzing these data sets. The relative accuracy of these methods is still being explored, but all of these methods are fast enough to be used on data sets of this size. A natural approach to species tree estimation for such data sets would be to run all these analyses (e.g., ASTRID, ASTRAL, SVDQuartets or SVDquest, and concatenation analyses) and then examine the resulting trees for common features. Furthermore, since concatenation analyses will tend to be the most computationally intensive, the other methods could be run first, and then concatenation analyses could be used if necessary, based on whether the evolutionary questions of interest are answered adequately using these analyses.

As n increases, the set of methods that are able to complete using standard resources (i.e., without supercomputers) decreases, and in particular concatenation analyses using maximum likelihood can become infeasible without substantial time on supercomputing platforms. The current research, described here, suggests that in these cases the better summary methods (notably ASTRAL) can provide good accuracy and can be fast enough (with low enough memory requirements) to complete within reasonable time frames for many data sets. The main effort in analyses using ASTRAL or ASTRID (or other summary methods) is the calculation of gene trees, which depends also on the calculation of multiple sequence alignments; these are both computationally and statistically challenging problems, but methods such as PASTA [505] (which coestimates alignments and trees) can be fast and highly accurate (assuming sufficient phylogenetic signal), even on data sets with several thousand sequences. Assembling the species tree from the estimated gene trees is then generally very fast, using in some cases a few CPU days of analysis but much less than the CPU years used by concatenation analyses.

For very large n, however, nearly all methods become either infeasible to use without extensive computational resources or without substantial modification. For example, concatenation analysis using maximum likelihood (if performed using RAxML or similarly accurate methods) becomes infeasible to run to completion without extensive computational resources (such as substantial memory), methods based on SVDQuartets need to be modified not to require the computation of all quartet trees, and even ASTRAL may not complete within reasonable time frames. As we noted in section 2.4, parallelism can speed up analyses but does not address the challenges of large numbers of species, as treespace increases exponentially with the number n of species. This is the context in which the divide-and-conquer strategies discussed in section 2.5 are likely to be very useful.

Addressing multiple biological sources of gene tree heterogeneity. When there are other biological sources for gene tree heterogeneity (e.g., gene duplication and loss, horizontal gene transfer, gene flow, etc.), then selecting a method is more complicated—in part because only a few studies have compared species tree estimation methods under multiple sources for discord. Four relatively recent studies [162, 418, 514, 713] provide some insight into the relative performance of methods under varying conditions. Davidson et al. [162] compared species tree estimation methods in the presence of both ILS and HGT. They found that both ASTRAL and Quartets MaxCut [706] (applied to the same set of quartet trees, derived from the input gene trees) had better accuracy than NJst and concatenation using RAxML, with the difference in accuracy increasing with the HGT rate. Solís-Lemus et al. [713] compared ASTRAL, ASTRID, concatenation, and PhyloNet [790] in the presence of gene flow and ILS and found ASTRAL generally had better accuracy than ASTRID, and ASTRID was better than concatenation. Interestingly, the maximum likelihood method in PhyloNet [790] had the best accuracy of all methods but was much more computationally intensive than the other methods. Finally, two other studies [418, 514] compared methods that estimate species trees from gene family trees (i.e., mul-trees) in the presence of ILS and GDL and found that ASTRAL-multi, FastMulRFS, and MulRF all had better accuracy than DupTree, ASTRID, and STAG [230]. Overall, these results suggest that ASTRAL may have some robustness to a variety of causes of gene tree discord in terms of empirical performance. Nevertheless, the studies comparing methods under a variety of conditions of discord are quite limited, and it is too early to draw any reliable conclusions.

2.7 Summary, Challenges, and Future Directions

The main focus of this chapter has been on large-scale species tree estimation in the presence of ILS, focusing on the computational requirements and statistical guarantees for several types of methods, including summary methods, coestimation methods, site-based methods, and concatenation analyses using maximum likelihood. We also presented some very recent results on species tree estimation in the presence of GDL, including some new theory establishing that FastMulRFS and ASTRAL-multi are statistically consistent under some stochastic models of GDL. These advances are likely to be surpassed in the near future, as new methods with even better accuracy and scalability are developed.

Since the most accurate methods can be computationally intensive when data sets have large numbers of species, we also presented recently developed divide-and-conquer approaches to scaling species tree estimation methods to large data sets, which are based on DTM techniques. Divide-and-conquer pipelines using DTM methods have been shown to enable statistically consistent estimates of the species tree (if properly designed), maintain accuracy, and reduce the computational effort in estimating species trees. Although these DTM methods have been tested only for species tree estimation in the presence of ILS, they should be adaptable to conditions with GDL. Thus, highly accurate ultra-large species tree estimation may be feasible using a combination of strategies. However, DTM methods are still very new and most likely there will be substantial improvements in their algorithmic design that affect accuracy and scalability to large data sets. Similarly, although no functional implementation currently exists for DACTAL, divide-and-conquer strategies that rely on supertree methods (rather than DTM methods) are also likely to be valuable tools for scaling species tree estimation methods to large data sets. Thus, biologists who are considering assembling ultra-large data sets may well benefit from future algorithmic developments and are encouraged to keep an eye out for new software that enable species tree methods to scale to large data sets.

Besides scalability, there are many other challenges to analyzing large multilocus data sets. The incidence of missing data is likely to increase on large data sets [649], and while some studies have suggested that the several species tree methods are robust to some patterns of missing data [819, 511, 543], this question has not been carefully evaluated on data sets with large numbers of species or with adversarial patterns of missing data. For example, the patterns of missing data that arise when combining data sets from different studies can negatively affect optimization-based methods (e.g., maximum likelihood heuristics) because of the creation of phylogenetic terraces [651]. Furthermore, some techniques for assembling large multilocus data sets (e.g., RADseq) may result in biased patterns of missing data that negatively impact multilocus species tree estimation [318]. Several different strategies are commonly employed to deal with missing data, including the removal of loci (that are missing too many species) and/or the removal of species (that are missing from too many loci) prior to species tree estimation. Whether or not such strategies are helpful depends on the model condition (e.g., the level of ILS) and the method [511]. There is also a great deal of recent literature exploring the impact of missing data and filtering strategies on biological data sets (e.g., [314, 731]), although the results of these studies can be difficult to interpret as the true species tree is unknown.

Another major challenge, at least for summary methods, is gene tree estimation error. The summary methods discussed here (ASTRAL and ASTRID) interpret gene tree discord as being due to ILS only. This is certainly not the case in practice, as gene trees computed for many recent biological studies are characterized by low bootstrap support values (table 1 in [511]). Many simulation studies have shown that the accuracy of summary methods is negatively affected by gene tree estimation error [505]. Furthermore, while summary methods have been shown to outperform concatenation analyses when the level of gene tree estimation error is sufficiently low and the level of ILS is sufficiently high, concatenation analyses (despite being statistically inconsistent) have been shown to outperform summary methods even under conditions with high ILS when the level of gene tree estimation error is sufficiently high [511]. As discussed in section 2.2.2.1, gene tree estimation error is expected to be a problem for many large multilocus data sets, so that summary methods may not always be appropriate for multilocus species tree estimation [264, 716].

Efforts to improve gene tree estimation in a multilocus setting would greatly enable the utilization of summary methods. Statistical binning [503] and its improvement, weighted statistical binning [55], can be seen as attempts to improve gene tree estimation (and thereby species tree estimation). Site-based methods may indeed provide powerful alternatives to summary methods, and more studies are needed to evaluate site-based methods, such as SVDQuartets, on large multilocus data sets. Coestimating the gene trees and species trees using *BEAST or similar techniques is likely to provide the best results (as well as estimates of uncertainty in the species tree); however, as discussed in this chapter, coestimation methods that can scale to large multilocus data sets have not yet been developed.

Given all of these challenges to large-scale species tree estimation, the quantification of uncertainty in large species trees is an important direction for future research. Computationally efficient methods for estimating branch support are greatly needed, especially methods that can address alignment uncertainty for site-based methods, gene tree uncertainty for summary methods, model misspecification, rogue taxa, and missing data. Another important direction for future research is phylogenetic network estimation, as hybrid speciation and horizontal gene transfer must be modeled as networks rather than trees. Unfortunately, estimating phylogenetic networks is far more computationally intensive than estimating species tree. Although the divide-and-conquer approaches presented here are not directly relevant to phylogenetic network estimation, other divide-and-conquer approaches have been developed (surveyed in [859]), and one of these techniques has been recently integrated into PhyloNet [749], a software suite for phylogenetic network estimation.

Finally, this chapter focused on the major methods in large-scale species tree estimation (and especially those for which statistical consistency has been established), but there are many other methods that can be used to estimate phylogenies from multilocus data sets. Three of the interesting ones to consider are Bayesian posterior probabilities (BPP) [602], revPoMo [662], and *guenomu* [165]. BPP is a Bayesian method for estimating the species tree from gene alignments that addresses gene tree heterogeneity due to ILS; as shown in [602], BPP has provided good accuracy on relatively small simulated and biological data sets. The next method, revPoMo, is an improvement on the earlier PoMo [164] method and is a maximum likelihood method that addresses polymorphism within a population. PoMo and revPoMo have not been established to be statistically consistent under the MSC, but these approaches address one of the main limitations of standard concatenation analyses (which do not consider polymorphism), and the accuracy of these methods in simulation studies is promising. The last of the methods, *guenomu*, is a Bayesian supertree method that is designed to address multiple sources of gene tree heterogeneity; here, too, the design is appealing, even though statistical consistency is not currently established.

2.8 Appendix: Big-O Analysis

An introduction to *Big-O running time analysis* can be found in many places (e.g., in Wikipedia, and also in [781]), but we provide a brief example here. Let f and g be two real-valued functions (i.e., f and g both map the real numbers to real numbers), and suppose for the sake of simplicity that both map positive real numbers to positive real numbers. Note that f is the name of the function and $f(n)$ is the value of the function on input value n.

We say that f is "Big-O" of g (written "f is $O(g)$") if there is a pair of constants C, C' such that $f(n) \leq Cg(n)$ for all $n \geq C'$. In other words, when n is large enough, then $f(n)$ is bounded from above by $Cg(n)$. Thus, Big-O ignores multiplicative or additive constants and provides an upper bound on the growth of a function.

Some examples may help clarify this definition. For example, if $f(n) = 5n + 1000$ and $g(n) = n^2$, then $f(n) \leq 6g(n)$ if $n \geq 1000$, and so f is $O(g)$. But we would also say that f is $O(h)$, where $h(n) = n$, since $f(n) \leq 6h(n)$ for large enough n. We can also express this by saying f is $O(n)$, which is easier to parse.

Now consider two functions $f(n) = 5n$ and $g(n) = n - 3$. Is it the case that f is $O(g)$? The answer is yes, since $f(n) \leq 6g(n)$ for n large enough. Note therefore that f is $O(g)$ in this case, even though $f(n)$ is always strictly bigger than $g(n)$. Also, $n^2 + 1000n$ is $O(n^2)$, but it is also $O(n^3)$ and $O(n^4)$. Thus, saying that a function f is $O(g)$ only expresses that $f(n)$ is asymptotically *bounded from above* by some constant times $g(n)$, and the key points are that it is asymptotic (so only depends on being true for large enough values of n) and that it is an upper bound.

Hence, when we talk about running time analysis using Big-O notation, we try to use the tightest and simplest upper bound we can. Thus, we will prefer to say that $f(n) = 3n^2 + 1000n$ is $O(n^2)$ rather than $O(n^3)$ (although both statements are true) because $O(n^2)$ is a tighter upper bound than is $O(n^3)$. Similarly, we will not want to say $f(n)$ is $O(3n^2)$ (even though that's true) because it is not as simple as saying $f(n)$ is $O(n^2)$. Thus, we aim for a tight upper bound that is simple to express, noting again that the definition of Big-O doesn't care about multiplicative or additive constants.

This definition of "Big-O" expresses the running time as a function $f(n)$, where n is the size of the input. For example, suppose that we have two different methods that both take an array of n integers as input. We will consider every operation (i.e., numerical operations, logical operations, reads, writes, etc.) as having the same cost (yes, this is not really true, but it simplifies the analysis and is how running times are generally analyzed). Then, if method A performs $n^2 + 3n + 10,000$ operations, we can say that method A has $O(n^2)$ running time (because the most important part of the running time is the n^2 part); similarly, if method B performs $1000n$ operations, then method B has $O(n)$ running time. From a "Big-O" perspective, it is easy to see that method B is more scalable than method A, since for large enough values of n the running time for B will be lower than the running time for A (just think about the difference between $1000n$ and n^2 when $n = 1,000,000$!). Importantly, asymptotic running time does not depend on experimental details (e.g., the language used to implement a method or the computer system used to run the analyses) and enables a comparison of methods based on the size of the input.

CHAPTER 3

Species Tree Estimation Using ASTRAL: Practical Considerations

Siavash Mirarab

3.1 Introduction

Understanding gene trees as entities evolving within species trees, the framework nicely summarized by Madisson [455] has given statisticians a powerful model to approach genome-wide phylogenetic reconstruction. Genome evolution can be understood using a hierarchical generative model (figure 3.1a): gene trees are first sampled from a distribution defined by a model of gene evolution and parameterized by the species tree; then, sequences are sampled from distributions defined by a model of sequence evolution and parameterized by the gene trees and other necessary parameters. The choice of the exact model of sequence evolution and the model of gene tree evolution defines the exact hierarchical model.

A leading model of gene evolution is the multispecies coalescent (MSC; see [175] and chapter 1 of this book). MSC models incomplete lineage sorting (ILS) and the resulting discordance between gene trees and the species tree (figure 3.1b). Note that I use terms *gene* and *locus* interchangeably to refer to a recombination-free region of the genome (not functional genes). The MSC model is widely adopted owing to its perceived biological realism and mathematical convenience. Under MSC, the species tree is identifiable from a distribution of gene trees [11], giving us hope to recover the species tree from gene trees.

Several approaches exist for inferring the species tree given multigene sequence data under MSC. Concatenating sequences from all loci and performing maximum likelihood (ML) inference under a model of sequence evolution (figure 3.2) amounts to ignoring the gene evolution component of the hierarchical model (figure 3.1) and is proved [624] not to be statistically consistent. This inconsistency, predicted earlier [385], has motivated the development of alternative MSC-based approaches.

Given the hierarchical nature of the model, the most statistically principled approach is to coestimate gene trees and the species trees as part of one joint inference (figure 3.2). Methods of coestimation have been developed, mostly using Bayesian Markov chain Monte Carlo (MCMC) to sample the distributions defined by the hierarchical model (e.g., [435, 298]) and have been shown in simulations to have good accuracy under the MSC model [56, 550]. These methods, however, need to sample a vast number of parameters: topologies of the species tree and all gene trees, their branch lengths, sequence evolution parameters (including rates of evolution), and population size.

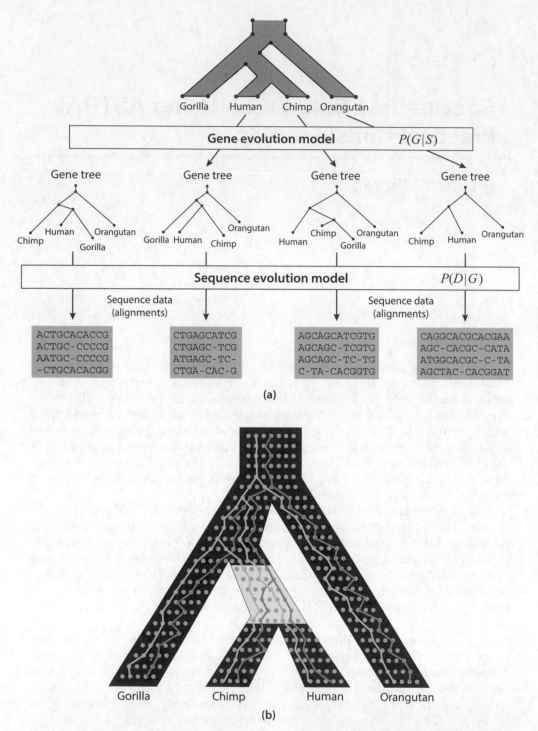

Figure 3.1. (a) The hierarchical model of genome evolution: the species tree parameterizes a model of gene tree evolution; gene trees, sampled from this model, parametrize a model of sequence evolution, which generates the sequences. (b) Tracing two lineages inside a species tree where each branch is a population. Dark grey lineages coalesce in ways that match the species tree topology. Light grey lineages fail to coalesce in the common ancestor of Human and Chimpanzee (light shaded population), giving the lineage from Chimpanzee a chance to coalesce with Gorilla before coalescing with Human—and creating incomplete lineage sorting (ILS). Gene tree = G, species tree = S, and data = D.

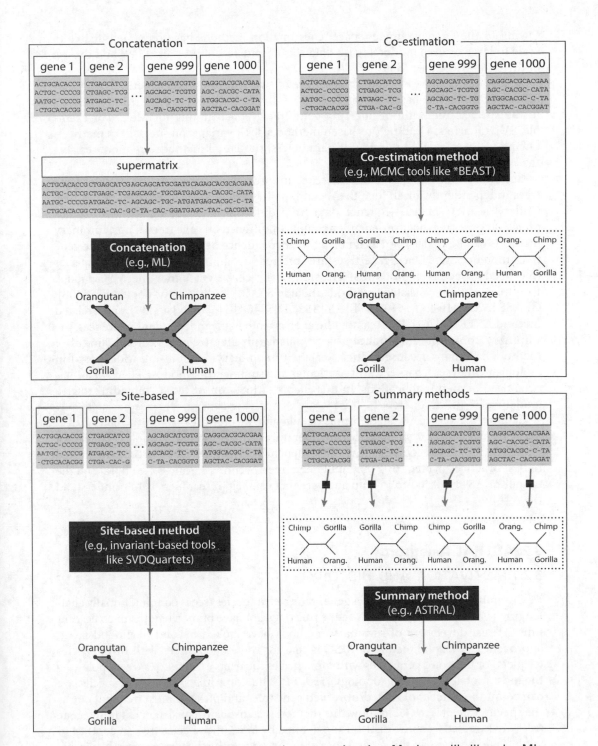

Figure 3.2. Four main approaches to species tree estimation. Maximum likelihood = ML; Markov chain Monte Carlo = MCMC.

Owing to the large parameter space, coestimation methods have remained unable to scale to large or even moderate-size data sets despite recent progress [549] and the use of divide-and-conquer [864].

Scalable alternatives to coestimation are of two types: summary methods and site-based methods. Site-based methods (e.g., [140, 98, 164]) go directly from gene data to the species tree, without inferring gene trees yet accounting for MSC. For example, SVDQuartets, a leading site-based method, uses invariants on site pattern matrices. Owing to their reduced number of parameters, site-based methods are more scalable than coestimation (chapter 2).

Summary methods divide the inference into two steps (figure 3.2); first, infer gene trees independently for all loci, then, combine these gene trees to get the species tree. Under the MSC model, sequence data from different genes are independent *conditioned* on gene trees but are dependent if not conditioned on gene trees. Thus, summary methods can be understood as ignoring the dependence between gene loci in the gene tree inference step. Once gene trees are inferred, combining them to infer a species tree needs specific methods that are statistically consistent under the MSC model. Examples of such consistent methods include STAR [442], BUCKy-population [405], GLASS [522], MP-EST [440], STELLS [816], DISTIQUE [656], NJst [439], and a related method ASTRID [761]. By breaking the analysis into many independent inferences, the summary approach can produce a very scalable pipeline (requires careful choices of methods). Perhaps because of their scalability, summary methods are widely used in biological analyses (chapter 2). In particular, the summary method ASTRAL [506] has been used in many publications. In this chapter, I focus on ASTRAL, intending to give guidelines to practitioners.

Section 3.2 overviews algorithmic details and theoretical properties of ASTRAL. Sections 3.3 and 3.4 summarize the literature on the accuracy and scalability of ASTRAL. Since the accuracy of ASTRAL depends on its input, section 3.5 is dedicated to best practices in preparing the input gene trees. Sections 3.6 and 3.7 elaborate on the output of ASTRAL and follow-up analyses that can help researchers better understand the results.

3.2 ASTRAL Algorithm

3.2.1 MOTIVATION AND HISTORY

Computing the probability of a gene tree given a species tree is computationally challenging [179], especially when the gene tree does not have branch lengths in coalescent units. Thus, developers of summary methods have looked beyond likelihood-based approaches. A helpful feature of MSC is that for rooted gene trees with three species (triplets) or unrooted gene trees with four species (quartets), the species tree topology is the most probable gene tree topology [559, 14]. Thus, on triplets/quartets of species, we can count the number of rooted/unrooted gene trees and pick the most frequent one as the species tree; it is trivial to show this method is statistically consistent assuming gene trees are sampled from the distribution defined by MSC on a species tree. In contrast to triplets and quartets, in the general case of more species, the species trees can be discordant with the most likely gene trees [176, 177], a condition known as the anomaly zone.

Several methods have extended the most-frequent-gene-tree method to more species by decomposing a data set of n species to all possible $\binom{n}{3}$ triplets or $\binom{n}{4}$ quartets. Larget et al. [405] suggested using Bayesian concordance factors [27] to compute the most frequent quartet tree for all possible choices of quartets and then combining the quartets using a quartet-joining method [453]. More recently, Sayyari and Mirarab [656] derived a consistent distance estimate between pairs of species based on how many times they are sisters among all possible quartets that include the two species of interest. Instead of finding the highest frequency gene tree, Liu et al. [440] defined the pseudolikelihood of the species tree by decomposing it into all possible triplets, computing the likelihood for each triplet, and combining the likelihoods by assuming independence. ASTRAL, too, decomposes gene trees to quartets.

The main insight behind ASTRAL is to realize that the solution to the following optimization problem is a consistent estimator of the species tree (easy to prove based on results of [12]). Let $Q(T)$ be the set of all quartet tree topologies induced by a tree T.

Maximum quartet support species tree (MQSST). *Given a set of k unrooted gene tree topologies G on (subsets of) n species, find the species tree T^* that shares the maximum total number of quartet trees with the set of gene trees. That is, find $T^* = arg\ max_T S(T)$ where*

$$S(T) = \sum_{G \in G} |Q(T) \cap Q(G)|. \tag{3.1}$$

MQSST has been studied even before its connection to MSC was realized. The problem is NP-hard in several variations [725, 347, 396], but heuristic solutions exist (e.g., [38]). One way to achieve scalability is to define a constrained version of the problem.

Constrained MQSST. *Solve the MQSST problem such that every branch (i.e., bipartition) of the species tree is drawn from a given set X of possible branches.*

Bryant and Steel [99] were the first to define this problem (to my knowledge), which they solved using dynamic programming in time that grows as $O(n^5 k + n^4 |X| + |X|^2)$. ASTRAL uses a dynamic programming algorithm similar (but not identical) to theirs [99], with an improved running time (I was unaware of their method at the time of my original publication). Crucially, solutions to constrained MQSST are consistent estimators under the MSC model [506].

3.2.2 ASTRAL ALGORITHM

ASTRAL has three published versions: ASTRAL, ASTRAL-II, and ASTRAL-III, and most recently a parallel implementation, ASTRAL-MP [838]. Below, when not otherwise specified, I discuss ASTRAL-III. Readers not interested in mathematical and algorithmic details can skip this section.

3.2.2.1 Weight Calculation and Dynamic Programming

A node in a binary (or multifurcating) unrooted tree T corresponds to a partition of leaves into three (or more) parts (figure 3.3a). Thus, a binary (multifurcating) tree can be represented as a set of tripartitions (multipartitions), one per node (figure 3.4). The ASTRAL algorithm is based on three insights:

1. The number of quartet trees shared between two trees equals half the sum of the number of quartet topologies shared among all pairs of tripartitions/multipartitions, one from each tree.

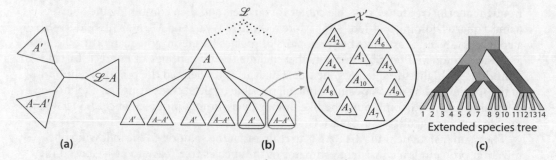

Figure 3.3. (a) An internal node in an unrooted tree creates a (tri)partition of leaves. (b) The dynamic programming recursively divides each cluster A into smaller clusters, drawing possible subsets from the set $\mathcal{X}$. (c) An extended species tree; species terminal branches are in dark grey; individuals are added as polytomies.

2. The number of quartet topologies shared between a tripartition and a multipartition can be computed efficiently *without* listing all $\binom{n}{4}$ quartet topologies.
3. Dynamic programming can be used to find a set of tripartitions that can be combined into a fully binary tree and in total have the maximum possible number of shared quartets with gene trees.

Let the set of species be $\mathcal{L}$. Let also $\mathcal{N}(T)$ be the set of internal nodes in a tree T, represented as multipartitions. Any (species) tree that includes a tripartition P as a node shares a certain number of quartet topologies with any (gene) tree that includes a multipartition M as a node; we let $QI(P, M)$ denote this quantity (figure 3.4). We define the *weight* of a tripartition as

$$w(P) = \frac{1}{2} \sum_{G \in \mathcal{G}} \sum_{M \in \mathcal{N}(G)} QI(P, M) . \tag{3.2}$$

Insight 1 asserts that $S(T) = \sum_{P \in \mathcal{N}(P)} w(P)$. Thus, we need to (1) compute $w(P)$ efficiently, and (2) find the tree with the maximum sum of $w(P)$ values. Zhang et al. [851] derived an efficient formula for QI. Given a multipartition $M = M_1 | \ldots | M_d$ (representing an internal node in a gene tree) and a tripartition $P = P_1 | P_2 | P_3$ (an internal node in a species tree), for $1 \leq i \leq d$ and $1 \leq j, k \leq 3$, let $I(i, j) = |M_i \cap P_j|$, $S(j) = \sum_{i=1}^{d} I(i, j)$, and $R(j, k) = \sum_{i=1}^{d} I(i, j) I(i, k)$. Let $\left(\begin{smallmatrix} 2 & 1 & 1 \\ 3 & 3 & 2 \end{smallmatrix} \right) = (h_{i,j})$ be a constant matrix. Then,

$$QI(P, M) = \frac{1}{2} \sum_{i=1}^{d} \sum_{j=1}^{3} \binom{I(i,j)}{2} \left(\left(S(h_{1,j}) - I(i, h_{1,j}) \right) \left(S(h_{2,j}) - I(i, h_{2,j}) \right) \right.$$

$$\left. - R(h_{1,j}, h_{2,j}) + I(i, h_{1,j}) I(i, h_{2,j}) \right) . \tag{3.3}$$

Computing this equation requires $\Theta(d)$ time given $I(j, i)$ values. ASTRAL-III uses a polytree data structure to represent gene trees such that each $I(j, i)$ can be computed in constant time. The data structure also compresses nodes that appear in multiple gene trees. Zhang et al. [851] showed how to compute $w(P)$ in $\Theta(D)$ where $D = O(nk)$ is the sum of the cardinalities of unique partitions observed in all gene trees.

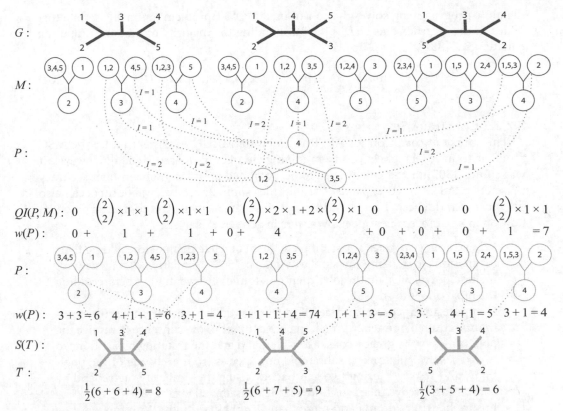

Figure 3.4. Insights behind ASTRAL. A set of gene trees *G* can be decomposed to a set of partitions *M* (tripartitions when input is binary), each corresponding to an internal node. A potential tripartition of the species tree *P* is scored against each gene tree partition *M* to compute the number of quartet topologies resolved identically by both ($QI(P, M)$). For each mapping from parts of *P* to *M* (six bijections are possible if *M* is a tripartition), we compute intersection sizes (*I*) between mapped parts (e.g., I_1, I_2, I_3). Then, the number of quartets shared by this mapping is $\binom{I_1}{2}I_2 I_3 + I_1\binom{I_2}{2}I_3 + I_1 I_2\binom{I_3}{2}$ and $QI(P, M)$ is the sum of this quantity over all mappings. Shown are all mappings with nonzero values (here, no more than one mapping has nonzero $QI(P, M)$). The weight $w(P)$ of tripartition *P* is the sum of all of its $QI(P, M)$ values. Each species tree (*T*) has $n-1$ tripartitions, and half the sum of the weights of its tripartitions give its total quartet score. Here, the topology in the middle has the highest score.

To maximize $S(T)$, ASTRAL recursively divides $\mathcal{L}$ into two smaller subsets (called *clusters*) until it reaches leaves (figure 3.3b). Each cluster is divided such that the total sum of the weights below it is maximized among allowable divisions. This recursive method solves the MQSST problem optimally if all ways of dividing a cluster *A* into $A' \subset A$ and $A \setminus A'$ are examined. But that approach has an exponential running time, hence, the need for the constrained MQSST problem. Assume we have defined a set $\mathcal{X}$ of allowable bipartitions for the species tree. Let $\mathcal{X}' = \{A : A|\mathcal{L} \setminus A \in \mathcal{X}\}$ and $Y = \{(C, D) : C \in \mathcal{X}', D \in \mathcal{X}', C \cap D = \emptyset, C \cup D \in \mathcal{X}'\}$. Kane and Tao [354] showed $|Y| < |\mathcal{X}'|^{1.726}$. We restrict the dynamic programming such that $(A', A \setminus A') \in Y$ (Figure 3.3b). Let $S^*(A)$ be the score for an optimal subtree on cluster *A*. Then, the following dynamic programming solves the constrained MQSST problem optimally in time that scales in the worst case as $O(D|\mathcal{X}|^{1.726})$, spending the majority of time in computing equation 3.2 [851]:

$$S^*(A) = \max_{(A', A\setminus A')\in Y} S^*(A') + S^*(A \setminus A') + w(A'|A \setminus A'|\mathcal{L} \setminus A). \tag{3.4}$$

3.2.2.2 Constraint Set

The sufficient condition for ASTRAL to be statistically consistent under the MSC model is to have all bipartitions from input gene trees in the set $\mathcal{X}$ [506]. However, it was shown [507] that $\mathcal{X}$ might need to be expanded in order to obtain high accuracy in practice. ASTRAL-III expands $\mathcal{X}$ using rules summarized below, but users can also directly expand the set. These heuristics rely on a similarity matrix computed based on how often a pair of species appear as sisters in gene tree quartets.

- When input gene trees are incomplete, first complete them before adding their bipartitions to $\mathcal{X}$ using the similarity matrix [507]. Similarly, when gene trees include polytomies, first resolve their polytomies in several ways before adding the bipartitions to $\mathcal{X}$ [851].
- Compute greedy consensus trees of gene trees with various thresholds (the minimum required frequency for adding bipartitions to the consensus), resolve the polytomies in the greedy consensus trees, and add the resulting bipartitions to $\mathcal{X}$. To resolve polytomies, subsample one species from each side of the polytomy, and resolve it using two approaches: using the similarity matrix and by computing greedy consensus trees on the subsampled taxa.
- Ensure heuristics do not add more than $O(nk)$ bipartitions. With this rule, we get running times that increase as $O(D(nk)^{1.726}) = O((nk)^{2.726})$.

3.2.2.3 Multiple Individuals

ASTRAL can easily be extended to inputs where more than one individual represents each species. Allman et al. [12] have introduced the concept of an extended species tree: start with the species tree, and for each species, add all individuals sampled from that species under it, creating polytomies when needed (Figure 3.3c). Rabiee et al. [595] have extended dynamic programming of equation 3.4 to compute the optimal extended species tree given gene trees with multiple individuals from some or all species (ASTRAL-multi). The dynamic programming is unchanged (treating individuals as taxon set $\mathcal{L}$), except for two modifications. (1) The boundary conditions need to change such that the algorithm stops as soon as a cluster equals the set of individuals of a species. (2) Set $\mathcal{X}$ needs to change such that each cluster has either all or none of the individuals of each species. Satisfying this condition required new methods for building $\mathcal{X}$. While it was developed to handle multiple alleles of the same gene, Legried et al. [418] have shown that ASTRAL-multi is statistically consistent under models of gene duplication and loss (GDL). Most recently, Markin and Eulenstein [469] showed that ASTRAL-multi is consistent under the DLCoal [604] model of ILS+GDL.

3.2.3 SUMMARY OF KNOWN THEORETICAL RESULTS RELATED TO ASTRAL

Consistency—general. All versions of ASTRAL give a statistically consistent estimator of the species tree if input gene trees are sampled randomly under the multispecies coalescent model (i.e., with no gene tree error, no sampling bias, and no model violations).

Consistency—missing data. Nute et al. [543] showed that ASTRAL remains statistically consistent when species are allowed to be missing from gene trees. For the exact version, key required assumptions are that the presence of a gene for a species should be independent of the gene tree topology and presence of other genes for that species. The default (constrained) version is also consistent if each clade of the species tree has a *nonzero* chance of having no missing data in each gene.

Inconsistency—estimated gene trees. Roch et al. [622] have proved that ASTRAL and other "reasonable" summary methods that use gene tree topology are statistically inconsistent if each gene has limited length and gene trees are computed using ML. Under specific conditions, they show ASTRAL and even partitioned ML fail because of long branch attraction even if there is *no gene tree incongruence*.

Inconsistency—Reticulation. Solís-Lemus et al. [713] have shown that ASTRAL can be statistically inconsistent under certain conditions when gene trees evolve on a phylogenetic network (thus, with a combination of ILS and gene flow).

Sample complexity. Shekhar et al. [674] have shown that the number of genes required by the exact version of ASTRAL to compute the correct species tree with high probability grows quadratically with the inverse of the shortest branch length and grows logarithmically with the number of species.

3.3 Accuracy

Versus concatenation. Simulation studies have indicated that the relative performance of concatenation and ASTRAL depends at least on two factors: the amount of true gene tree discordance (ILS) and the amount of gene tree estimation error (e.g., [506, 507, 268, 511]). For example, simulations [507] found ASTRAL to be more accurate than concatenation using ML (CA-ML) when true gene tree discordance (i.e., level of ILS) was high (figure 3.5a) or when gene tree error was relatively low (figure 3.5b). In contrast, CA-ML was more accurate when either the true gene tree discordance was low or when the discordance was moderate or high but the gene tree error was also high. The two methods had similar accuracy when ILS levels were moderate, *and* gene tree error was also moderate; for example, when normalized [620] (Robinson-Foulds) distance between true and estimate gene trees was between 20% to 40% (figure 3.5b). Moreover, a separate experiment [162] found ASTRAL outperforms CA-ML in the presence of both ILS and horizontal gene transfer (HGT) (figure 3.5f).

In practical terms, when gene tree discordance is very low or when gene tree error is expected to be high, CA-ML may be preferable to ASTRAL whereas in other scenarios ASTRAL is preferable. Because neither method universally dominates the other, it seems wise to use both methods and compare the results. One way to decide is to simulate data emulating real data and compare methods. For example, Giarla and Esselstyn [268] found that for their data set of tree shrews, both ASTRAL and CA-ML produced one wrong branch in simulations that emulated the real data, but only CA-ML had high bootstrap support for the wrong branch. Ballesteros and Sharma [44] showed in simulations that ASTRAL could recover the correct branch even when a branch does not appear in *any* of the input gene trees.

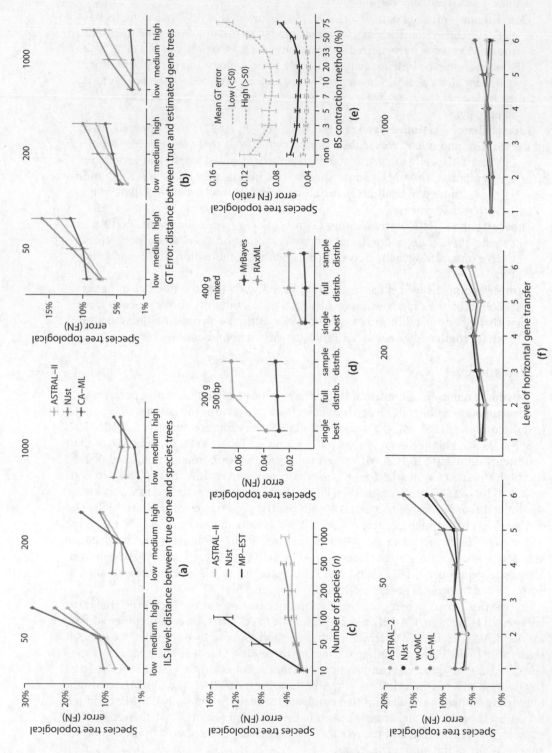

Figure 3.5. Legend on next page.

Versus other summary methods. Several simulation studies compare summary methods [e.g., 506, 507, 656, 511, 761], including some that do not involve developers of ASTRAL [268, 44]. Overall, the accuracy of ASTRAL has compared favorably with alternative ILS-based summary methods such as NJst [439], ASTRID [761], MP-EST [440], wQMC [38], as well as consensus and supertree methods such as greedy consensus, MulRF [132], and MRP [597]. For example, ASTRAL outperformed NJst by small but consistent margins in ILS simulations [507] (figure 3.5a–c) and in simulations with HGT [162] (figure 3.5f) and was essentially tied with ASTRID in [511] and [761]. DISTIQUE was close to ASTRAL but not any better [656]. ASTRAL dominated MP-EST [506, 507], especially with large numbers of species (figure 3.5c). Giarla and Esselstyn [268] showed ASTRAL outperforms MulRF in terms of topological accuracy on simulations that match their real data set of tree shrews. Beyond accuracy on conditions that seek to emulate real data, Shekhar et al. [674] have compared ASTRAL with NJst in idealized cases in terms of data requirements: the number of error-free genes required to recover the correct species tree with high accuracy. Their simulations showed mixed patterns: out of three true species trees tested, they found ASTRAL required fewer genes in two cases (with only short branches), and NJst required fewer genes in the third case (with long basal branches).

Figure 3.5. Accuracy of ASTRAL in simulations. Shown is the species tree error, defined as the proportion of branches in the true tree missing from the estimated trees; i.e., False Negative (FN) rate. (a–c) Simphy simulations by [507] comparing concatenation (CA-ML) and summary methods. (a) The level of incomplete lineage sorting (ILS) is set to low, medium, or high by adjusting tree height (10^7, 2×10^6, 5×10^5 generations) and affects mean Robinson-Foulds (RF) distance between true species trees and true gene trees (9%, 34%, 68%). Speciation rate: 10^{-7}; number of genes: 50, 200, or 1000 (boxes); 201 species. (b) Replicates in (a) are categorized into three sets based on the mean normalized RF distance between true gene trees and gene trees estimated by FastTree: $[0, 25]$%, $(25, 40]$%, $(40, 100]$%, corresponding to low, medium, and high gene tree (GT) error. (c) Error versus the numbers of species for medium levels of ILS (2×10^6 height) and speciation rate 10^{-6} with 1000 genes. (d) Mammalian-like simulations from [506] with 37 species with 200 gene (g) trees (500bp) or 400 gene trees (mix of 500bp and 1000bp). Gene trees estimated using RAxML or MrBayes. Input: *single best* tree per gene (ML for RAxML or maximum credibility for MrBayes); *full distribution* per gene (200 bootstrapping [BS] replicates for RAxML or a sample of 200 trees for MrBayes); *sample distributions* to get a single tree per gene and repeat 200 times to report their consensus (aka MLBS). MrBayes results are unpublished. (e) Simulations with 100 species and moderately high ILS (46% RF between true gene trees and species trees) by [851]. Contracting branches with ≤20% BS support (threshold of contraction shown on x-axis) from best ML trees reduces error. Dividing replicates into high and low gene tree error shows that contraction helps mostly in the high error case. (f) 50 species simulations by [162], comparing accuracy of methods in presence of horizontal gene transfer (HGT). All six model conditions (x-axis) have ILS but differ in level of HGT, ranging from no HGT (1) to very high (6). Thus, true gene tree discordance varies: ~33% (1–3), ~45% (4), ~55% (5), and ~70% (6). CA-ML is the least robust and ASTRAL is the most robust to HGT.

Other results. Simulation studies have tested the robustness of ASTRAL to factors such as HGT (figure 3.5f), gene flow [713], gene tree error [55], and missing data [511] (discussed later). Rabiee et al. [595] studied the relative impact of increasing the number of loci or the number of individuals per species on ASTRAL accuracy and found more loci to be far more beneficial. Beyond simulations, researchers have also compared ASTRAL with other methods on empirical data (e.g., [268, 682, 221, 493, 731, 675]). Since the ground truth is not known on real data, these results are harder to interpret and cannot be easily summarized without loss of important nuance. Referring the reader to these publications, I note that overall, the performance of ASTRAL on real data has been positive.

3.4 Running Time

The running time of ASTRAL has improved through its three versions, both in theoretical guarantees of worst-case asymptotic running time and in empirical measures. ASTRAL-I had guaranteed polynomial running time but is the slowest version. ASTRAL-II did not guarantee polynomial running time but was faster than ASTRAL-I. The current version, ASTRAL-III, has an asymptotic worst-case running time of $O(D(nk)^{1.726})$, which itself is $O((nk)^{2.726})$ (recall that D is the sum of degrees of *unique* nodes in input gene trees). The theoretical running time, thus, is a function of n, k, and the amount of gene tree discordance, which controls both the search space and D. In practice, on simulated data sets [851], the empirical running time of ASTRAL-III seems to increase with $n^2 k^2$. Thus, for example, a researcher planning to double the number of gene trees can expect a fourfold increase in the running time. Similarly, other things equal, increasing the number of species should roughly quadruple the running time.

Example numbers may be instructive. Zhang et al. [851] report that ASTRAL-III took roughly 16 hours on average for a data set with very high ILS, 48 species, and 16,000 genes (so, a large-k scenario) and roughly 9 hours on a data set with moderate ILS, 5000 species, and 1000 genes (so, a large-n dataset). For a comparison of the running time of ASTRAL to other methods, see chapter 2.

ASTRAL-MP. Yin et al. 2019 [838] have recently developed a parallel version of ASTRAL for CPU (multicore and vectorization) and GPU that can analyze very large data sets. This version, called ASTRAL-MP, speeds up runs by up to $150\times$ compared with ASTRAL-III, especially for data sets with large numbers of gene trees. On a data set with 10,000 species, 1000 gene trees, and moderate ILS, ASTRAL-MP takes between 5 to 32 hours (11 hours on average) given a single GPU and 24 cores with AVX2. On a real insect transcriptomic data set with 144 taxa and 1478 genes, each with 100 bootstrapped gene trees ($k = 147,800$ in total), ASTRAL-MP with four GPUs and 24 cores finished in 35 hours.

3.5 Input to ASTRAL: Practical Considerations

The ideal input to coalescent-based methods is a set of perfectly aligned orthologous regions present in all genomes, with each region small enough to avoid recombination but large enough to have a strong phylogenetic signal and with regions distributed randomly across the genome and placed far enough apart to make them fully unlinked. For summary methods, gene trees are ideally estimated under models of sequence evolution that are correct (but not overly parameterized) by means of consistent methods that utilize the data efficiently (i.e., have optimal sample complexity). Satisfying all these requirements is hard, if not impossible. Thus, phylogenomic projects seeking to use summary methods like ASTRAL face many practical choices.

In practice, phylogenomic analyses include many steps before arriving at gene tree and species tree estimation (figure 3.6). From sample preparation to sequencing, assembly, and annotation to orthology detection and multiple sequence alignment (MSA), the pipeline includes steps that are far from trivial [574]. Each step is error prone, and some steps (e.g., orthology detection, and MSA) seek to solve computational problems that are incidentally best solved with the knowledge of the phylogeny.

Several groups have discussed error propagation through steps of the pipeline and the impact on the species tree [e.g., 562, 503, 264, 574, 511]. We can aspire to move away from a pipeline approach and toward a unified statistical inference, with full join modeling of uncertainty [742]. Since this end-to-end coestimation remains unavailable currently and likely impractical in the near future, we are left having to deal with pipelines, which requires awareness of errors and making an effort to mitigate their effect. Below, I discuss best practices that have emerged from published work in preparing the input to ASTRAL.

3.5.1 GENE TREE ESTIMATION

3.5.1.1 Gene Tree Uncertainty

The standard input to ASTRAL is ML gene trees inferred under standard models of sequence evolution. Restricting ourselves to ML, several options are available.

bestML: The most straightforward choice is to use the tree with the best likelihood found by a heuristic ML method. The bestML input is the most natural approach but ignores gene tree uncertainty.

Contracted bestML: Each gene tree is bootstrapped, and support values are computed for bestML trees. Then, branches with extremely low support in bestML trees are contracted, and the resulting multifurcating trees are used as input to ASTRAL (one per gene).

MLBS: Multilocus bootstrapping (MLBS) seeks to model uncertainty by performing bootstrapping for each gene. Then, these bootstrapped gene trees are used to create several inputs to ASTRAL (with or without gene resampling); running ASTRAL on each input set produces a set of outputs, which are then summarized using methods such as greedy consensus to generate a final consensus result.

ALLBS: All replicate bootstrapped gene trees are combined to form a single input to ASTRAL.

At least two simulation studies [504, 506] have shown that MLBS or ALLBS have lower accuracy than simply using bestML, except perhaps when only a small number of genes are available (see figure 3.5d). Sayyari and Mirarab [657] have provided an explanation. The set of bootstrapped gene trees show a higher level of gene tree discordance than do the set of bestML gene trees. The increased discordance is not biological but is a result of the lowered phylogenetic signal in bootstrapped gene alignments. This increased level of gene tree error, as we previously saw (figure 3.5b), can reduce the accuracy of ASTRAL.

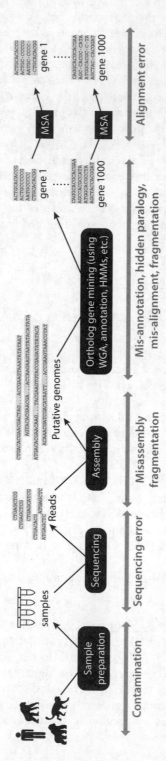

Figure 3.6. The phylogenomics pipeline. To get the typical input to coalescent-based methods, a set of gene alignments, many steps have to be taken, and each step is prone to errors that can propagate. Whole-genome alignment = WGA, hidden Markov models = HMM, and multiple sequence alignment = MSA.

Contracted bestML, in contrast to MLBS, can improve accuracy. The important (if somewhat counterintuitive) point to remember is that only collapsing branches with *extremely low* support improves accuracy, and contracting other branches can *increase* error. Zhang et al. [851] have shown that collapsing branches with BS below 5%–20% can improve accuracy by a substantial margin (figure 3.5d) and that the improvements in accuracy are higher when input gene trees have higher levels of error and when more gene trees are available. For example, for input gene trees with the mean error above 50%, the ASTRAL tree enjoys a 25% reduction in error (from 0.114 to 0.085) after contracting gene tree branches with support below 10%. Thus, contracting *very* low support branches can increase accuracy substantially. Aggressively collapsing branches with support <50% or <75% (i.e., only keeping high support branches) can substantially reduce the accuracy (figure 3.5d). Future research should explore smarter algorithms for collapsing low support branches.

Arcila et al. (2017) have suggested inferring gene trees constrained to include a set of predefined undisputed clades chosen by the researcher. They show promising results on several data sets using ASTRAL applied with such constrained gene trees. However, the method can also remove some of the real discordances among gene trees (as opposed to noise). As pointed out elsewhere [502], this approach runs several theoretical risks, including biasing result in unexpected ways.

3.5.1.2 Inference Tools and Models

The choice of the gene tree inference tool may be consequential, and published simulations have compared FastTree [586] and RAxML [720]. Despite earlier results [432], [659] have found using simulated and empirical data that FastTree *can* be less accurate than RAxML in inferring gene trees (under limited conditions they test), and the increased gene tree error leads to less accurate ASTRAL trees. Thus, using best available ML methods (ideally with multiple starting trees) is preferable to faster tools in biological analyses. Moreover, effects of misspecified sequence evolution models have been discussed for phylogenomics in general (e.g., [575, 344]), but to my knowledge have not been studied for ASTRAL. We should expect that systematic model misspecification can lead to biases in estimated gene tree distributions, which can lead to errors in the species tree.

In unpublished simulations, I have compared the ML method RAxML and the Bayesian method MrBayes [334] on a mammalian-like simulated data set [506]. Like ML, the output of MrBayes is used in three ways: using a single maximum credibility tree per gene, using a large sample of trees per gene (akin to ALLBS), and repeatedly sampling single trees from gene tree distributions produced by MrBayes (akin to MLBS). Interestingly, unlike ML and BS, the use of Bayesian distributions removes the sensitivity to the mode of input so that all three types of input perform similarly (figure 3.5d). These results also show a small advantage in using MrBayes compared with RAxML when both are used in their best setting (i.e., a single tree per gene). These preliminary results warrant more studies in the future.

3.5.2 FILTERING OF DATA

A vexing problem in phylogenomics is that curating sequence data using visual inspection is impossible. Thus, methods for *detecting* errors automatically, perhaps in downstream steps, are also needed. Empirical studies often employ several mostly ad hoc methods for filtering erroneous data [574], typically relying on a mix of visual inspection of (parts of) data and automatic error detection tools. While the extent of the negative effect of errors in input on the output ASTRAL tree is not fully understood, efforts to minimize such errors seem necessary. However, tampering with data to remove error can also remove signal and introduce bias—and thus warrants caution and careful study.

3.5.2.1 Filtering of Leaves from Gene Trees

One way of detecting abnormalities is to examine the estimated gene trees that show unexpected patterns. For example, Wickett et al. [798] rooted gene trees and detected and removed branches from gene trees with extremely long root to tip distance compared with the other species. Visual inspection of gene trees is also what Gatesy, Springer, and colleagues have used in several of their published criticism of previous phylogenomic studies [264, 681, 716, 718]. A well-studied automated approach for detecting species with unstable positions in individual gene trees is rogue taxon detection (e.g., [2, 793]). Rogue taxon detection methods tend to identify the same species (usually those on long branches) on many genes [460]. Since removing the same taxon from many genes reduces taxon occupancy, rogue taxon removal may prove problematic. The effect of rogue taxon removal on ASTRAL, to my knowledge, has not been tested.

TreeShrink. Mai and Mirarab [460] developed an automatic method called TreeShrink to find suspicious patterns of branch length in gene trees (figure 3.7a). TreeShrink tries to shrink the diameter (i.e., the maximum total branch length between any two leaves) of each gene tree successively by removing species. Then, for each species in each gene tree, TreeShrink computes a *signature* (figure 3.7b), which quantifies its impact on the diameter of the gene tree. Finally, it examines the *distribution* of signatures of each species across all gene trees and detects outliers in these distributions using a simple heuristic. Outliers is a case in which a species has an uncharacteristically large signature (e.g., is on a long branch) in a gene tree compared with the rest of gene trees. Since outliers are defined using a distribution across genes for a single species, a taxon with high signatures in most genes will not be detected as an outlier in those genes (outgroups tend to be like this). In contrast, a taxon with low signatures in most genes but high signatures in a handful of genes will be detected as an outlier for those high-signature genes. Once the abnormally long branches are detected, TreeShrink removes specific species from specific genes but does not remove the entire gene. Mai and Mirarab [460] showed on several biological data sets that TreeShrink reduces the pairwise discordance among gene trees beyond random removal of taxa (figure 3.7c) and also often beyond methods such as RogueNaRok [2]. Moreover, it avoids removing any specific species from too many genes.

3.5.2.2 Filtering of Entire Gene Trees

A more extreme form of filtering is to remove entire gene trees. Several criteria for removing entire genes have been proposed and tested (e.g., [493, 314, 136, 318, 448, 80]). Some of the criteria have to do with missing data and will be discussed later, while others relate to error and uncertainty in gene trees. Molloy and Warnow [511] give a recent summary of the literature and also, to my knowledge, give the only study that evaluates the accuracy of ASTRAL in simulation in response to removing full genes. Their simulations show that removing genes with high gene tree estimation error can improve accuracy for low levels of ILS but *reduces* accuracy for moderate to high levels

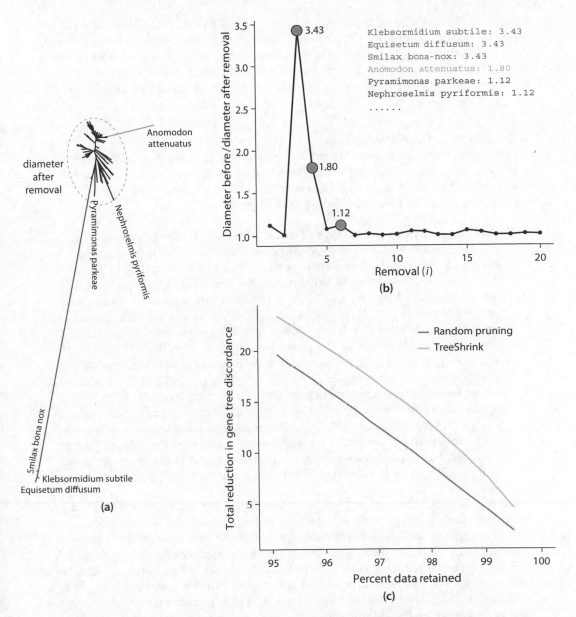

Figure 3.7. **TreeShrink** on [798] data set. (a) A gene tree with abnormally long branches. (b) TreeShrink identifies three (dark grey) tips that increase the diameter 3.4x and are likely erroneous, followed by another (light grey) that may also be problematic. (c) TreeShrink reduces discordance substantially more than random filtering. Y-axis: *reduction* in gene tree discordance defined as mean pairwise matching splits distance between gene trees.

of ILS. Nevertheless, even for ILS levels for which removing genes helps accuracy, on real data we do not have direct access to gene tree error to decide what genes to remove. Instead, we have to resort to filtering by other factors, such as bootstrap support. Since these proxies do not perfectly correlate with gene tree error, the positive impact of filtering may further diminish. Neither Molloy and Warnow [511] nor any other simulation study tested filtering by proxies for ASTRAL. Lanier and Knowles [404] and

Li et al. [438] have studied the question for STEM and MP-EST and have observed little or no reason for filtering. On empirical data sets, the conclusions have been mixed, with some studies (e.g., [314, 493, 448]) recommending removal of gene trees and others finding no evidence that filtering helps [136, 80]. Overall, there is little evidence in the literature suggesting that removing entire genes because of lack of support is helpful to ASTRAL analyses.

3.5.2.3 Filtering of Missing Data

The most common type of filtering is in response to missing data. For summary methods, two types of missing data exist—*missing genes* and *fragmentary data* (*type I* and *type II* in the parlance of Hosner et al. [314]). The two types have different consequences and have inspired two types of filtering.

Type I (missing genes). This type of missing data occurs when a gene is entirely missing for some of the species but is present in others. The only suggested filtering for these types of missing data is to remove them. Molloy and Warnow [511] found no evidence in simulations that removing genes with this type of missing data helps accuracy. Their results are in agreement with several empirical studies (e.g., [136, 314]) that also saw no benefit in filtering genes.

Type II (fragmentation). When a species includes a gene, but only partially, it introduces missing data in the gene tree inference step. Both Hosner et al. [314] and Sayyari et al. [659] showed that the presence of fragmentary data can be problematic, confirming earlier observations (e.g., [798, 716]). They show fragmentary sequences increase gene tree error, which translates to increased species tree error. Sayyari et al. [659] suggested a simple yet effective solution: remove species with fragmentary sequences from gene alignments before inferring the gene tree. The optimal level of filtering depends on the data set; Sayyari et al. [659] defined species with less than half of the total alignment length as fragmentary while Wickett et al. [798] used a one-third threshold.

To summarize, removing loci because of missing species is not recommended, but removing specific species from loci because of fragmentation is recommended. Note that filtering fragmentary data replaces type II missing data with type I. That such a trade-off helps accuracy, once again, underscores the negative effect of gene tree error and the benefit in reducing error—even if this reduction adds to missing data. Note that the distinction between types I and II is not relevant for concatenation. There is no reason to think that removing fragmentary data from a concatenation analysis could help accuracy, as it only adds missing data.

Results showing that removing type I missing data fails to help accuracy do not imply missing genes are harmless. As shown in simulations [543], missing genes can increase error in ASTRAL trees, especially when the number of genes is low, the amount of ILS is very high, or when entire clades tend to be missing. Thus, missing data can hurt accuracy, but filtering low occupancy loci is not a solution.

Recently, Gatesy et al. [262] added a twist. For a set of empirical data, given two alternative species trees, they computed the difference in quartet score of the two trees *for each gene* and called it partitioned coalescence support (PCS). Genes with extremely high PCS for either alternative tree *tend to* be a lot more complete than other genes. One problematic observation is that in some cases, ASTRAL trees change if only a couple of these high PCS genes are removed. Thus, in the presence of uneven taxon occupancy, results may be driven by a handful of genes, an observation that makes sense given the rapid growth of the number of quartets as trees become larger. These results suggest that perhaps, contrary to common wisdom, having genes with similar levels of occupancy is more important than avoiding missing data. Future work should further explore the implications of these results.

3.6 ASTRAL Output

3.6.1 SPECIES TREE TOPOLOGY AND ITS QUARTET SCORE

ASTRAL outputs the tree with the maximum quartet score among all trees within its search space (defined by $\mathcal{X}$). Since ASTRAL limits the search space (unless run with -x), it is possible that other trees with better quartet scores exist. In simulations, increasing the search space beyond the default used in ASTRAL-II or -III (e.g., by adding all bipartitions from the true tree) rarely even improves accuracy, though it occasionally improves the quartet score slightly [507, 851].

ASTRAL is a statistically consistent estimator under the MSC model given gene trees sampled from the true distribution defined under MSC. However, ASTRAL does not use a parametric model and is not tied to likelihood under the MSC model. As such, ASTRAL can be considered a nonparametric estimator. As it has been long argued [312], absent access to the correct model, reliance on nonparametric methods can be beneficial. Thus, ASTRAL (and other nonparametric methods like NJst/ASTRID) may be more robust than parametric methods (e.g., MP-EST). ASTRAL is a natural estimator for any model for which the most likely gene tree matches the species tree for quartets. Consistent with this expectation, several exciting new results have emerged that show ASTRAL is statistically consistent under some models [623] of random HGT [162], GDL [418], and even the DLCoal model of GDL+ILS [469]. Moreover, Zhang et al. [852] were able to extend ASTRAL to handle paralogs. These results, obtained years after widespread use of ASTRAL, are a testament to the power of nonparametric estimation.

Along with the tree topology, ASTRAL outputs its quartet score, which is the number of quartet trees in gene trees that are present in the species tree. We normalize the absolute value by the total number of quartet trees in input gene trees (e.g., $k\binom{n}{4}$ if there is no missing data) to give a more interpretable score. For example, a quartet score of 0.8 means that 80% of quartet trees in input gene trees are in the output tree. Thus, the normalized quartet score can be used as a measure of the amount of gene tree discordance. However, the score has to be interpreted with care as gene tree error is likely to reduce the quartet score.

3.6.2 BRANCH LENGTHS IN COALESCENT UNITS

ASTRAL estimates coalescent units (CU) lengths of all internal branches and of terminal branches corresponding to species with multiple individuals. True CU branch lengths are proportional to the number of generations spanned by the branch and inversely proportional to the population size [177]. Coalescent unit is important in MSC modeling because Coalescent unit length is what identifies the amount of topological discordance. Shorter branches lead to more discordance, especially when adjacent to each other [173]. For a quartet, if the length of the only internal branch is d in CU, the probability of a gene tree matching the species tree is $\theta = 1 - \frac{2}{3}e^{-d}$, and the probability of the two alternative topologies is $\frac{1-\theta}{2} = \frac{1}{3}e^{-d}$ (figure 3.8a). The proportion of gene

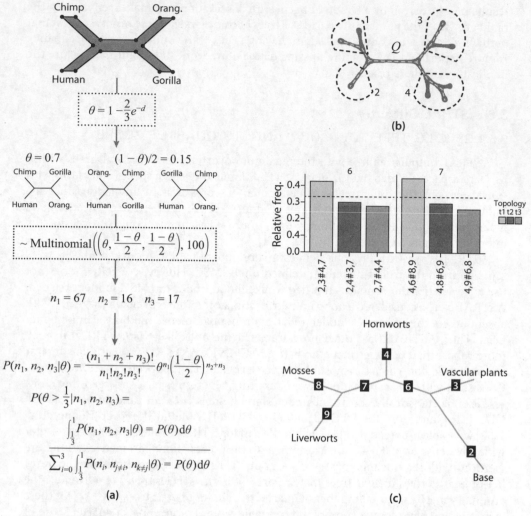

Figure 3.8. Per-branch quartet frequencies. (a) The probability of quartet gene tree topologies $(\theta, \frac{1-\theta}{2}, \frac{1-\theta}{2})$ is a function of the coalescent units (CU) branch length d. Observed counts of topologies n_1, n_2, n_3 follow a multinomial distribution, allowing us to compute the likelihood term $P(n_1, n_2, n_3|\theta)$ as shown. Assuming a prior $P(\theta)$, we can compute the posterior probability $P(\theta > \frac{1}{3}|n_1, n_2, n_3)$, as shown. (b) Branch $Q = 12|34$ can be rearranged in two ways: 13|24 and 14|24. Choosing a leaf from each group 1...4 gives a quartet *around* branch Q (e.g., white dashed). (c) Visualizing quartet support using Disco-Vista. Bars shows quartet frequencies around two internal branches (labeled 6 and 7) of the tree shown in below. Light grey bars correspond to the frequency of the main resolution shown in the tree and dark grey bars are for alternative resolutions, identified with branch lables matching the tree. (d) Accuracy of ASTRAL branch length in simulations [657] on true (gt) and estimated gene trees (alignments length: 250bp to 1500bp). Image copyright © 2016 Oxford University Press. (e) Accuracy of local posterior probability (localPP) in the same simulations. For a threshold p, branches of the ASTRAL tree and the two rearrangements around each of them ($3(n-3)$ in total) are categorized into true positives (correct, support $\geq p$), true negatives (correct, support $< p$), false positives (incorrect, support $\geq p$), false negatives (incorrect, support $< p$), and

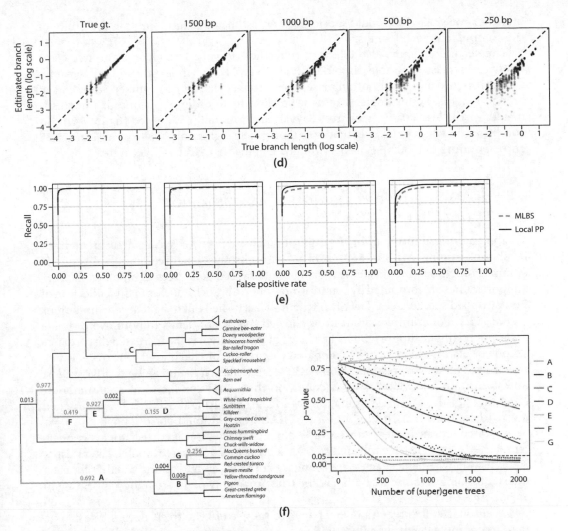

(d)

(e)

(f)

Figure 3.8. (*continued*) recall and false positive rates are computed. Exploring *p* produces the receiver operating characteristic curve, where a higher line means more true positives for each false positive rate. LocalPP dominates MLBS. (f) Testing the polytomy null hypothesis for seven recalcitrant branches (A–F) in the ASTRAL species tree computed from 2022 supergene trees of an avian data set. With more genes, p-values drop for some branches but not for others [658]. Branch E (position of enigmatic species Hoatzin) seems to be best explained by a hard polytomy.

trees that exhibit a quartet topology (f) estimates θ or $\frac{1-\theta}{2}$; thus, when $f > \frac{1}{3}$, we can use $-\ln \frac{3}{2}(1-f)$ to estimate the CU branch length.

ASTRAL exploits this observation to compute CU branch lengths (with simplifying assumptions) using a fast algorithm for computing mean quartet frequencies "around" each branch (figure 3.8b). Sayyari and Mirarab [657] have shown that despite assumptions, ASTRAL CU branch lengths are accurate *when* a sufficiently large number of true gene trees are used (figure 3.8d). ASTRAL CU branch lengths, however, suffer from two issues, which limit their usability in practice. An obvious shortcoming is that terminal branches for single-individual species lack an estimated length, limiting the utility of the computed branch lengths.

The second difficulty is the lack of robustness to gene tree error. Gene tree error tends to increase gene tree discordance; as ASTRAL branch lengths are only a function of discordance (and nothing else), gene tree error results in underestimation of branch lengths. For example, lengths were close to an order of magnitude underestimated for the least strong gene trees Sayyari and Mirarab [657] tested (figure 3.8d). In conditions in which gene trees had even moderately high resolution (e.g., 60% mean BS corresponding to 1500 bp genes), estimated lengths were relatively accurate.

3.6.3 BRANCH SUPPORT USING LOCAL POSTERIOR PROBABILITY (LOCALPP)

The traditional method for obtaining branch support for species trees is multilocus bootstrapping (MLBS) [668]. MLBS first bootstraps gene trees and then runs the summary method in replicate runs using bootstrapped gene trees as input. In the end, support values are computed by counting how often a branch appears in this collection of bootstrapped species trees. The MLBS method has turned out to have severe limitations (e.g., [683]). For example, simulations showed that MLBS tends to both overestimate and underestimate support [504]. The heart of the problem is the increased discordance among bootstrapped gene trees, compared with maximum likelihood gene trees (which themselves tend to overestimate conflict). Recall that each locus can be relatively short and lacking in informative sites, a condition that is not conducive to accurate bootstrapping [240]. To address limitations of MLBS, Sayyari and Mirarab [657] designed a way to compute branch support for ASTRAL trees without bootstrapping.

As stated repeatedly, if a branch in an estimated species tree is correct, for every quartet selected around it, the probability θ of observing the species tree topology is at least $\frac{1}{3}$. Thus, asking whether a species tree branch is correct is akin to asking whether for each quartet topology around that branch, the true probability of that topology appearing in a gene tree is at least $\frac{1}{3}$.

Given quartet frequencies in an error-free sample of gene trees, Bayes's rule can be used to compute the probability that $\theta > \frac{1}{3}$. More formally, assume the CU species tree is generated under the Yule process. Stadler and Steel [719] have shown that under this assumption, the branch lengths will be exponentially distributed. This observation helped [657] derive a prior distribution for the θ parameter. By (conditional) independence of gene trees, the number of times the three quartet topologies are observed (n_1, n_2, n_3) are drawn from a multinomial distribution with parameters $\theta, \frac{1-\theta}{2}, \frac{1-\theta}{2}$, and $\sum_1^3 n_i = k$ tries. Thus, the likelihood of our observations f_1, f_2, f_3 for a given θ can be computed using the multinomial probability mass function (see figure 3.8a). Since we have both the prior and the likelihood, we can compute the posterior probability (PP) of θ given f_1, f_2, f_3 using Bayes's rule. Recalling that a branch is correct as long as $\theta > \frac{1}{3}$, the PP of correctness of a branch is obtained by marginalizing over all $\theta > \frac{1}{3}$. Luckily, the prior according to the Yule model and the likelihood by the multinomial model are conjugate, allowing us to compute an analytical equation for the PP of four species.

Exact calculation of PP is not similarly easy for more than four species. However, with several simplifying assumptions, we can fall back to the case of a quartet. One assumption is *locality*: in computing the support for a branch, we assume that all four branches around it are correct, enabling us to consider only three rearrangements around the branch (figure 3.8b). Because of this assumption, this measure of support is called localPP. Sayyari and Mirarab [657] show in simulations that localPP is more accurate than MLBS when gene trees are inferred using short loci (i.e., from gene trees with relatively high error) and matches MLBS when gene trees are highly accurate (figure 3.8e). Moreover, localPP does not require bootstrapping gene trees and therefore is much faster than MLBS. Since its introduction, localPP has been adopted by many studies.

3.7 Follow-up Analyses and Visualization

Several analyses can follow the species tree inference to gain additional insights. These follow-up analyses can be performed on any tree, whether computed by ASTRAL or not. ASTRAL can perform the following analyses and compute localPP for any tree given to it using the -q option.

3.7.1 TESTS FOR POLYTOMIES

A central question in systematics is whether a particular branch is resolvable given the present data or, more broadly, at all. Branches that cannot be resolved are removed, resulting in polytomies. Polytomies are called *hard* when the multifurcation is biological, and no amount of data should be able to resolve it, or *soft* when the present data cannot resolve the relationships owing to lack of power. Sayyari and Mirarab [658] introduced a frequentist approach for testing the null hypothesis that a given branch in the tree has length zero and thus should be contracted. Note that under the null hypothesis, the quartet frequencies around a branch are $\frac{1}{3}$ for all three resolutions around the branch. A simple chi-squared test can be used to test this null hypothesis. However, the failure to reject the null is not the acceptance of null; thus, when the null hypothesis is not rejected, we replace the branch with a polytomy, but we cannot say if it is a soft or a hard polytomy. Sayyari and Mirarab [658] showed in simulations that the method successfully controls the false positive rate and is powerful in rejecting the null, given sufficient genes. The method also showed intriguing patterns when applied to the base on Neoaves (figure 3.8f), indicating that some (but not all) recalcitrant branches should perhaps be replaced with polytomies. The test for polytomies is implemented in ASTRAL and can be invoked using the option -t 12 (see ASTRAL documentation).

We note that tests such this should be applied on a *preselected* set of branches and may require multiple hypothesis testing using methods such as FDR [63]. However, note that unlike the standard applications of statistical tests, here we typically expect very few polytomies and thus frequent rejection of the null. Thus, to avoid losing power, it is probably best to specify in advance the branches for which a polytomy null hypothesis is reasonable and correct only with respect to those p-values. This process adds subjectivity, a problem often affecting frequentist tests.

3.7.2 PER BRANCH QUARTET SUPPORT (MEASURE OF DISCORDANCE)

Phylogenomic studies are often interested in the amount of discordance *per branch* of the species tree. The computation of branch length and localPP in ASTRAL is contingent on first computing, for each branch, its quartet support. Note that around each branch of an unrooted tree, many quartets are defined (figure 3.8b) that map to that branch and only that branch (there are $n - 3$ to $(\frac{n}{4})^4$ such quartets per branch).

The quartet support of a species tree branch with respect to gene trees (with no missing leaves) is the proportion of times that quartets around the branch are resolved identically to the species tree in the gene trees. When there are missing data, the definition becomes more tricky because several normalization schemes become possible. We use the following definition. First, we discard all genes that do not fully include *any* of the quartets around the branch. Then, for each gene, we compute what portion of its quartets supports each topology, and we compute the mean of these values over all genes. Thus, we get a number between 0 and 1 for each quartet topology around the branch.

The quartet score of a branch can be used as a measure of discordance around the branch. ASTRAL can output quartet scores for all branches of a given tree (-t 1 and -t 8). Several points are worth mentioning:

- Values close to $\frac{1}{3}$ point to very high levels of discordance. However, given a large number of gene trees, quartet scores that have relatively small divergences from $\frac{1}{3}$ (e.g., 40%) can lead to high localPP. Also, remember that discordance includes both true discordance and the effects of gene tree error.
- Under ILS, one expects quartet scores of the second and third topologies to be identical. When the two frequencies diverge substantially, ILS assumptions are violated, either during gene tree estimation (e.g., due long branch attraction) or because other biological sources of discordance (e.g., paralogy) also exist. Both cases warrant extra caution in interpretation.
- In rare occasions, a branch of the ASTRAL tree has a quartet score below $\frac{1}{3}$. This can happen for several reasons, but in all cases, the branch should be considered unresolved (will have a localPP of 0).

DisvoVista. The best way to interrogate the quartet scores produced by ASTRAL is to visualize them for branches of interest. Sayyari et al. [660] have developed a tool called DiscoVista to visualize quartet scores around important branches (and also produce other visualizations of discordance). For example, in figure 3.8c, I summarize quartet scores around two focal branches of a plant tree [798]; helpfully, DiscoVista collapses large groups into individual nodes for better visualization.

3.8 Conclusion

I reviewed the relatively substantial body of knowledge available in the literature on the ASTRAL method, including best practices for using it. I hope the reader comes away with these messages:

- ASTRAL is a statistically consistent method of species tree estimation given inputs sampled with no error under the MSC model. More generally, it is consistent under any model for which the quartet that matches the species tree is expected to occur with the highest frequency.
- ASTRAL is extremely scalable and can analyze many thousands of species.
- ASTRAL, like other summary methods, can be sensitive to gene tree estimation error, a problem that is alleviated but not eliminated if *extremely* low support branches in gene trees are contracted.

- ASTRAL has performed well in terms of accuracy in simulation analyses compared with other summary methods. The performance with respect to concatenation depends on the amount of discordance and phylogenetic signal in input genes.
- ASTRAL's native localPP is a better method of computing support than is multilocus bootstrapping.
- On real data, care is needed for preparing the input to ASTRAL, in particular, to avoid negative effects of fragmentary data. However, extensive gene tree filtering is not recommended.
- ASTRAL is statistically inconsistent under models of gene evolution that include gene flow. However, it has shown high accuracy under simulations with high levels of (randomly distributed) HGT.

CHAPTER 4

Species Tree Estimation Using Site Pattern Frequencies

David L. Swofford and Laura S. Kubatko

4.1 Introduction

Methods for estimating species trees can be divided into several classes, as described in chapter 1. The advantages and disadvantages of each class of methods are widely appreciated. For example, summary methods are known to be computationally efficient, often requiring only seconds or minutes of computation once gene trees have been estimated for each of the individual loci. However, these methods fail to account for uncertainty in the estimated gene trees, and thus their performance is dependent on the quality of the gene tree estimates used as input. On the other hand, Bayesian coestimation methods are based on probabilistic models linking the observed sequence data directly to the species tree and associated parameters, typically by explicitly incorporating the multispecies coalescent (MSC) model. These have the advantage of being fully model based, thus enabling estimation of associated model parameters. However, owing to the complexity of the underlying models, coestimation methods typically rely on Markov chain Monte Carlo (MCMC) methods for statistical inference and thus require extensive, and sometimes prohibitive, computation in order to carry out inference.

An alternative to summary methods and MCMC-based coestimation methods are methods that utilize site pattern frequencies as input to estimate the species tree under the MSC directly. In this chapter, we describe two such methods. The first, SVDQuartets, is a method for estimation of the species tree topology that does not rely on likelihood computation for estimation, thus enabling coalescent-based inference to be carried out on large-scale genomic data in a computationally efficient manner. We describe the theory underlying SVDQuartets, give some details concerning the algorithms used for estimation, and provide examples of its implementation in PAUP*. The second method we describe is designed to estimate parameters associated with a fixed species tree topology, such as the speciation times, based on a composite likelihood approach. We describe how the composite likelihood is computed as well as its use in a Bayesian context to derive statistically consistent estimators of the speciation times. Both procedures are implemented in the PAUP* software, allowing the analyses to be carried out with user-friendly software. We provide recommendations for use of these methods and carefully describe their strengths and weaknesses. Finally, we use the genome-scale data on gibbon species analyzed by Carbone et al. [109] and Shi and Yang [676] to demonstrate the performance of the methods.

4.2 Estimation of the Species Tree Topology Using SVDQuartets

4.2.1 THEORETICAL BASIS

As its name implies, SVDQuartets is a method for species tree inference that is based on the examination of quartet relationships. Thus, we begin by describing the model that SVDQuartets assumes for the relationships among quartets (collections of four taxa) under the coalescent model. Consider four species numbered 1 through 4 and note that there are three possible unrooted phylogenies relating these four species, as shown in figure 4.1. To derive the theory underlying SVDQuartets, we make the following assumptions. First, we assume that our data consist of a collection of aligned sites from the genomes of the four species under consideration. We assume that, conditional on the species tree, each site evolves independently of the other sites in the data set. Thus, each site has its own underlying gene genealogy that arises under the MSC model on the species tree. We call this data type *coalescent independent sites* (CIS). Though we argue later that SVDQuartets can be properly applied to linked sites sharing the same underlying gene tree, the theory underlying the method is motivated by the setting of CIS data, and we focus on that data type in this initial description.

Next, we assume that sequence data arise along these gene trees according to one of a set of commonly used nucleotide substitution models that includes the GTR+I+G model or any submodel thereof. We further assume that no gene flow occurs following speciation and that no processes other than the coalescent contribute to variation among gene trees; that is, we assume that no horizontal transfer, gene duplication and loss, or gene conversion has occurred. The model underlying SVDQuartets *does* allow for species trees that violate the molecular clock, and it allows for variation in the effective population sizes along the branches of the species tree.

Given species 1 through 4, we can consider the probability of observing a particular configuration of nucleotides at the tips of the tree. Let X_H denote the nucleotide observed at a particular site for species H, and let p_{ijkl} be the probability that species 1 has nucleotide i, species 2 has nucleotide j, species 3 has nucleotide k, and species 4 has nucleotide l, that is,

$$p_{ijkl} = P(X_1 = i, X_2 = j, X_3 = k, X_4 = l). \tag{4.1}$$

We call $ijkl$ a site pattern, and we refer to the collection of probabilities $\{p_{ijkl} : i, j, k, l \in \{A, C, G, T\}\}$ as the site pattern probability distribution. For any particular unrooted quartet tree, we can arrange these $4^4 = 256$ site pattern probabilities in the form of a 16×16 matrix, called a *flattening matrix*, for which the rows correspond to possible nucleotides for two of the species that form a cherry in the tree and the columns correspond to possible nucleotides for the other two species. As an example, the flattening matrix below corresponds to the leftmost tree in figure 4.1:

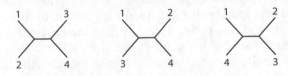

Figure 4.1. The three possible quartet trees for four taxa.

$$
Flat_{12|34} =
\begin{pmatrix}
 & [AA] & [AC] & [AG] & [AT] & [CA] & \cdots & [TT] \\
[AA] & p_{AAAA} & p_{AAAC} & p_{AAAG} & p_{AAAT} & p_{AACA} & \cdots & p_{AATT} \\
[AC] & p_{ACAA} & p_{ACAC} & p_{ACAG} & p_{ACAT} & p_{ACCA} & \cdots & p_{ACTT} \\
[AG] & p_{AGAA} & p_{AGAC} & p_{AGAG} & p_{AGAT} & p_{AGCA} & \cdots & p_{AGTT} \\
[AT] & p_{ATAA} & p_{ATAC} & p_{ATAG} & p_{ATAT} & p_{ATCA} & \cdots & p_{ATTT} \\
[CA] & p_{CAAA} & p_{CAAC} & p_{CAAG} & p_{CAAT} & p_{CACA} & \cdots & p_{CATT} \\
\cdots & \cdots & \cdots & \cdots & \cdots & \cdots & \cdots & \cdots \\
[TT] & p_{TTAA} & p_{TTAC} & p_{TTAG} & p_{TTAT} & p_{TTCA} & \cdots & p_{TTTT}
\end{pmatrix}
$$

In the above matrix, the $(3,2)$ entry, p_{AGAC}, is the probability of observing nucleotide A for species 1, nucleotide G for species 2, nucleotide A for species 3, and nucleotide C for species 4, for example.

Chifman and Kubatko [141] showed that when the flattening matrix corresponds to the true species tree (i.e., when the species used for the rows and columns of the matrix reflect a true split in the species tree), then the rank of the flattening matrix is 10 when the molecular clock holds for the GTR+I+G model and all submodels. When the flattening matrix does not correspond to a split in the true tree, then the rank of this matrix is 16. Long and Kubatko [446] extended this result to the case in which the molecular clock is no longer required and for which effective population sizes are allowed to vary along the tree under the generalized time reversible (GTR) model and all submodels. These results are important in two ways. First, they establish that the species tree is identifiable from sequence data under the MSC, which is necessary to prove consistency of likelihood-based inference methods (see [783]). Second, they provide the theoretical basis for the SVDQuartets method, which we now describe.

Given an empirical data set consisting of either a collection of SNPs or of alignments for multiple loci, we use the observed site pattern frequencies in the data to estimate the flattening matrices corresponding to each of the three unrooted four-taxon trees in figure 4.1. For each of these matrices, we then compute a measure, which we call the SVDScore, that gives the distance to the nearest rank 10 matrix. As a specific example, consider again the leftmost tree in figure 4.1 and suppose that the $Flat_{12|34}$ matrix has been estimated by substituting the observed frequency of each site pattern for the true probabilities. For this matrix, we then carry out singular value decomposition. Letting $\hat{\sigma}_j$ represent the jth singular value, we define the SVDScore as

$$
SVDScore := \sqrt{\sum_{i=11}^{16} \hat{\sigma}_i^2}, \tag{4.2}
$$

that is, the SVDScore is the square root of the sum of squares of the 11th through 16th singular values. The Eckhart-Young theorem [214] establishes that this is the distance to the nearest rank 10 matrix under the Frobenius norm. When the flattening matrix corresponds to the true quartet relationships found in the species tree, the true values of the 11th to the 16th singular values are 0, and thus their estimated values are expected to be small. What this means in practical terms is that small values of the SVDScore indicate that the quartet relationship under consideration is likely to be that found on the

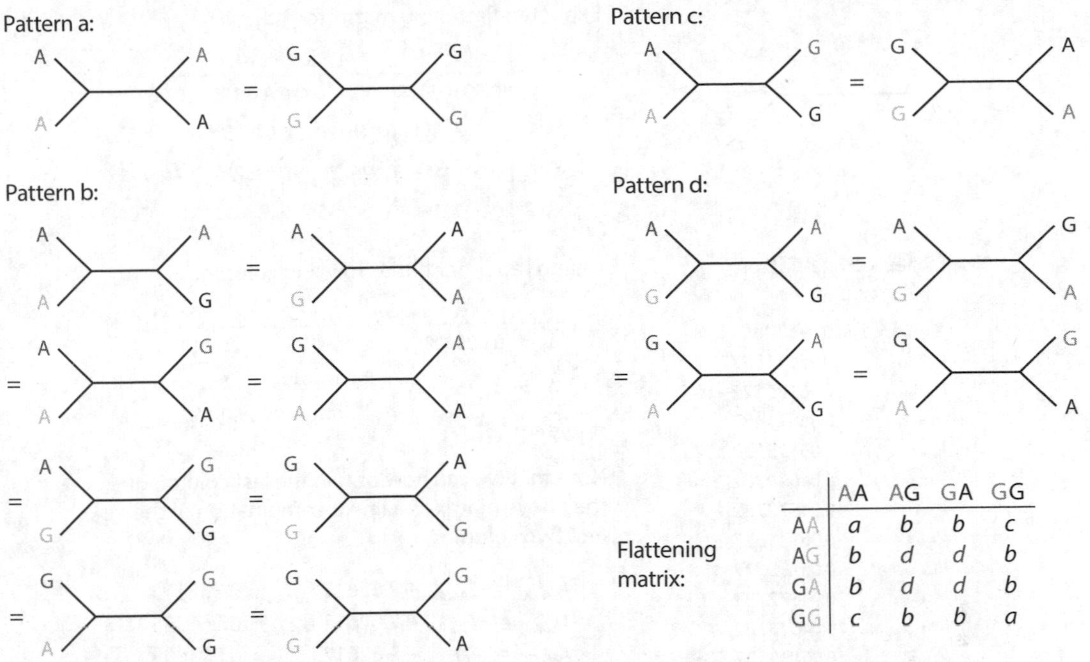

Figure 4.2. Possible site patterns under the simplified model considered in our example, for which only nucleotides *A* and *G* are possible. In each tree shown, shading indicates species (dark grey = species 1, light grey = species 2, black = species 3, intermediate grey = species 4) and nucleotides at the tips of tree indicate a possible site pattern. For this simplified model, all possible site patterns can be reduced to four classes in such a way that patterns within a class have the same probability. For example, "Pattern a" (top left) includes the case in which all species have either nucleotide *A* or nucleotide *G*. These two patterns are equally probable under the model assumed here. The flattening matrix corresponding to this tree is shown in the bottom right panel. Note that the middle two columns and middle two rows in the flattening matrix are identical to one another.

true species tree, while larger values indicate lack of support for that particular quartet relationship on the true species tree.

4.2.1.1 Example: Rank Reductions for Flattening Matrices

To make the ideas introduced in this section more concrete, we provide a specific example that provides some intuition for the matrix rank results discussed above. To simplify the problem, we consider the gene tree (rather than the species tree) setting, and we suppose that there are only two possible nucleotides that could be observed at the tips of the tree, *A* and *G*. We assume that mutations between these two nucleotides occur at equal rates and that the nucleotides are equally frequent. Under this model, there are four distinct classes of site patterns as depicted in figure 4.2, such that all patterns in the same class have the same probability of occurring on the tree under this model. The flattening matrix that results is shown in the bottom right panel of figure 4.2. It is easy to see that the middle two columns and middle two rows of this matrix are identical. Thus, although the flattening matrix has four rows and four columns (and thus its rank

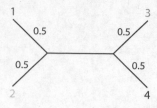

Expected flattening matrix for 1,2|3,4

	AA	AG	GA	GG
AA	0.093008	0.061355	*0.061355*	0.068115
AG	0.061355	0.046728	*0.046728*	0.061355
GA	*0.061355*	*0.046728*	0.046728	*0.061355*
GG	0.068115	0.061355	*0.061355*	0.093008

Expected site-pattern frequencies

p_{AAAA}	0.09300841
p_{AAAG}	0.06135527
p_{AAGA}	0.06135527
p_{AAGG}	0.06811487
p_{AGAA}	0.06135527
p_{AGAG}	0.04672782
p_{AGGA}	0.04672782
p_{AGGG}	0.06135527
p_{GAAA}	0.06135527
p_{GAAG}	0.04672782
p_{GAGA}	0.04672782
p_{GAGG}	0.06135527
p_{GGAA}	0.06811487
p_{GGAG}	0.06135527
p_{GGGA}	0.06135527
p_{GGGG}	0.09300841

Delete redundant third row and column…

	AA	AG	GG
AA	0.093008	0.061355	0.068115
AG	0.061355	0.046728	0.061355
GG	0.068115	0.061355	0.093008

Note that we can now obtain the last column of the above matrix as a linear combination of the first two columns:

$$f_{AA,GG} = -f_{AA,AA} + 2.62617\, f_{AA,AG} = 0.068115$$
$$f_{AG,GG} = -f_{AG,AA} + 2.62617\, f_{AG,AG} = 0.061355$$
$$f_{GG,GG} = -f_{GG,AA} + 2.62617\, f_{GG,AG} = 0.093008$$

∴ *matrix has only two linearly independent rows and columns; rank is 2*

Figure 4.3. Continuation of the example in figure 4.2 to consider a specific numerical example. The example tree with branch lengths shown in the top left can be used to compute the 16 possible site pattern probabilities listed in the left column. Arranging these in the flattening matrix and removing the redundant third column shows that the last column can be expressed as a linear combination of the first two, leading to a further reductions in the rank of the flattening matrix to rank 2.

could be as large as four), one of those columns is redundant, leading to a reduction in the rank of the matrix by at least one.

We now consider a numerical version of this example to show that the rank is reduced even further. In figure 4.3, the tree considered in this example is assigned branch lengths, which are then used to compute the probabilities of each of the 16 possible site patterns, as shown in the left column. These site patterns can be arranged into the flattening matrix, as demonstrated in figure 4.2, and then the redundant third row and third column can be removed. It can additionally be noted that the probabilities given in the last column can be obtained from those in the first two columns (and similarly for the rows, though this is not shown in the figure), and so the rank of the matrix is again reduced by one. Thus, although the maximum possible rank of the flattening matrix is four, the actual rank is two because the entire matrix can be reproduced once the first two columns of the matrix are known.

Thus far, we have provided some intuition for why the matrix rank is reduced for flattening matrices that correspond to the true tree. However, the question about whether similar rank reductions occur for the two alternative flattening matrices has not been

	AA	AG	GA	GG
AA	a	b	b	c
AG	b	d	d	b
GA	b	d	d	b
GG	c	b	b	a

	AA	AG	GA	GG
AA	a	b	b	d
AG	b	c	d	b
GA	b	d	c	b
GG	d	b	b	a

(a) Flattening matrix for 1,2|3,4   **(b)** Flattening matrix for 1,3|2,4

Figure 4.4. Flattening matrices for (a) the original example tree and (b) the tree in which species 1 and 3 and species 2 and 4 are grouped together. Shading refers to species labels as defined in figure 4.2, and matrix entries refer to the site pattern probabilities from figure 4.2. Note that the matrix in (b) has no duplicate rows or columns.

addressed. To examine this, consider the tree that groups species 1 and 3 and species 2 and 4 together (see figure 4.4). Note that the flattening matrix for this tree can be obtained by rearranging the entries of the matrix for the original tree, as shown in part (b) of figure 4.4. With these rearrangements, no columns (or rows) are duplicates of one another, and it can be shown that none are linear combinations of the others, either. Thus, the rank of the flattening matrix for this alternative tree is 4.

This is a significant result, and one that forms the basis of the theory underlying SVDQuartets: flattening matrices corresponding to the true tree show reductions in their ranks, while flattening matrices corresponding to incorrect trees (i.e., trees that did not give rise to the observed data) do not. Though our example here was simplified to the case of gene trees for a model with only two observed nucleotides and equal mutation rates, the principles hold directly for the more complicated case of data arising under the MSC. The only changes are that the full set of nucleotides are considered, leading to 256 site patterns and flattening matrices that are 16×16. For a quartet that reflects the true species-level relationships, the rank of the corresponding flattening matrix is 10 (rather than 2, as in our example), while the rank for an incorrect tree is 16 (rather than 4, as in our example). These results have been proven mathematically and hold for any substitution model that fits the assumptions of the GTR+I+G model when the molecular-clock assumption is satisfied. Even if the clock assumption is violated, the result still holds for all submodels of GTR that assume equal rates among sites. For GTR submodels including invariable sites or other mixtures of rates, the rank is still reduced for the true tree, but by a lesser amount.

One reason that this result is so important is that calculating the rank of a small matrix is trivial in terms of computation time and can be completed in fractions of a second. In practice, however, the flattening matrix must first be estimated from data, and the resulting numerical calculation of the matrix will always give a rank of 16, even for the true tree. This is because sampling error in the estimates of the entries of the flattening matrix will result in the middle two columns in our example being *similar* but not *identical*. This is why the SVDScore in equation 4.2 is used; it allows us to measure how similar a matrix is to being of a certain rank, even when it is not precisely the desired rank. This score has well-developed mathematical theory behind it and has been used in other phylogenetic contexts as well [13, 15]. It can also be rapidly computed and has been shown to be useful in differentiating between trees [140, 256, 447].

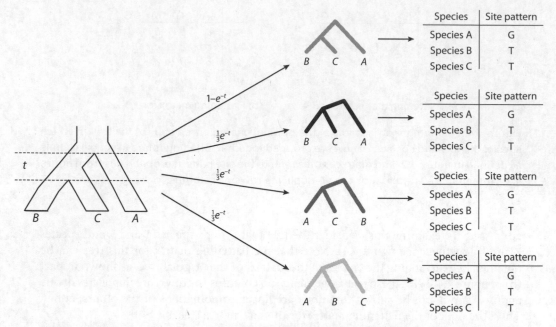

Figure 4.5. The data generation process for a species-level phylogeny with three species. The species tree is shown in the leftmost panel. The length of the branch between speciation events is t. The second panel shows the four possible gene tree histories that are compatible with the species trees, along with their probabilities (labeled on the arrows). Note that the first two gene tree histories share the same topology but differ in the timing of the coalescent event uniting species B and C (see chapter 1 for details). The third panel shows that site pattern GTT can arise from any of the four histories.

4.2.2 ACCOUNTING OF INCOMPLETE LINEAGE SORTING IN SVDQUARTETS

In the example above, gene trees were used to gain intuition about how the reduction in matrix rank arises for the flattening matrix corresponding to the true tree. In the full SVDQuartets method, however, the reduction in matrix rank is achieved for a species tree under a data model based on the multispecies coalescent. In this section, we provide a short, nontechnical description of how the coalescent process is incorporated into SVDQuartets to accommodate incomplete lineage sorting (ILS; see chapter 1 for a description of ILS and how it is related to gene trees).

To illustrate the model, consider a species tree with three taxa, as shown in the left panel of figure 4.5. Recall from chapter 1 that the coalescent process can be used to specify the probability of each of the four possible gene tree histories that can arise from this species tree. These are shown in the middle panel of figure 4.5 with their probabilities labeled on the arrows that lead from the species tree. The coalescent model also specifies a probability distribution on the lengths of the branches within these gene tree histories. These probabilities distributions are not shown here as they don't provide any insight into the model (the interested reader can consult [383, 601]). Along each gene tree history, sequence data arise in the same way as described in the gene tree example above to generate site patterns (third panel in figure 4.5). When a single site pattern is assumed to have arisen along each gene tree, a CIS data set is generated. When multiple site patterns arise from a fixed gene tree and data from many such gene trees are observed, the data are referred to as multilocus data.

The previous paragraph described the mechanism by which data are generated, but how is this data generation mechanism incorporated into SVDQuartets? The key idea is that the entries in the theoretical flattening matrix (i.e., the one that contains the true site pattern probabilities rather than those estimated from the data) are computed under the model in figure 4.5. For example, consider the probability associated with the observed site pattern GTT. We can see from figure 4.5 that there are several "paths" by which we could observe this site pattern: it could have arisen from any of the four gene tree histories in the second panel of the figure. Thus, to compute its probability, we need to sum over the probability that it was generated from each of the gene trees histories, weighting by the probability that the particular gene tree history was generated by the species tree. Branch lengths are handled similarly, except that we must integrate over the branch length distribution because the branch lengths are continuous. Chifman and Kubatko [141] used the Mathematica software to compute these integrals for the two possible rooted four-taxon species trees. They obtained analytical expressions that could then be used to prove mathematically that the rank of the flattening matrix is reduced for the true tree under the coalescent model. Thus, although the data used for SVDQuartets are often presented as a concatenated data matrix, it is important to remember that SVDQuartets does formally model ILS arising from the multispecies coalescent.

4.2.3 SPECIES TREE INFERENCE: QUARTET SAMPLING AND ASSEMBLY

In the previous section, we described the analysis used by SVDQuartets for trees with only four taxa. We now describe the method used by SVDQuartets as implemented in PAUP* (http://paup.phylosolutions.com) for inferring the species tree. To begin, the data must be input into PAUP* in Nexus format. Optionally, the Nexus file also contains a taxpartition command that can be used to map sequences to species in the case in which multiple individuals are sampled for one or more of the species under consideration. The analysis is run using the svdq command in PAUP*.

By default, the software will consider all possible combinations of four sequences, each selected from a distinct species, if this number is not too large. If this number is too large, then samples of four sequences may be randomly selected. PAUP* allows this step to be parallelized via multithreading, as each quartet can be evaluated independently of the others. SVDQuartets can easily evaluate all possible quartets for up to 100 taxa, and this number can be pushed to 200 or more taxa depending on the speed of the computer, the number of processor cores available, and the length of time the user is willing to wait. Unfortunately, there is not a straightforward answer to the question of how many random quartets should be sampled for problems that are too large for exhaustive quartet sampling. In general, the more the better, but the number needed to obtain an accurate estimate of the species tree topology will depend on several data-specific characteristics, such as the informativeness of the inferred quartet trees and the extent to which they conflict with one another. We recommend that users experiment with increasingly large numbers of sampled quartets and assess stability in the inferred species tree across these runs.

For each quartet, three flattening matrices, one corresponding to each of the unrooted four-taxon trees in figure 4.1, are estimated, and the SVDScore defined in equation 4.2 is computed for each. The three scores are compared, and the tree that produces the smallest score is retained as that inferred for the quartet under consideration. The result after considering all quartets (or a random sample of quartets) is a list of quartet trees that is then used as input to a quartet assembly algorithm. More information on the process of quartet assembly is given in section 4.2.4. The result of the assembly process is an estimate of the unrooted species tree topology. The procedure used by SVDQuartets is shown in figure 4.6.

4.2.4 ALGORITHMIC DETAILS

A complication that arises when calculating the SVDScore (equation 4.2) involves the numerical accuracy and stability of the singular value decomposition. A number of algorithms exist for computing the SVD of a matrix, and care must be taken to select an appropriate one; for example, accuracy of SVD algorithms is tied, in part, to minute details of floating-point arithmetic characteristics on the processor being used. Our usage in SVDQuartets differs from many applications of SVD in that accurate estimation of the smallest entries in the vector of singular values is important.

We use routines from the highly regarded LAPACK Fortran library [25], which are the only ones we evaluated that do not crash or exhibit other numerical problems for some inputs. The primary method used for singular value decomposition in PAUP* is the DGESDD routine from LAPACK (which is also used by R, MATLAB, and other numerical software). However, we have discovered that the singular values returned by this method can be inaccurate for sparse flattening matrices containing many zero entries (i.e., when many of the 256 possible site patterns for four taxa are not observed). Very small singular values can occur in this case, and sometimes values that should be exactly zero are instead returned as small positive numbers. The numerical inaccuracy can be great enough to cause the wrong quartet topology to be preferred or one of the topologies to be chosen as best when in fact all three topologies should have equal scores.

Several modifications to the code have been made to improve accuracy of computing the singular values of the flattening matrix under these conditions. First, rows and columns containing all-zero entries are removed prior to performing the SVD (which does not affect the estimated rank of the matrix). In addition to improving the accuracy of DGESDD, this matrix reduction sometimes makes SVD calculation entirely unnecessary, as the rank of the matrix cannot be greater than min(number of rows, number of columns). Second, when all three of the quartet topologies have very similar SVDScores, an additional SVD is performed using the slower, but more accurate, LAPACK routine DGEJSV; the singular values returned by DGEJSV are then used to recalculate the scores for the quartet. Third, if the scores for all three topologies for a quartet are equal (within a tolerance for floating-point round-off error), the quartet is simply discarded. In earlier versions of PAUP*, one of the three resolutions was kept, and the decision as to which one was chosen was affected by the numerical inaccuracy described above.

A second critical issue is the method used to assemble the results from individual quartets into an overall tree. This assembly requires solving the NP-complete maximum quartet consistency (MQC) problem [347]. Many heuristics have been proposed to approximate solutions to MQC. We tested several of them and found the Quartets MaxCut (QMC) [706] and Quartet FM (QFM) [608] to outperform the alternatives: we chose to implement QFM for PAUP*. We developed a novel implementation of QFM based on the algorithmic descriptions in Reaz et al. [608] that runs much faster than their original implementation. We also modified certain aspects of the algorithm that improved its effectiveness (to be published elsewhere).

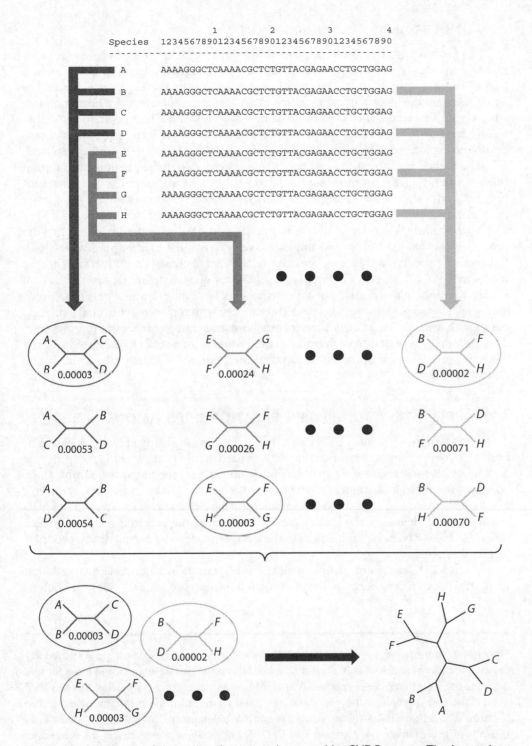

Figure 4.6. Schematic of the estimation procedure used by SVDQuartets. The input data set consists of eight species, labeled A through H. The first step is to consider quartets of species and compute the SVDScore for each of the three possible unrooted topologies for these quartets. The figure shows three example quartets, indicated by arrows of

4.2.5 UNCERTAINTY QUANTIFICATION

Thus far, we have described the process by which SVDQuartets can be used to estimate the species tree topology. When obtaining a phylogenetic estimate at any level, it is also important to obtain a measure of uncertainty in the phylogenetic estimate. For SVDQuartets, the most natural measure of uncertainty is obtained by bootstrapping, which is implemented in PAUP* in two different forms. If the input data to be used in SVDQuartets are SNPs, then the proper bootstrap procedure is to sample sites at random with replacement from the input aligned SNP data. For each bootstrap replicate, SVDQuartets is used to estimate a species tree as described in the previous section, and these bootstrap species trees are summarized with a consensus tree. The bootstrap support for partitions not appearing in the consensus tree is also reported, and the species trees estimated for each bootstrap replicate can optionally be written to an output file.

The procedure is similar in the case of multilocus data, with the exception that the process by which the bootstrap samples are drawn uses the method of Seo [668]. To illustrate the process in this case, consider a data set that consists of M loci with n_i sites for locus i, $i = 1, 2, \ldots, M$. To obtain a single bootstrap sample in this case, M loci are selected at random with replacement from the set of loci in the sample. For each sampled locus, say j, a sample of n_j sites is selected at random with replacement from the original data for that locus. These data then form a bootstrap data set from the original data. This process is repeated to obtain the desired number of bootstrap data sets, and the analysis and summary of the bootstrap data then proceeds as described above in the case of SNP data.

4.2.6 APPLICATION TO SPECIES RELATIONSHIPS AMONG GIBBONS

As an example, we'll apply SVDQuartets to a data set consisting of five species of gibbons [109, 676]. The data set for coding regions used in [676] consists of 11,323 genes for a total of 2,264,600 sites for six species: *Hylobates moloch* (one individual sampled); *H. pileatus* (one individual sampled); *Symphalangus syndactylus* (two individuals sampled); *Hoolock leuconedys* (two individuals sampled); *Nomascus leucogenys* (two individuals sampled); and *Homo sapiens* (one sequence was used as the outgroup). Note that each individual contributes two sequences to the data set, and so the overall data set consists of 17 sequences.

We ran SVDQuartets by sampling all possible quartets, with uncertainty measured using 100 bootstrap replicates in the multilocus bootstrap procedure. The entire analysis

---

Figure 4.6. (*continued*) different shades of grey. For example, the dark gray arrow indicates selection of species *A*, *B*, *C*, and *D*, and of the three possible unrooted trees for this quartet, the one that places species *A* and *B* together (circled in dark grey) has the lowest SVDScore. This tree is then retained and used as input for the quartet assembly step (bottom of the figure). A similar process is shown for quartets consisting of species *E*, *F*, *G*, and *H* (intermediate grey arrow) and *B*, *D*, *F*, and *H* (light grey arrow). The black dots indicate that all possible quartet samples, including those not shown here, would need to be evaluated in this way. The bottom of the figure indicates that the inferred quartet trees from the first step are used as input for a quartet assembly procedure that then produces the estimate of the species tree (bottom right).

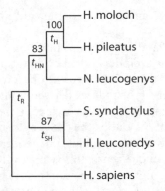

Branch	Estimate	Standard error	95% HPD interval
t_H	1.8892	0.0412	(1.8140, 1.9822)
t_{HN}	2.5×10^{-10}	4.86×10^{-9}	$(0, 1.19 \times 10^{-89})$
t_{SH}	0.0822	0.0195	(0.0419, 0.1193)
t_R	10.60	0.2616	(10.11, 11.12)

Figure 4.7. Species tree estimated by SVDQuartets (left) and maximum a posteriori composite-likelihood MAP_{CL} estimates of branch lengths in coalescent units (right) for the gibbon data of Carbone et al. [109]. Numbers above the internal nodes are bootstrap support values, and notation below the branches indicates the label of the branch used to report the estimated length in the table at the right. Note that SVDQuartets estimates an unrooted tree; the tree displayed here has been rooted using the human sequence as an outgroup. Highest posterior density = HPD.

took 8.23 seconds on a desktop machine, and the estimated tree is shown in figure 4.7. Shi and Yang [676] analyzed these data using the MCMC-based method Bayesian Phylogenetics and Phylogeography (BPP) and found that, across 10 independent runs, two distinct trees were identified as the maximum a posteriori (MAP) tree (one was found in seven of the ten runs, with the other found by the remaining three). Interestingly, the two trees found by BPP differ from the tree inferred by SVDQuartets in the placement of *N. leucogenys*, with it being placed sister to the clade containing *S. syndactylus* and *Ho. leuconedys* in the majority of the runs and sister to the entire group in the remaining runs. The somewhat low bootstrap support for the clade containing *N. leucogenys* (i.e., bootstrap support of 83) indicates that the data are not strongly informative about placement of this species, which is reflected in the BPP analysis. Further analysis of these data (see section 4.3.4) indicates very short intervals between speciation events in the evolutionary history of this species.

4.2.7 PROPERTIES OF SVDQUARTETS

In this section, we review the assumptions underlying SVDQuartets and describe its properties. In doing so, we describe scenarios under which SVDQuartets should and should not be applied.

4.2.7.1 Assumptions and Misconceptions

Above, SVDQuartets was introduced for CIS data, that is, data for which each site is generated from its own underlying gene tree, which in turn has been generated from the species tree via the MSC. The theoretical derivations described above assume this data structure. It is also directly applicable to SNP data sets because constant site patterns (which are included in CIS data but would not be included in SNP data) do not impact the reduced rank results on which the method is based (e.g., in our example above, the two middle columns in the flattening matrix that were equal to one another did not involve any of the constant site patterns).

One can also use SVDQuartets to analyze multilocus data. To gain some intuition for why this works, recall again that for CIS data each site is an independent observation arising from the species tree under the MSC. The method then works by using observed site pattern frequencies to estimate theoretical site pattern probabilities derived under the model. What changes when moving to multilocus data is that now each gene tree provides information about the distribution of site patterns under the MSC by contributing many *correlated* sites; that is, sites within a given locus are not independent because they come from the same underlying gene tree. In one sense, this is an advantage: we get more information about each gene tree. But on the other hand, this is not quite as good as having an equivalent number of *independent* sites, as these will carry more information that is directly relevant about the species tree. Nonetheless, having increased information about each locus is helpful in obtaining the overall phylogenetic estimate. When the number of loci is large, so that we have sufficient data to capture the variation in the underlying gene trees, then the multilocus estimate will perform very well, and better, in fact, than an estimate obtained from using only one SNP from the same number of loci. For this reason, we recommend that multilocus data be used when they have been collected. For example, we do not suggest that RADseq data be filtered to provide only one SNP per locus; rather, all sequence data collected should be used in the analysis.

The description above highlights a key requirement for SVDQuartets to perform well: the amount of data must be sufficient to enable accurate estimation of the flattening matrix corresponding to the true tree so that the SVDScore will be able to "detect" the reduction in matrix rank. The number of sites needed depends to some extent on the underlying species tree, particularly on the probability with which sites that are informative about the reduced rank are generated. It also depends on whether CIS or multilocus data are used. Application of SVDQuartets to data collected for only a handful of loci is not likely to result in accurate species tree estimates, even if all of the loci are quite long, because such data do not provide adequate information about variation among gene trees. On the other hand, several thousand CIS may be sufficient for some species trees since each site is directly informative about the species tree. In both cases, however, it should be noted that trees with short internal nodes may require large sample sizes as short species tree branches result in shallow gene trees, which in turn produce data with a relatively low proportion of variable sites. Though this situation is challenging for all inference methods, it is particularly difficult for SVDQuartets when the number of sites is small.

A strength of SVDQuartets is that it is valid for a wide range of substitution models, although the model need not be specified explicitly. In addition, there is no requirement to estimate model parameters or to compute likelihoods. These properties have led some researchers to consider this a nonparametric or "model-free" method, but this is not accurate: the theory on which SVDQuartets is based assumes both the MSC and a standard substitution model, and features of the probability distributions arising from these models are precisely the information that is used for inference. Another misconception about SVDQuartets is that it is a "concatenation" method, an idea that appears to have originated from the input format for the analysis being a concatenated data matrix for which gene boundaries need not be specified. However, it is important not to confuse the data format with the underlying model. Recall that SVDQuartets assumes a model for which each site has its own underlying gene tree. Thus, the reason that gene boundaries need not be provided is because each site is assumed to be its own gene, independently sampled from the species tree. This is in direct contrast to concatenation methods that assume that all sites have a single underlying gene tree.

Finally, SVDQuartets has been referred to as a summary method for species tree inference ([676], page 172) because the data used as inputs are counts of the 256 possible site patterns for each quartet. However, SVDQuartets is distinct from summary methods that estimate gene trees for each locus as a first step and subsequently use these estimated gene trees as input to estimate the species tree. Further, we do not view SVDQuartets as a summary method in any sense, as counts of the observed site patterns are a valid presentation of the input data in the same way that a multiple sequence alignment (MSA) is a valid presentation of the input data. In fact, an MSA could also be viewed as a summary, in the sense that the raw read count data obtained in the sequencing reaction have been summarized to produce a single sequence to represent each taxon.

4.2.7.2 Statistical Consistency

One property of a phylogenetic estimator that is often used to compare methods is that of statistical consistency. A statistically consistent estimator is one that becomes increasingly likely to be equal to the truth as the amount of data increases. Recent work [783] has shown that SVDQuartets provides a statistically consistent method for estimating the species tree. This result builds on two key ideas. First, entries of the flattening matrix are consistently estimated, leading to consistent estimation of the SVDScore. Consistent estimation of the SVDScore ensures that the quartet trees used as input to the quartet assembly algorithm are all correct when the data set size is sufficiently large. Second, it is well-known that quartet relationships uniquely identify a tree [667], and thus an input data set consisting of true quartet relationships will result in the correct species tree estimate. Importantly, this consistency result holds for **both** multilocus and CIS data, and thus SVDQuartets is an accurate method of species tree estimation for large data sets, whether multilocus or CIS data are used. A detailed simulation study is included in [783], including comparisons with the Bayesian method BPP, for four-taxon problems.

4.2.7.3 Missing Data and Data Filtering

Phylogenomic data commonly contain a very high proportion of missing data. SVDQuartets has two options for handling missing data. The first option is to ignore the missing data. If this option is selected, then the complete data set is considered for each set of four taxa that are sampled. If there are sites for which some data are missing for one of the four taxa under consideration, then those sites are not used in computing the SVDScore for that quartet. In this way, each quartet evaluation uses the maximum amount of nonmissing data possible. The second option is to impute the missing data in the following way. Each site in the sequences contributes a count of 1 to the relevant site pattern frequency estimate, so if data are missing for one or more sites, then this count could be distributed over possible site patterns. For example, consider a site at which taxa A, B, and C have nucleotide T, but data at this site for taxon D is missing. Since the missing nucleotide could be either A, C, G, or T, we increment the counts of the patterns $TTTA$, $TTTC$, $TTTG$, and $TTTT$ each by some fractional amount. This amount could be chosen based on the observed proportion of each nucleotide in the input data or based on some assumptions about the relative proportions of nucleotides. For example, if all nucleotides are assumed to be equally frequent, then each of the four site patterns listed above could be given a count of 0.25. In this way, missing data is accommodated by using the information available at this site (i.e., taxa A, B, and C have nucleotide T) while incorporating the information about the uncertainty in the nucleotide at that site for taxon D.

Large phylogenomic data sets are often fairly informative about the species tree, even when the proportion of missing data is large. In other words, what often matters is how much data you *have*, rather than how much data you *don't* have. Large data sets will necessarily often have a high percentage of missing data yet still lead to relatively certain estimates for many relationships in the species tree. For this reason, we do not recommend that users filter their data prior to analysis with SVDQuartets. Subjective filtering can lead to biases in the retained sites, and the computational efficiency of SVDQuartets, coupled with its intuitive approach for handling missing data, make its application to all sampled data the recommended approach.

4.2.8 RECOMMENDATIONS FOR USING SVDQUARTETS

When selecting a method for analyzing data with the goal of estimating the species tree, it is important to consider the characteristics of the data in relation to the requirements of possible inference methods. At the core of the SVDQuartets method is the need to estimate the site pattern probabilities accurately for each of the quartet trees, so the primary requirement in terms of data is that the data are sufficient to obtain reliable estimates of these probabilities. In general, this means that the performance of the method will improve as the amount of data increases. It also means that when the overall data set size is small (either in the number of loci or the number of sites or both), the method may not perform as well. We suggest that users exercise extreme caution when applying the method to fewer than 20 loci or fewer than 5000 SNPs. On the other hand, a strong advantage of SVDQuartets is that it is computationally efficient for very large data sets. We have successfully applied the method for millions of sites (see, e.g., the gibbon data analysis in section 4.2.6) and hundreds of taxa.

One should always run SVDQuartets in conjunction with an appropriate bootstrapping approach (i.e., either the standard or the multilocus bootstrap). It is also important to note that SVDQuartets returns an unrooted estimate of the species tree, and hence rooting with an outgroup or some other appropriate method must be applied if a rooted estimate of the species tree is desired. An advantage of SVDQuartets is that it is not necessary to specify a substitution model, and the method remains valid if there is variation in the substitution model across the input sites when the molecular clock holds, provided that the models governing all partitions are submodels of GTR+I+G. In practice, settings in which there is extensive heterogeneity in the models that apply to different data partitions may require large sample sizes for accurate estimation of the species tree. SVDQuartets can also be applied in settings in which the molecular clock is violated and when there is variation in effective population sizes throughout the tree [446]. SVDQuartets is also robust to the presence of gene flow between sister taxa [445].

4.3 Estimation of Speciation Times

In this section, we describe a method that uses site pattern frequencies to estimate speciation times for a fixed species phylogeny. The method uses the likelihood for

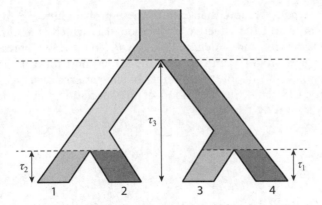

Figure 4.8. Example symmetric species tree with speciation times τ_1, τ_2, and τ_3.

four-taxon trees to construct a composite likelihood for species trees of arbitrary size. Maximization of this composite likelihood leads to estimators that are statistically consistent and asymptotically efficient and that can be computed efficiently in practice. In the sections below, we describe the theoretical basis for computing these estimators, methods for quantifying uncertainty in the estimates, and recommendations for their use. We note that although the estimates are based on quartets and use site pattern frequencies as input data, they are distinct from the SVDQuartets method described in the previous section. In particular, they can be applied to a species tree topology that has been estimated by **any** method.

4.3.1 THEORETICAL BASIS

As mentioned in the previous section, there are 256 possible site patterns that can arise on a four-taxon species tree. Chifman and Kubatko [141] showed that for the rooted symmetric and asymmetric species trees, these 256 site patterns can be reduced to 9 and 11 classes, respectively, in such a way that site patterns within each class occur with equal probability under the MSC with the JC69 substitution model. For example, it is clear that for the JC69 model, $p_{AAAA} = p_{CCCC} = p_{GGGG} = p_{TTTT}$ for any tree, while for the leftmost tree in figure 4.1, $p_{AAGG} = p_{AACC} = \cdots = p_{GGTT}$. In addition to defining these classes for both symmetric and asymmetric trees, [141] derived analytic expressions for the probability of a site pattern in each class. These expressions can be used to compute the likelihood for site pattern frequency data on a fixed species phylogeny by means of the multinomial distribution, as described below.

We consider the symmetric species tree shown in figure 4.8 in defining the relevant notation; the asymmetric case is analogous. This species tree includes three speciation times, labeled τ_1, τ_2, and τ_3. For this tree, the nine distinct classes of site patterns can be specified by

$$p_{xxxx}, p_{xxxy} = p_{xxyx}, p_{xyxx} = p_{yxxx}, p_{xyxy} = p_{yxxy}, p_{xxyy},$$

$$p_{xyxz} = p_{yxxz} = p_{xyzx} = p_{yxzx}, p_{xxyz}, p_{yzxx}, p_{xyzw}, \tag{4.3}$$

where x, y, z, and w are states $\in \{A, C, G, T\}$ observed in each of four distinct taxa. For example p_{xxxx} includes the patterns $p_{AAAA} = p_{CCCC} = p_{GGGG} = p_{TTTT}$. We label the classes above 1–9 and denote the vector of 9 probabilities corresponding to these classes by $\mathbf{p} = (p_1, p_2, \ldots, p_9)$. We note that each of these probabilities is a function of both the speciation times and the effective population size, which is assumed to be constant throughout the tree and is denoted here by θ. The precise expressions for these probabilities are given in [141], and we refer the reader there for further details.

Let $\mathbf{Y} = (y_1, y_2, \ldots, y_9)$ denote the number of site patterns observed in each of the classes. Then for CIS data, the likelihood of the speciation times and the effective population size is given by

$$L(\tau_1, \tau_2, \tau_3, \theta | \mathbf{Y}) \propto \prod_{j=1}^{9} p_j^{y_j}. \tag{4.4}$$

We note again that the p_j are functions of the τ's and θ, though we have suppressed that notation here for ease of exposition. This likelihood is used to form the basis of our inference procedure. For example, this likelihood (or equivalently, its logarithm) can be maximized to obtain maximum likelihood estimates of the parameters in the case of a fixed four-taxon species tree.

While this procedure is straightforward in the four-taxon case, expressions analogous to the p_is cannot be efficiently computed for five or more taxa, prohibiting the use of this likelihood framework directly for species trees with more than four tips. Therefore, as an alternative to computing the true likelihood, we use instead the *composite likelihood* for estimation of the parameters for species trees of arbitrary size. In order to form the composite likelihood, we first decompose the set of taxa into all possible quartets. We then compute the four-taxon likelihood given by equation 4.4 for each of the quartets. Finally, we combine the four-taxon likelihoods by taking their product. Let Q be the number of unique quartets that can be obtained by sampling one lineage from each of four distinct species. Any quartet i induces a subtree on the full tree containing three internal nodes u_i, v_i, and w_i corresponding to nodes 1–3 in equation 4.4. This subtree may be either symmetric or asymmetric, with either $n(i) = 9$ or $n(i) = 11$ distinct site patterns, respectively. Rewriting the likelihood for quartet i as $L_i(\tau_{u_i}, \tau_{v_i}, \tau_{w_i}, \theta | \mathbf{Y})$, corresponding to equation 4.4, the composite likelihood is given by

$$\ell(\tau_1, \tau_2, \ldots, \tau_R, \theta | \mathbf{Y}) \propto \prod_{i=1}^{Q} L_i(\tau_{u_i}, \tau_{v_i}, \tau_{w_i}, \theta | \mathbf{Y}) = \prod_{i=1}^{Q} \prod_{j=1}^{n(i)} p_j^{y_j}. \tag{4.5}$$

The parameter values for which this function is maximized are called maximum composite likelihood estimates (MCLEs). See figure 4.9 for a depiction of the process of computing the composite likelihood for a five-taxon tree.

Note that the procedure above entails multiplication of the quartet likelihoods, even though they are not independent. If the quartet likelihoods were independent of one another, then multiplying them together would lead to computation of their joint probability. However, this will not be the case when there is dependence among the quartets, which is why a likelihood formed in this way is termed a *composite likelihood* (or *pseudolikelihood*). To see the lack of independence, refer to figure 4.9, which shows the five quartet trees that can be obtained from the species tree in the top left. Comparing quartets 1 and 2, for example, we observe that they share three taxa in common and that both include the same set of branch lengths. Thus, they use overlapping sets of data to compute their quartet likelihoods, and they share a set of parameters (the

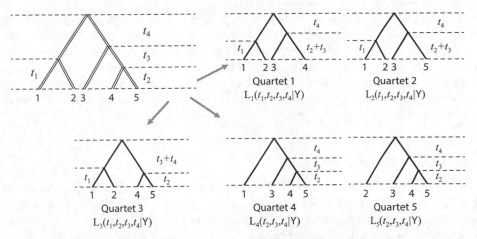

Figure 4.9. Schematic representation of the computations of the composite likelihood for a five-taxon species tree (top left) with branch lengths t_1, t_2, t_3, and t_4. Arrows point to the five possible quartet trees that are obtained by removing one taxon from the species tree. Branches for each quartet tree are relabeled appropriately, and the likelihood based on the multinomial likelihood is computed for each, as indicated by the terms below each tree. To compute the composite likelihood, the terms below the five quartet trees are multiplied.

three branch lengths on their quartet trees), so they are clearly not independent of one another. Similar comparisons can be made among all pairs of quartets.

Intuitively, the composite likelihood is useful for inference because values of the parameters that maximize the true likelihood are likely to lead to large values for each of the quartet likelihoods as well. Thus the composite likelihood contains information about which values of the parameters are likely to provide reasonable explanations of the data. Indeed, there is a rich literature on the use of composite likelihood approaches for inference in situations in which computation of the full likelihood is infeasible, including theoretical results about the statistical properties of estimators obtained using composite likelihood approaches (see, e.g., [763] for a recent review). Peng et al. [572] have shown that the values of the speciation times that maximize the composite likelihood in equation 4.5 are statistically consistent and asymptotically normally distributed by proving that the results of [33] can be applied in this case. This enables implementation of a computationally efficient method for obtaining estimates of these parameters with good statistical properties.

Instead of using composite likelihood directly, however, we have found it preferable to estimate parameter values via Bayesian maximum a posteriori (MAP) estimation. In addition to allowing incorporation of prior knowledge into the estimate, weighting the (composite) likelihood by the priors improves the computational efficiency and stability of the optimization algorithms by reducing the flatness of the optimality surface in regions of the parameter space that have very low likelihood (see below). Prior distributions are placed on both θ and the total tree height $h = \tau_R$ (R is the index of the root node). If $f_\theta(\theta)$ and $f_h(\tau_R)$ represent the probability density functions for these prior distributions, then the log of the posterior density is proportional to

$$\log g(\tau_1, \tau_2, \ldots, \tau_R, \theta | \mathbf{Y}) = \log f_\theta(\theta) + \log f_h(\tau_R) + \sum_{i=1}^{Q} \log L_i(\tau_{u_i}, \tau_{v_i}, \tau_{w_i}, \theta | \mathbf{Y}) \quad (4.6)$$

(g is an unnormalized posterior density because it lacks a marginal likelihood term). We refer to the parameter values that optimize this function as MAP_{CL} estimates, with the subscript "CL" signifying that a composite-likelihood term is used in equation 4.6 rather than a true likelihood. Note that PAUP* allows frequentists to omit the prior terms in equation 4.6 to obtain MCLEs instead, although for reasons explained below, larger amounts of data may be needed to avoid numerical difficulties during the optimization.

4.3.2 ALGORITHMIC DETAILS

The goal is to obtain estimates of the speciation times and the effective population size that maximize the posterior density function g in equation 4.6. Several algorithmic issues arise in the development of a computationally efficient estimator that performs well, and we briefly describe those here.

A key concern is that the value of the effective population size, θ, has a large impact on the estimates of the speciation times, but the composite likelihood surface is relatively flat with respect to the effective population size parameter. Thus routines for numerical optimization are extremely sensitive to the starting point chosen for θ, making it important to choose a good initial value for this parameter. To address this complication, a grid search is first carried out in conjunction with the moment-based estimators for the speciation times given in [384] to determine an interval that provides loose lower and upper bounds on the optimal θ value. The starting θ value is then optimized numerically using a two-step procedure in which the bracket is first narrowed via a golden section search, then polished using a derivative-free one-dimensional optimizer. Initial values for the speciation times are determined using the moment-based estimators computed at the starting value chosen for θ.

We then use multidimensional optimization techniques to search numerically for values of the speciation times and effective population size parameters that maximize the log posterior density. This optimization involves reparameterization of the speciation-time parameters in order to enforce the constraint that a node cannot be older than its parent, as well as variable transformations to eliminate bounds, facilitating efficient unconstrained optimization. Partial derivatives of function g with respect to the transformed parameters can be computed efficiently, allowing use of gradient-based optimizers; we use the quasi-Newton L-BFGS algorithm.

Full details of the PAUP* implementation are provided in the supplemental material to [572], which interested readers can consult for more information.

4.3.3 UNCERTAINTY QUANTIFICATION

As was the case when estimating the species tree topology using SVDQuartets, it is important to provide a measure of the uncertainty associated with the MAP_{CL} estimates. While the application of results from [33] provides an explicit expression for the asymptotic variance, we have found this variance estimator to be unstable in practice. Thus, we recommend instead that bootstrapping be used to estimate the variance. The implementation of the MAP_{CL} method in PAUP* includes options for both the standard and the multilocus bootstrap, and we have found that these methods perform well in practice [572].

4.3.4 APPLICATION TO SPECIES RELATIONSHIPS AMONG GIBBONS

To demonstrate the performance of the MAP_{CL} estimators, we return to the gibbon data analyzed with SVDQuartets in the section 4.2.6. Using the tree estimated by SVDQuartets (figure 4.7), we obtain estimates of the speciation times as well as their variances using the PAUP* command: qage patProb=exactJC taxpartition= gibbonspecies loci=lociset bootstrap=multilocus;. The estimates are shown in figure 4.7. We note that the MAP_{CL} procedure provides estimates that are both fast and accurate—this analysis took 10.79 seconds on a desktop machine and compares favorably to the estimates provided by BPP using a much longer run time, as reported in [572]. In addition, the MAP_{CL} estimates were found to be much more robust to the choice of prior distribution than those obtained by BPP [572].

4.3.5 RECOMMENDATIONS FOR USING COMPOSITE LIKELIHOOD ESTIMATORS OF THE SPECIATION TIMES

As described above, the MAP_{CL} estimators assume the JC69 model under the MSC. We have used simulations (not shown here) to demonstrate that the method can be somewhat sensitive to this assumption. To handle cases in which more general substitution processes may be operating, PAUP* includes an option to compute the MAP_{CL} estimation procedure under more complex substitution models. The primary difficulty involved in using more general models is that closed form expressions for the site pattern probabilities are no longer available, and thus these site pattern probabilities must be approximated numerically. The implementation of this step in PAUP* is very accurate but requires some additional computation time. Nonetheless, estimation using the MAP_{CL} procedure will generally be much more efficient than the corresponding Bayesian approach.

The MAP_{CL} estimates will benefit from larger data sets in the same way that SVDQuartets does. That MAP_{CL} is statistically consistent means that these estimates become increasingly accurate as the sample size increases. As was the case for SVDQuartets, there is little computational cost associated with increasing the number of sites in the data, as counting the number of each type of site pattern is a rapid procedure that need be done only once since the species tree is fixed in this case. Thus, we recommend the MAP_{CL} estimates when the number of sites in the data set, whether multilocus or CIS, is large.

4.4 Conclusion and Future Work

Statistical methods for inferring species trees and associated parameters using site pattern frequencies provide computationally efficient, model-based approaches with provable statistical properties, including consistency. The methods described here, SVDQuartets and MAP_{CL} estimation of speciation times, have been implemented in PAUP* in a user-friendly interface, making them widely accessible. We have provided a tutorial that can be downloaded and/or viewed that replicates the analyses for the gibbon data that are included here. The tutorial is available at https://phylosolutions.com /tutorials/svdq-qage/.

Site pattern probabilities have also been applied to other problems in species-level phylogenetic inference. For instance, they are used to form both the HyDe [77] and

ABBA-BABA [208] statistics that allow for identification of species that have arisen via hybridization. They have also been used to identify the root position of a species tree [753]. Because they enable rapid computations under the MSC model, methods based on site pattern frequencies are promising approaches for computationally efficient species tree inference for large-scale data sets.

We note, however, that good performance of methods based on site pattern frequencies depends crucially on having obtained a large sample. For CIS, this means that many independent sites are needed, while for multilocus data, it means that the number of loci should be reasonably large (certainly more than ~50, but ideally several hundred to a thousand or more). Our view is that methods based on site pattern frequencies offer a strong alternative approach to Bayesian methods in cases in which the data are too large to allow Bayesian inference in a reasonable time. When data set sizes are smaller, Bayesian approaches can be expected to perform better and offer the advantage of a full characterization of the posterior distributions of interest. However, as sequencing advances continue to outpace the development of computational tools that can efficiently analyze the wealth of available data, methods based on site pattern frequencies will be invaluable in carrying out efficient inference in a theoretically justified framework.

CHAPTER 5

Practical Aspects of Phylogenetic Network Analysis Using PhyloNet

Zhen Cao, Xinhao Liu, Huw A. Ogilvie, Zhi Yan, and Luay Nakhleh

5.1 Introduction

Several evolutionary processes acting within and across species can give rise to non-tree-like evolutionary histories. These processes include admixture, gene flow, hybridization, hybrid speciation, introgression, and horizontal gene transfer (see section 5.7 for details related to PhyloNet and these processes). When any of these processes occurs, the evolutionary history (of the species or subpopulations) is more adequately represented by a network, rather than a tree, structure. In our context, a phylogenetic network takes the shape of a rooted, directed, acyclic graph [528], so that the single root represents the most recent common ancestor of all taxa at the leaves, and every internal node represents either divergence or reticulation.

As reticulation events occur in evolutionary settings—often among very closely related species [461], it is important to account simultaneously for *incomplete lineage sorting* (ILS), which also arises under similar conditions. In other words, for accurate inference of reticulate evolutionary histories, gene tree heterogeneity should not be explained solely by reticulation; rather, the potential for ILS to have occurred must be accounted for as well. This is a fundamental assumption in PhyloNet [790], which is a software package that implements a variety of methods for phylogenetic network inference while accounting for both reticulation and ILS.

For parsimonious inference where a network is inferred to minimize a measure of conflict with a set of gene trees, the *minimizing deep coalescence*, or MDC, criterion [455, 456, 748] was extended so that it works on species networks [840]. Inference based on this criterion requires a set of gene trees, one per locus (multiple trees per locus could also be used in order to account for uncertainty), and a network with a given number of reticulations that optimizes the MDC criterion (the lower its value, the more optimal the network) is sought.

As inference based on the MDC criterion does have its limitations in terms of statistical consistency [750] and does not allow for estimating evolutionary parameters beyond the topology of the network, method development turned to statistical inference. To accomplish likelihood-based inference, the multispecies coalescent [177] was extended to the multispecies network coalescent, or MSNC [841, 842]. This extension allowed for deriving the probability distribution of gene trees in the presence of both ILS and reticulation. Heuristic search techniques were developed for maximum likelihood estimation of phylogenetic networks, under the MSNC using gene tree estimates as the data [842]. As likelihood calculations on phylogenetic networks proved computationally very challenging, a set of techniques was first developed to speed up the exact likelihood calculations [844], and then a pseudolikelihood measure of phylogenetic networks was devised to allow for faster maximum (pseudo-)likelihood inference [843].

A major challenge facing the aforementioned inference techniques was determining the true number of reticulations, that is, the issue of model complexity. The MDC criterion does not penalize model complexity, and adding more reticulations almost always results in "better" networks under the criterion. Maximum likelihood estimation faces the same issue, and attempts at employing information criteria such as the Akaike Information Criterion (AIC) and Bayesian Information Criterion (BIC) did not provide satisfactory results [842]. We illustrate this issue in section 5.4 and discuss how to deal with it in practice when doing either maximum parsimony or maximum likelihood inference. To address the challenge of model complexity more systematically, a Bayesian approach was pursued so that model complexity is accounted for in the prior that is defined on the model. Bayesian inference of phylogenetic networks from gene trees was first devised by [789], and then to enable estimation of more parameters and to handle gene tree uncertainty more systematically, methods for Bayesian inference from multiple sequence alignment data and bi-allelic marker data were devised [787, 861]. To speed up computations from bi-allelic markers, a pseudolikelihood based technique for bi-allelic marker data was developed [860].

Last but not least, as phylogenetic network inference is computationally very expensive (could be orders of magnitude more expensive than tree inference), initial steps have been taken toward scalable phylogenetic network inference using divide-and-conquer techniques [859].

All of the aforementioned inference methods have been implemented in PhyloNet [790], in addition to other features for comparing and analyzing phylogenetic networks [749]. In this chapter, we do not review the mathematics or algorithmic techniques behind the methods, as those were reviewed recently [229]; rather, we illustrate the use of these methods on a simulated data set in the hope that the reader will learn how to use the various methods in PhyloNet, how to interpret the results, and what the capabilities and limitations of the methods are. The rest of the chapter is organized as follows. In section 5.2, we review what a phylogenetic network depicts and how to interpret it, and in section 5.3, we discuss the nature of the algorithmic techniques employed by PhyloNet for phylogenetic network inference under the various criteria. In section 5.4, we illustrate the application of the various inference methods on a simulated data set and discuss how to interpret the results and how to use the methods with caution, especially with respect to the number of reticulations. Furthermore, we report on the running times of the various analyses so as to give an idea to the user about the computational requirements of the various inference methods in PhyloNet. In section 5.5 we discuss how to analyze large data sets with the current methods in PhyloNet. Given that a set of phylogenetic networks can be inferred, we discuss in section 5.6 how to summarize them. In section 5.7, we discuss the issue of evolutionary processes underlying the umbrella term *reticulation* and how they cannot be distinguished by methods in PhyloNet (or any method, we believe). We also discuss how PhyloNet can be used to analyze data with polyploids. We end with conclusions in section 5.8.

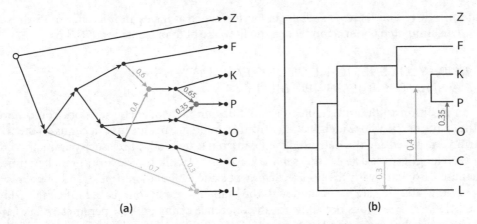

Figure 5.1. "Reading" a phylogenetic network. (a) The shape, or topology, of a phylogenetic network is a rooted, directed, acyclic graph. The (unique) root is open circle node.
The reticulation nodes are colored in light, dark, and intermediate greys. (b) Depicting the
same network in terms of a backbone tree (the solid black lines) and a set of reticulation
edges (the light, dark, and intermediate greys). The inheritance probabilities are shown
on the reticulation edges.

5.2 Reading and Interpretation of a Phylogenetic Network

The phylogenetic networks that PhyloNet deals with take the shape of rooted,
directed, acyclic graphs (to distinguish them from other types of phylogenetic network
[520]). Figure 5.1a gives an example of a phylogenetic network on seven taxa (e.g.,
species): C, F, K, L, O, P, and Z. What distinguishes a phylogenetic network from a phylogenetic tree is the presence of *reticulation nodes* in the former. A reticulation node has
two parental nodes, and its presence denotes that some of the genetic material in the
node's descendants are inherited from one parent while other genetic material in
the node's descendants are inherited from the other parent. For example, the phylogenetic network of figure 5.1a depicts a scenario in which some loci in P's genome share
a most recent common ancestor with loci from O's genome, and other loci in P's genome
share a most recent common ancestor with loci from K. Roughly speaking, the percentages of loci belonging to these two categories are 35% and 65%, as denoted by the *inheritance probabilities* that label the two *reticulation edges* associated with that reticulation
node.[1]

For ease of visualization, a phylogenetic network can be drawn as a tree, which we
refer to hereafter as the "backbone tree," and a set of reticulation edges, each of which
connects two edges of the backbone tree, as shown in figure 5.1b. In this case, the user
could interpret the backbone tree as the species tree and the reticulation edges as gene
flow events. However, we cannot overemphasize the point that this is only a visualization
decision; graph theoretically, and as far as PhyloNet methods are concerned, the two
seemingly different phylogenetic networks in figure 5.1 are exactly the same (more on
this issue in section 5.7).

A comment about the meaning of a reticulation edge is in order. For actual
biological species and populations, a reticulation edge abstracts an epoch of gene
flow of unknown duration, but likelihood-based species network methods make the
simplifying assumption that horizontal gene transfer occurs through instantaneous

intermixture. This issue, as well as the performance of Bayesian inference in PhyloNet on data simulated under a model of gene flow, are both presented in [787].

5.2.1 PHYLOGENETIC NETWORK PARAMETERS AND THEIR IDENTIFIABILITY

In addition to the inheritance probabilities, the phylogenetic network could have other parameters whose identifiability from an inference perspective depends on the criterion employed and the data used by the inference method. Three such parameters are branch lengths in coalescent units that are associated with the network's branches, times that are associated with the nodes (divergence and reticulation times), and population mutation rates that are associated with the branches. Parsimony-based inference using the MDC criterion does not allow for the inference of any of these parameters. Statistical inference based on gene tree topologies mainly allows for estimating branch lengths in coalescent units. Statistical inference from the sequence data directly allows for estimating all of these parameters (assuming, for example, the generation times are known). However, the identifiability of parameter values depends on the data being used, as we now illustrate.

Figure 5.2a shows a simple three-taxon phylogenetic network with a single reticulation node where the time associated with the reticulation node is 0.0358. With the method MCMC_SEQ in PhyloNet, which implements Bayesian inference from sequence alignment data [787], the time associated with the reticulation node is unidentifiable when only one individual is sampled from species K, but its estimation improves significantly when two individuals (or alleles) are sampled from species K, as shown in figure 5.2b and 5.2c, respectively. The sampled values for the time range between 0 and 0.2 have similar frequencies when a single individual from K is sampled, indicating no confidence in this value. When two individuals are sampled, the distribution of the sampled values changes drastically and peaks around the correct value, providing high confidence in the estimated value. However, even if multiple alleles were sampled from K, if the hybridization event is very old, then with high probability these alleles would coalesce into a common ancestor more recently than the hybridization event and, consequently, provide the same signal as a single allele. In other words, the ages of speciation and hybridization events play a critical role in the identifiability of hybridization times.

Examples of indistinguishability under the MSNC were discussed in [841], and a more thorough discussion, including with respect to what set of trees characterize a network under the MSNC, was presented in [560, 863, 862].

5.3 Heuristic Searches, Point Estimates, and Posterior Distributions, or, Why Am I Getting Different Networks in Different Runs?

The inference methods listed above and implemented in PhyloNet can be divided into two categories. The first category consists of solutions to optimization problems—problems defined by a solution, or solutions, that optimize a certain criterion. Belonging to this category are InferNetwork_MP, InferNetwork_ML, InferNetwork_MPL, and MLE_BiMarkers, which search for networks that optimize the MDC criterion, likelihood and pseudolikelihood from gene trees, and likelihood from bi-allelic markers, respectively. The second category consists of MCMC_GT, MCMC_SEQ, and

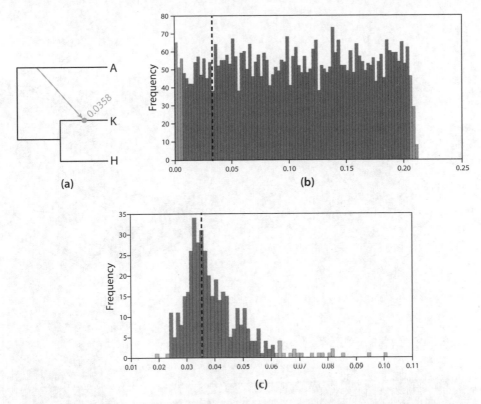

Figure 5.2. Identifiability of the network's parameters. (a) A three-taxon phylogenetic network. Shown in light grey is the time, in coalescent units, associated with the reticulation node. Histograms of the values sampled for the reticulation node time using mcmc_seq when (b) a single allele per taxon is sampled, and (c) when two alleles per taxon are sampled. The dotted vertical lines correspond to the true value (0.0358).

MCMC_BiMarkers, which use Markov chain Monte Carlo (MCMC) simulations to approximate the posterior distribution of the networks and their parameters from gene tree, sequence alignment, and bi-allelic marker data, respectively. Methods in the second category do not optimize a criterion.[2]

All problems that these methods are designed to solve are intractable—no exact algorithms for solving these problems exist in practice. By their nature, heuristics and approximations are not guaranteed to solve problems optimally, though well-designed ones perform very well in practice in terms of both accuracy and computational efficiency. When run from different starting points, these methods could return different solutions (e.g., networks) on the same data set. We illustrate why this happens and how to handle the seemingly different results from different runs on an example data set involving trees in figure 5.3.

For the results in the figure, we simulated a sequence alignment of length 300 bp, where the first 150 bp evolved down the tree (D,((A,C),(B,E))) and the last 150 bp evolved down the tree (D,((A,E),(B,C))). Starting the search for the maximum likelihood estimate twice from the same starting tree could take the search along two different paths, one that reaches a local optimum that is not the maximum likelihood

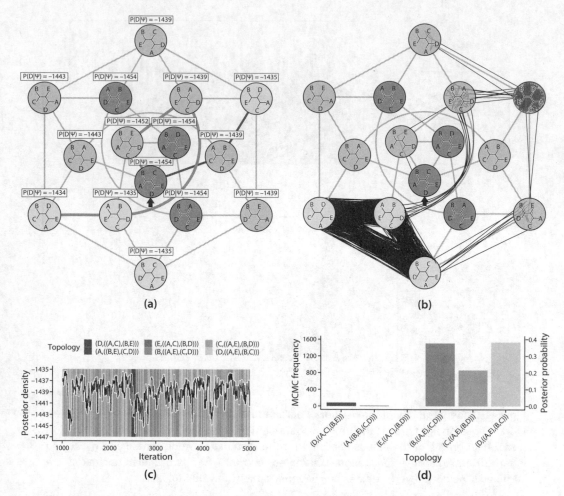

Figure 5.3. Heuristic search versus MCMC sampling. (a) Two possible hill climbing paths for five-taxon unrooted tree topologies based on nearest-neighbor interchange (NNI). Phylogenetic likelihoods, $P(D|\Psi)$ are for the optimal branch lengths for a given topology. Edges between topologies represent possible NNI moves. The light grey path begins from the randomly chosen starting topology (indicated with a solid arrow) and improves the likelihood until a local optimum is reached. The dark grey path begins from the same starting topology but instead by chance improves the likelihood until the global optimum is reached. Topologies are shaded in a spectrum spanning dark grey (lowest likelihoods) to light grey (highest likelihoods). (b) Markov chain Monte Carlo (MCMC) for unrooted trees with multiple optima. A 5000-iteration MCMC random walk (black path) is used to estimate the posterior distribution over topologies and branch lengths, with a uniform prior over both, by NNI operators and by random changes to branch lengths. The walk is initialized from a starting topology (black arrow). Grey lines show changes to tree topologies that are possible using NNI. (c) The posterior density appears well sampled, but the minor topology mode is sampled only once during the chain. (d) The number of MCMC samples is proportional to the estimated posterior probability of each topology, a mathematical feature of MCMC.

estimate, while the other does indeed reach the global optimum that is the maximum likelihood estimate (figure 5.3a).

In the case of Bayesian MCMC, a random walk starts from a certain tree and traverses the space of trees, not in search of an optimum but to "collect" information on the posterior distribution over all the possible trees. As figure 5.3b shows, the random walk visits the different trees but "circulates" among three topologies more than the other twelve topologies. The posteriors of the trees visited are plotted as in figure 5.3c, and then the marginal probability distribution on tree topologies is summarized from the samples, as in figure 5.3d. While one could return the tree (D,((A,E),(B,C))) as the one having the highest marginal probability (the light grey bar is the highest in figure 5.3d), the power of Bayesian analysis is that the totality of the results shown in the figure provide more information than just the optimal point. For example, the results in figure 5.3d show that the confidence level does not exceed 40% for any of the possible 15 trees. Furthermore, it shows that two trees have comparable probabilities, while a third one is close to them.

Back to network inference, any search strategy begins with one or more starting networks, proposes changes to the network or networks to generate a new candidate or set of candidate networks, and through some evaluation chooses which networks to retain or discard. This is true of the simplest strategy (basic hill climbing) or more complex strategies such as stochastic hill climbing, simulated annealing, or genetic algorithms. A starting network may be either randomly generated or it may be a point estimate derived from some other network. A popular way to generate a starting network is using some kind of maximum parsimony algorithm; for example, MCMC_GT uses a starting network with 0 reticulations (i.e., a tree) inferred using the MDC algorithm. The point of using a nonrandom starting network is to find the optimal network in less time by beginning the search closer in tree space to the optimum.

Beginning from the starting network, modifications are proposed using "operators" that sample from a distribution of networks that is conditional on the current network; this distribution can be considered the "neighborhood" of the current network. For the MCMC search strategies the mathematical properties of the proposal distribution are important, but for point estimate search strategies they are irrelevant; as long as the operators can efficiently traverse the space of networks from the start to the optimum, they are good operators. At some point a stopping condition is reached, and the best network sampled during the search reported as the optimal network. The danger of these search strategies is that if a stopping condition causes a search to terminate before the optimal network is sampled, a suboptimal network will be reported as the best network.

How are modifications accepted or rejected? The simplest criterion is to accept modifications that improve the optimality criterion (likelihood or parsimony) and reject those which do not. An algorithm that uses this criterion is called "hill climbing", because it attempts to climb up "hills" in the parameter space. A stopping criterion often used for hill climbing is the number of proposals since any proposal was accepted. If say 1000 modifications are proposed without any being accepted, it is likely that the search has reached *an* optimal point (not necessarily global), and further proposals will also be rejected.

However, just because we can be fairly certain when an optimal point is reached, it is much harder to be certain when the *global* optimum has been reached. In the space of networks (or other parameters) there may be several "hills" of good scores (either high likelihoods, posterior densities, or small parsimony scores). Because basic hill climbing will not climb down a hill, if the current network is sampled from one of the *local* optima far from the global optimum, the global optimum may be beyond the neighborhood of the current state, and the search will get stuck on top of the hill (as illustrated in the example of figure 5.3).

If a hill-climbing algorithm is run multiple times with different starting seeds, a different series of proposals will be made for each run. While one or more runs may be stuck in local optima, hopefully at least one will by chance reach the global optimum and report it as the best network. Therefore, when using a simple hill-climbing method, users must execute it multiple times with different starting seeds. The network with the best score across all of these runs should be reported as the best point estimate.

5.4 Illustration of the Various Inference Methods in PhyloNet

For the network figure 5.1, we generated 100 gene trees with two individuals per species using the program ms [329]. The exact ms command is

```
ms 14 100 -T -I 7 2 2 2 2 2 2 2 -es 0.5 4 0.35 -ej 0.6 7 4 -es 0.864 2
0.29999999999999993 -ej 1.0368 5 8 -es 1.7915903999999998 8 0.4 -ej
3.7150418534399994 6 8 -ej 5.349660268953598 8 2 -ej 6.419592322744317 10
3 -ej 11.093055533702179 4 2 -ej 15.973999968531139 2 3 -ej
19.168799962237365 3 9 -ej 50.0 9 1
```

We used the program Seq-gen [598] to simulate sequence data. We set the population mutation rate at 0.02, base frequencies of A, C, G, and T at 0.2112, 0.2888, 0.2896, and 0.2104, respectively, and the transition probabilities at [0.2173, 0.9798, 0.2575, 0.1038, 1, 0.2070]. The exact Seq-gen command is

```
seq-gen -m GTR -s 0.01 -f 0.2112, 0.2888, 0.2896, 0.2104  -r 0.2173,
0.9798, 0.2575, 0.1038, 1.0, 0.207  -l 1000
```

We removed the outgroup Z and ran the phylogenetic network inference methods InferNetwork_MP, InferNetwork_ML, InferNetwork_MPL, and MCMC_GT based on true gene trees and inferred gene trees from IQTREE [536]. In terms of the accuracy of gene tree estimates, 72 out of 100 gene trees inferred by IQTREE are topologically identical to true gene trees. The average error of gene tree branch lengths, which is computed as the rooted branch score distance [392] between each pair of inferred and true gene trees normalized by the branch length of the true gene tree, is 6.58%.

For each inference method, the maximum number of reticulations is set to be 0, 1, 2, 3, and 4, respectively. The number of processors used is eight, and all other parameters are the default settings. We refer to the tree drawn with black lines (the network without the three arrows) in figure 5.1 as the backbone tree.

5.4.1 INFERENCE UNDER THE MDC CRITERION

Phylogenetic networks inferred under the MDC criterion (the InferNetwork_MP command) from the true gene tree topologies are shown in figure 5.4. As the figure shows, the optimal tree (i.e., network with no reticulation nodes) inferred is the backbone tree of the true network. Furthermore, as the number of reticulations allowed was increased, the three reticulation events were identified one by one, until the true network was inferred (figure 5.4d).

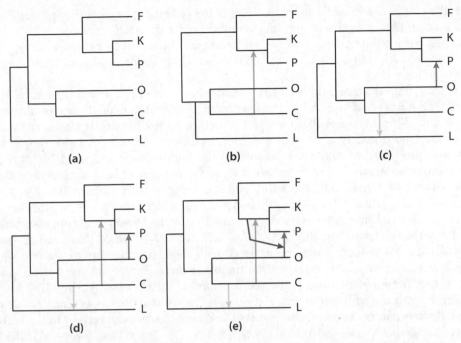

Figure 5.4. Inference results using the minimizing deep coalescence (MDC) criterion on the true gene tree topoloies. (a)–(e) The optimal networks inferred when the maximum number of reticulations was set to 0, 1, 2, 3, and 4, respectively. The MDC scores of the five networks in (a)–(e) are 363, 276, 175, 67 and 63, respectively.

Three points are worth highlighting based on the results of figure 5.4. First, while the reticulation between C and (K,P) was correctly inferred when the maximum number of allowed reticulations was set at 1, the placement of L is incorrect (figure 5.4b). There could be two explanations for this error in the network. One is that the network shown has a better score under the MDC criterion than the one with the correct placement of L. The second is that the network with the correct placement of L has a better score under the MDC criterion but that the heuristic for inferring the optimal network with a single reticulation was not run long enough to find it. To answer this question, we compared the MDC score of the network in figure 5.4b and the MDC score of the network that differs only in correcting the placement of L so that it matches that in the true network. We found that the network in figure 5.4a, with the reticulation from C to (K,P), has an MDC score 278, which means the network with the wrong placement of L has a better score under MDC. However, the difference in the MDC score between the two networks is only 2, an indication that the MDC criterion, in this case, cannot place L with high confidence. It is important to note here that this type of analysis cannot be done with real data, as the true network is unknown. However, if the biologist has an evolutionary hypothesis in mind (e.g., some modification to the inferred network), they can compute the MDC score of that network and compare it to the one inferred by the method (the DeepCoalCount_network command).

Second, observe that the network with two reticulations (figure 5.4c) is not simply the network with one reticulation (figure 5.4b) plus an additional reticulation. In fact, the two networks are disjoint in terms of the reticulation events they model. This observation illustrates the important point that under the MDC criterion, an optimal network with $k + 1$ reticulations cannot necessarily be obtained by first inferring an optimal network with k reticulations and then searching for one additional reticulation to add to it.

Third, while the true number of reticulations is three, the method inferred a network with four reticulations when the maximum number of reticulations allowable was set at four. This illustrates that the MDC criterion by itself is not sufficient to determine the true number of reticulations. One technique to use in practice, yet is not foolproof under all conditions, is to inspect the improvement to the MDC score as more reticulations are inferred. In this case, as the number of reticulations identified went from 0 to 4, the MDC score dropped according to the sequence 363, 276, 175, 67, and 63, respectively. Translating these MDC scores into subsequent rates of change (subtracting two consecutive numbers and dividing by the larger) yields improvements in the MDC scores of roughly 24%, 37%, 62%, and 6%. Notice that the improvement in the MDC score from the network with three reticulations to the one with four reticulations is much smaller than the other three improvements, a good indication that three reticulations is the correct number in this case. As the MDC score decreases with the addition of reticulations, and to avoid artificial percentages due to small denominators[3], one can also inspect the rate of change with respect to the MDC score of the tree: $(363 - 276)/363 \approx 24\%$, $(363 - 175)/363 \approx 52\%$, $(363 - 67)/363 \approx 82\%$, and $(363 - 63)/363 \approx 83\%$. This inspection makes it even clearer that the fourth added reticulation is very unlikely to be a real one but rather a reflection of the fact that the MDC score cannot get worse by adding more reticulations to the same network.

As gene trees are estimated from data in practice, we repeated the analysis but with the data consisting of gene tree topologies estimated using IQTREE and rooted at outgroup Z. The results are shown in figure 5.5. The main observations here are that the method inferred the correct network with three reticulations, despite the error in the gene tree estimates, and that the order in which the reticulation events were inferred is not consistent with that in the case of the true gene trees. The placement of L is now wrong when the method inferred a network with two reticulations, and the fourth "fake" reticulation is different from the one identified before. However, there are no generalizable patterns here. All these observations could be affected by the quality of the estimated gene trees.

The improvements to the MDC scores are now $(384 - 270)/384 \approx 30\%$, $(384 - 175)/384 \approx 54\%$, $(384 - 100)/384 \approx 74\%$, and $(384 - 88)/384 \approx 77\%$. We observe that in this case as well the criterion points to three as the correct number of reticulations.

Having illustrated MDC, especially that it worked perfectly on this data set, it is important to note that inference under the MDC criterion is not statistically consistent in the case of species trees [750]. Therefore, there would be cases in which the wrong network is inferred under the MDC criterion. However, we have observed that the criterion does very well in general [840].

5.4.2 MAXIMUM LIKELIHOOD INFERENCE

Phylogenetic networks inferred under maximum likelihood (the InferNetwork_ML command) from the gene tree topologies are shown in Figure 5.6.

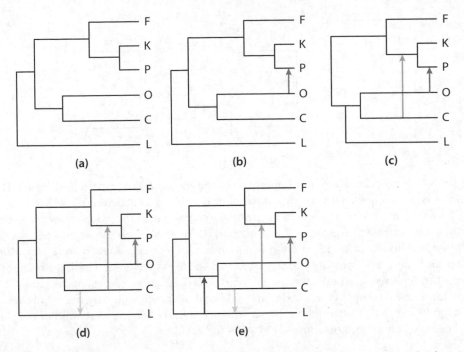

Figure 5.5. Inference results using the MDC criterion on the gene tree topologies estimated by IQTREE. (a)–(e) The optimal networks inferred when the maximum number of reticulations was set to 0, 1, 2, 3, and 4, respectively. The MDC scores of the five networks are 384, 270, 175, 100, and 88, respectively.

Given that the likelihood calculations are very expensive when gene tree topologies are used as input [862, 860, 229], analyses in which the maximum number of reticulations was greater than two were infeasible within the time allotted (we report on actual running times of all methods in section 5.4.5). For analyses in which the number of inferred reticulations was capped at two or fewer, inference based on the true and estimated gene tree topologies resulted in the same networks, which is a reflection of the method's robustness to gene tree error (up to a certain level).

The MDC criterion is not a multispecies coalescent criterion per se; rather, it is a simple criterion that aims at minimizing conflict between the input gene trees and the species phylogeny. On the other hand, maximum likelihood inference in this case *explicitly* makes use of the multispecies (network) coalescent. The results shown in figure 5.6, and even more so in figure 5.7 where branch lengths of the gene trees are used, highlight the big impact of gene flow on species *tree* inference, an issue that has been discussed and analyzed before [862, 787, 229, 108].

When a tree is inferred under maximum likelihood (that is, the maximum reticulation is set at 0), the tree (figure 5.6a) does not match the true backbone tree. The reason for this is that under the multispecies coalescent, to explain in a tree model the clades that group O and P genes together because of gene flow (the dark grey arrow in the true network), O is grouped with P and K. Similarly, owing to gene flow between C and (K,P) (the medium grey arrow in the true network), C is grouped with the clade the contains P and K before F is. This illustrates the issues that could arise when a species tree is inferred despite the fact that gene flow had occurred and is congruent with previous

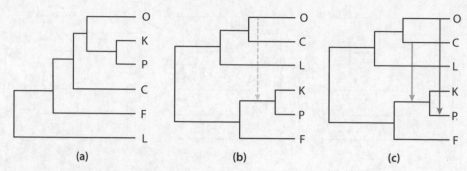

(a) (b) (c)

Figure 5.6. Maximum likelihood inference results on gene tree topologies. (a)–(c) The optimal networks inferred when the maximum number of reticulations was set to 0, 1, and 2, respectively. While the same network topologies were inferred from the true gene trees as well as from the gene trees estimated by IQTREE, the network parameters, as well as their likelihoods, differed. The log likelihoods of the three networks in (a)–(c) when using the true gene trees are −790, −686, and −653, respectively. The log likelihoods of the three networks in (a)–(c) when using the IQTREE gene tree estimates are −792, −719, and −635, respectively. Maximum likelihood inference with maximum number of reticulations set at 3 and 4 did not finish within 24 hours. The log likelihood of the true network is −388 for true gene trees and −624 for IQTREE.

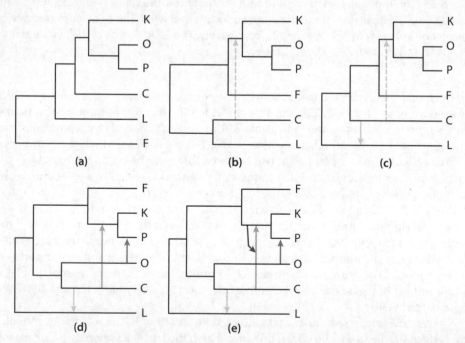

(a) (b) (c)

(d) (e)

Figure 5.7. Maximum likelihood inference results on the true gene trees (topologies and branch lengths). (a)–(e) The optimal networks inferred when the maximum number of reticulations was set to 0, 1, 2, 3, and 4, respectively. The log likelihoods of the five networks in (a)–(e) are −11925, −6292, −3492, −1642, and −1494, respectively. The log likelihood of the true network (topology, branch lengths, and inheritance probabilities) is −1408.

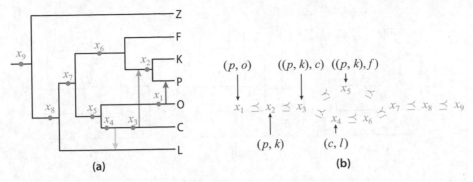

(a)   (b)

Figure 5.8. The constraints that coalescent times impose when a species tree is inferred in the presence of gene flow. (a) Coalescent events can occur at any internal branch in the phylogenetic networks, including the points $x_1, \ldots, x_9$. The backbone tree defines a partial order on those nine points: $x_1 \preceq x_6$, $x_2 \preceq x_5$, $x_3 \preceq x_4$, $x_4 \preceq x_6$, $x_5 \preceq x_7$, $x_6 \preceq x_7$, $x_7 \preceq x_8$, and $x_8 \preceq x_9$. The three reticulation events impose additional order. For example, owing to gene flow, the smallest possible time at which x_1 could occur is smaller than or equal to the smallest possible time at which x_2 could occur because alleles from P and O coalesce before alleles from P and K coalesce, thus adding $x_1 \preceq x_2$. (b) The partial order defined by the network and smallest possible coalescent times of $x_1, \ldots, x_9$.

results showing that the species tree methods ASTRAL and NJst infer erroneous clades under ILS and gene flow [713, 445]. As reticulations are allowed during inference, the clades, as well as the true reticulations, start to be recovered correctly, as shown in figure 5.6c.

As we mentioned, this issue is even more pronounced when the maximum likelihood inference utilizes the branch lengths (effectively the coalescent times) in the gene trees. As figure 5.7a shows, the inferred tree, while very different from the true backbone tree, makes much sense when the information on coalescent times is taken into consideration. The earliest possible (looking backward in time from the leaves toward the root) coalescent event can happen between alleles from O and P because of gene flow between the two species (the dark grey arrow in the true network). The second earliest possible coalescent event can happen between alleles from P and K, and so on. These temporal constraints defined by the phylogenetic network and the fact that a coalescence between two alleles from two species cannot occur before the divergence of these two species from their most recent common ancestor (MRCA) or the most recent episode of gene flow between them define a partial order as illustrated in figure 5.8. This partial order explains the species tree inferred and shown in figure 5.7a. However, as the permitted number of reticulations is increased, the tree and network structures are fully and correctly inferred, as shown in figure 5.7d. As with inference based on the MDC criterion, maximum likelihood without a penalty for model complexity cannot determine the number of reticulations, as illustrated in the network of figure 5.7e.

Finally, results based on gene trees with estimated topologies and branch lengths are similar to results based on true gene tree topologies and branch lengths, as shown in figure 5.9. However, while the method infers the true phylogenetic network topology with three reticulations, the branch lengths are different from those of the true one. In fact, the likelihood of the true network (true topology, true branch lengths, and true inheritance probabilities) given the estimated gene trees is 0, the reason being that the

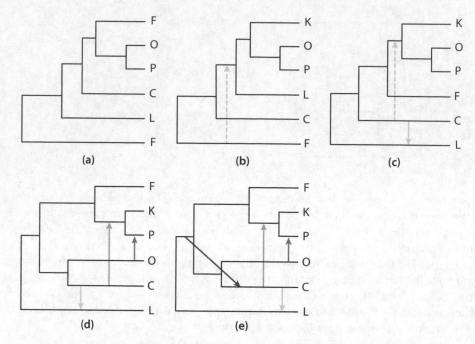

Figure 5.9. Maximum likelihood inference results on the gene trees (topologies and branch lengths) estimated by IQTREE. (a)–(e) The optimal networks inferred when the maximum number of reticulations was set to 0, 1, 2, 3, and 4, respectively. The log likelihoods of the five networks in (a)–(e) are -15683, -9232, -5977, -4044, and -3653, respectively. The true network (true topology, branch lengths, and inheritance probabilities) is not viable given the gene tree estimates (see text).

coalescent times are underestimated for some of the gene trees, forcing the network to a nonviable model (e.g., if two species A and B diverged from their MRCA at time t, and alleles from A and B coalesce at time τ; having $\tau < t$ is not a viable setting). The issue of underestimating the coalescent times in the gene trees, as well as the implications thereof on species tree estimation, was discussed in [172]. Unless one is certain of the gene tree branch lengths or coalescent times, we recommend against inference that utilizes gene tree branch lengths, despite that it is computationally much less demanding than the inference based on gene tree topologies alone.

5.4.3 MAXIMUM PSEUDOLIKELIHOOD INFERENCE

Phylogenetic networks inferred under maximum pseudolikelihood (the Infer Network_MPL command) from the true and inferred gene tree topologies are shown in figure 5.10 and figure 5.11, respectively.

As the pseudolikelihood of a network is efficiently computable, analyses now were run with up to four reticulations allowed. Using this method we obtained more accurate results than maximum likelihood, particularly in the case of an inferred species tree in which maximum pseudolikelihood seems to be robust to the presence of gene flow. Furthermore, it appears easier to identify the correct number of reticulations using pseudolikelihood. When estimated gene trees were used, the pseudolikelihood for the

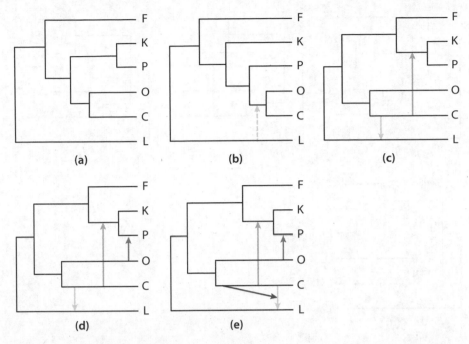

Figure 5.10. Maximum pseudo-likelihood inference results on the true gene tree topologies. (a)–(e) The optimal networks inferred when the maximum number of reticulations was set to 0, 1, 2, 3, and 4, respectively. The log pseudolikelihoods of the five networks in (a)–(e) are −13449, −11992, −10791, −10314, and −10311, respectively. The log pseudolikelihood of the true network (topology, branch lengths, and inheritance probabilities) is −10,341.

best network with three reticulations was identical to that with four reticulations. When the true gene trees were used, the log pseudolikelihood of the best networks with three and four reticulations found differed by only 3.

We also inferred phylogenetic networks under maximum pseudolikelihood from bi-allelic markers (the MLE_BiMarkers command), and the results are shown in figure 5.12. Except for the species tree in figure 5.12a and the incorrect placement of L in figure 5.12c, the results are very good, with the true network inferred when the maximum number of reticulations is set to 3. It is worth mentioning that when the inference was run with the maximum number of reticulations set to 4, the search still terminated with the network with three reticulations, shown in figure 5.12d, as no network with four reticulations had a better pseudolikelihood. This is almost the same pattern we observed above with maximum pseudolikelihood using gene tree topologies, for which the network with four reticulations has almost the same pseudolikelihood as the one with three reticulations.

5.4.4 BAYESIAN INFERENCE

PhyloNet has three functionalities that allow for Bayesian inference of phylogenetic networks: MCMC_GT infers phylogenetic networks using Bayesian MCMC from gene tree topologies [789], MCMC_BiMarkers works directly on bi-allelic marker data

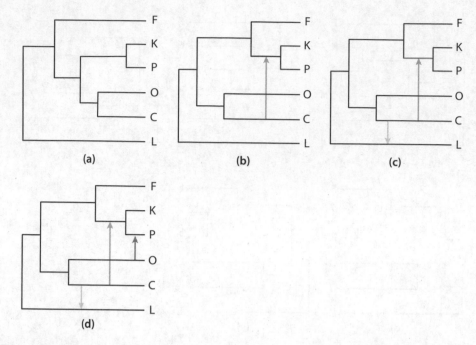

Figure 5.11. Maximum pseudolikelihood inference results on the gene tree topologies estimated by IQTREE. (a)–(c) The optimal networks inferred when the maximum number of reticulations was set to 0, 1, and 2, respectively. (d) The optimal network when the maximum number of reticulations was set to 3 and to 4. The log pseudolikelihoods of the four networks in (a)–(d) are −14582, −13339, −12540, and −12079, respectively. The log pseudolikelihood of the true network (topology, branch lengths, and inheritance probabilities) is −13,707.

[861], and MCMC_SEQ infers phylogenetic networks from sequence alignment data [787]. The former two are computationally very demanding and would not converge within 24 hours on the simulated data set used here. For MCMC_GT, the computational demands are similar to those of InferNetwork_ML when gene tree topologies alone are used. For MCMC_BiMarkers, the analytical computation of the integration over all gene trees becomes infeasible for certain phylogenetic networks, as illustrated in [860, 229]. Therefore, we analyzed the data set using MCMC_SEQ, which, unlike the other two, does not compute the joint integration of species network and gene tree histories entirely analytically; instead, it samples gene trees with their coalescent times and calculates the density function of gene trees. The results are shown in figure 5.13.

We make three observations. First, similar to the results based on maximum likelihood when the true gene trees with their branch lengths are used (figure 5.7), the species tree (when the analysis is constrained as to not allow any reticulations) has incorrect groupings of the taxa. The explanation for this is the same as that illustrated above with figure 5.8. Second, as the number of reticulations allowed is increased, the inference converges onto the correct network. Third, even when the maximum number of reticulations allowed was set at 4, the method inferred the network with the correct number

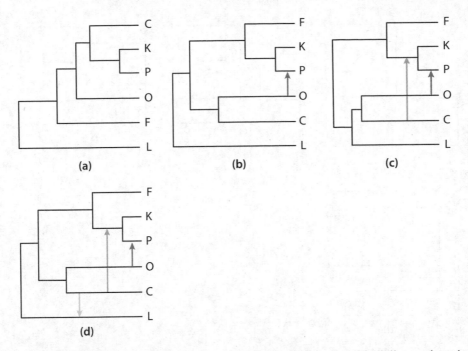

Figure 5.12. Maximum pseudolikelihood inference results using bi-allelic marker data. (a)–(d) The optimal networks inferred when the maximum number of reticulations was set to 0, 1, 2, and 3, respectively. The log pseudolikelihoods of the five networks in (a)–(d) are −677309, −676093, −675271, and −675165, respectively.

of reticulations (three), demonstrating the ability of Bayesian inference to determine the correct number of reticulations by imposing a prior distribution on the number of reticulations.

5.4.5 RUNNING TIME

While phylogenetic analyses take nontrivial amounts of computational time to perform, generally it will take up the minority of time required for a study, which requires fieldwork, sample preparation, and so on. Previous work has shown that Bayesian inference of species trees from sequences scales with the size of the data set employed roughly following a power law [550]. The CPU time required to achieve a posterior density effective sample size (ESS) of 200 on our test data set is consistent with a power law for a given number of individuals per species when inferring a species tree by setting the maximum number of reticulations to 0 (figure 5.14).

Low ESS values may be the result of a chain that is still in its burn-in phase, or a chain that is stuck in a local optima, or a chain that has reached the mode of the stationary distribution but has a high autocorrelation time. After concatenating the posterior samples from two independent chains for each combination of data set and maximum number of reticulations, chains that had not finished burning in or were stuck in different local optima will cause ESS to be very low (we used less than 30 as a rule of thumb). The calculated time to 200 ESS, extrapolated from these very small ESS values, will not be reliable.

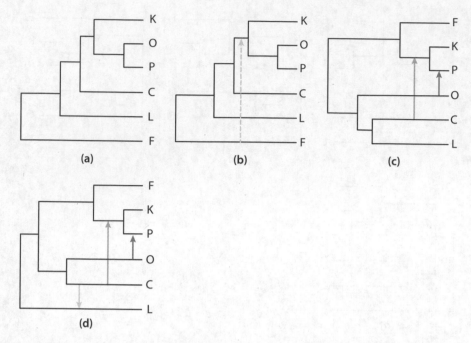

Figure 5.13. Bayesian inference results on the sequence alignment data. (a) The tree that appeared in 99.7% of the collected samples when the maximum number of reticulations was set to 0. (b) The network that appeared in 100% of the collected samples when the maximum number of reticulations was set to 1. (c) The network that appeared in 100% of the collected samples when the maximum number of reticulations was set to 2. (d) The network that appeared in 100% of the collected samples when the maximum number of reticulations was set to 3 or to 4.

Judging from the more reliable extrapolated times, increasing either the maximum number of reticulations permitted or the number of individuals per species increases the required CPU time. The time required for the 50×1 (loci $\times$ individuals per species) analyses were similar to the 25×2 analyses, and the 100×1 times were similar to 50×2, suggesting the total number of sequences in a data set is the key variable determining required time (figure 5.14).

Table 5.1 shows the running times of the various network inferences we discussed above on NOTS (Night Owls Time-Sharing Service), which is a batch scheduled high-throughput computing (HTC) cluster. If the inference exceeded 192 CPU hours, it was terminated.

5.5 Analysis of Larger Data Sets

Statistical inference of phylogenetic networks is computationally very demanding as the computational requirements of likelihood calculations on networks can be orders of magnitude larger than those on trees [860, 229]. Currently, analyses of larger data sets in terms of the numbers of loci and taxa can be done in PhyloNet in one of three ways:

1. **Maximum pseudolikelihood inference:** Inference using maximum pseudo-likelihood from gene tree topologies (the InferNetwork_MPL command). The

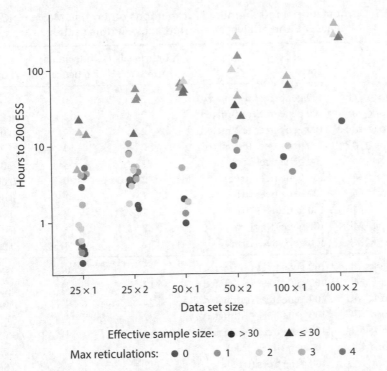

Effective sample size: ● > 30   ▲ ≤ 30
Max reticulations: ● 0   ● 1   ● 2   ● 3   ● 4

Figure 5.14. Running time requirements of MCMC_SEQ for the analysis reported in section 5.4.4. Two chains were run for each combination of data set and maximum reticulations, then concatenated after removing 50% burn-in before computing effective sample size (ESS). The time to a posterior density ESS of 200 was extrapolated from the ESS of the concatenated chains. The CPU time was halved to account for the time taken on the sample removed as burn-in. The full 100 locus data set was segmented into four 25 locus and two 50 locus data sets. Data set size is given as number of loci by number of individuals per species. Max reticulations refers to the maximum number of reticulations permitted.

advantage of this approach is that computing the pseudolikelihood of a network candidate is very efficient (can be done in seconds or minutes even on networks with 200 taxa). However, this approach does not circumvent the problem of searching the enormous space of phylogenetic networks. Therefore, maximum pseudolikelihood inference of very large networks can suffer in practice from getting stuck in local optima. Furthermore, while pseudolikelihood performed well on the data set above in terms of determining the number of reticulations, this is not the case in general.

2. **Tree-based augmentation:** Inference of a species tree and then augmenting it into a phylogenetic network (the InferNetwork_MPL -fs or Infer Network_MDC -fs command). The advantage of this approach is that there are very efficient methods for inference of species trees, and then the number of ways to add k reticulations to the tree is polynomial in n. The disadvantages of this approach are that the start tree could be wrong, thus resulting in a wrong network. Furthermore, most criteria for evaluating the network candidates are computationally very demanding, leaving mainly the MDC and pseudolikelihood criteria to use. In other words, this can be viewed as a very

Table 5.1. Computational requirements for methods other than MCMC_SEQ on the data set and analyses reported above in this section.

Method	Loci	Individuals per species	Maximum # reticulations	CPU hours
InferNetwork_MP	100 gene trees (estimated)	2	0	0.0024
InferNetwork_MP	100 gene trees (estimated)	2	1	0.0023
InferNetwork_MP	100 gene trees (estimated)	2	2	0.0055
InferNetwork_MP	100 gene trees (estimated)	2	3	0.0055
InferNetwork_MP	100 gene trees (estimated)	2	4	0.0247
InferNetwork_MP	100 gene trees (true)	2	0	0.0002
InferNetwork_MP	100 gene trees (true)	2	1	0.0014
InferNetwork_MP	100 gene trees (true)	2	2	0.0035
InferNetwork_MP	100 gene trees (true)	2	3	0.0052
InferNetwork_MP	100 gene trees (true)	2	4	0.0098
InferNetwork_ML	100 gene trees (estimated)	2	0	0.0477
InferNetwork_ML	100 gene trees (estimated)	2	1	0.7828
InferNetwork_ML	100 gene trees (estimated)	2	2	8.6983
InferNetwork_ML	100 gene trees (estimated)	2	3	51.5806
InferNetwork_ML	100 gene trees (estimated)	2	4	38.5744[b4]
InferNetwork_ML	100 gene trees (true)	2	0	0.0362
InferNetwork_ML	100 gene trees (true)	2	1	0.5652
InferNetwork_ML	100 gene trees (true)	2	2	2.5620
InferNetwork_ML	100 gene trees (true)	2	3	90.8990
InferNetwork_ML	100 gene trees (true)	2	4	15.8162[b2]
InferNetwork_ML	100 gene trees[a] (estimated)	2	0	1.2064
InferNetwork_ML	100 gene trees[a] (estimated)	2	1	91.3122
InferNetwork_ML	100 gene trees[a] (estimated)	2	2	77.7539[b1]
InferNetwork_ML	100 gene trees[a] (true)	2	0	0.6156
InferNetwork_ML	100 gene trees[a] (true)	2	1	75.4391
InferNetwork_ML	100 gene trees[a] (true)	2	2	170.5937[b1]
InferNetwork_MPL	100 gene trees (estimated)	2	0	0.0046
InferNetwork_MPL	100 gene trees (estimated)	2	1	0.0077
InferNetwork_MPL	100 gene trees (estimated)	2	2	0.0156
InferNetwork_MPL	100 gene trees (estimated)	2	3	0.0256
InferNetwork_MPL	100 gene trees (estimated)	2	4	0.0335
InferNetwork_MPL	100 gene trees (true)	2	0	0.0046
InferNetwork_MPL	100 gene trees (true)	2	1	0.0075
InferNetwork_MPL	100 gene trees (true)	2	2	0.0177
InferNetwork_MPL	100 gene trees (true)	2	3	0.0297
InferNetwork_MPL	100 gene trees (true)	2	4	0.0408
MCMC_GT[c]	100 gene trees (estimated)	2	0	32.0089
MCMC_GT[c]	100 gene trees (estimated)	2	1	76.8422
MCMC_GT[c]	100 gene trees (true)	2	0	22.1422
MCMC_BiMarkers	10,000 bi-allelic markers			not finished[d]

Table 5.1. (*continued*)

Method	Loci	Individuals per species	Maximum # reticulations	CPU hours
MLE_BiMarkers	10,000 bi-allelic markers	1	0	0.0773
MLE_BiMarkers	10,000 bi-allelic markers	1	1	0.2532
MLE_BiMarkers	10,000 bi-allelic markers	1	2	0.4390
MLE_BiMarkers	10,000 bi-allelic markers	1	3	2.1085
MLE_BiMarkers	10,000 bi-allelic markers	1	4	11.3406
MLE_BiMarkers	10,000 bi-allelic markers	2	0	0.1954
MLE_BiMarkers	10,000 bi-allelic markers	2	1	1.5781
MLE_BiMarkers	10,000 bi-allelic markers	2	2	9.9110
MLE_BiMarkers	10,000 bi-allelic markers	2	3	not finished[d]
MLE_BiMarkers	10,000 bi-allelic markers	2	4	not finished[d]

[a] Topologies only, no branch lengths.
[b1] Time for only one run rather than all five runs; exceeded 192 CPU hours for the second run.
[b2] Time for only two runs rather than all five runs; exceeded 192 CPU hours for the third run.
[b4] Time for only four runs rather than all five runs; exceeded 192 CPU hours for the fifth run.
[c] Finished all 1100 samples. The remaining cases have fewer than 100 samples (burn-in) within the time limit of 192 CPU hours.
[d] Within the time limit of 192 CPU hours.

constrained version of number 1 above, where the search space consists of only augmentation of a given tree into a network. The merits and limitations of this approach were recently studied in [108].

3. **Divide-and-conquer:** Inference based on a divide-and-conquer approach (the NetMerger command). This approach is based on the recently developed divide-and-conquer method of [859], in which the set of taxa is divided into overlapping three-taxon subsets, trinets (networks on three taxa) are inferred, and then these trinets are merged into the full network. A major advantage of this method is that it avoids searching the space of large phylogenetic networks, as only the space of trinets is searched (in parallel for each three-taxon subset). Furthermore, model complexity is handled in the inference of the trinets; combining them together need not deal with it.

The networks in figure 5.1 and figure 5.2 are in fact sampled from the larger network shown in figure 5.15a, whose inference is infeasible via direct application of InferNetwork_MP, InferNetwork_ML, MCMC_GT, MCMC_SEQ, MCMC_BiMarkers, or MLE_BiMarkers.[4] Maximum pseudolikelihood inference using the InferNetwork_MPL command with the gene trees estimated by IQTREE and setting the maximum number of reticulations at 3 yields the correct network that is shown in figure 5.15a. The log pseudolikelihood of the network is -158498.

For the tree-based augmentation approach, we ran ASTRAL [850] on the gene trees (estimated by IQTREE) to infer a backbone tree. ASTRAL inferred the correct backbone tree; that is, the tree obtained from the network in figure 5.15a by removing the three arrows. The method then searched for augmentation of the tree into a network in four different ways: searching for the optimal network under the pseudolikelihood criterion of [843] and under the MDC criterion of [840], with the number of reticulations set at 3 and at 5 for each of the two criteria. Augmentation into networks based on both

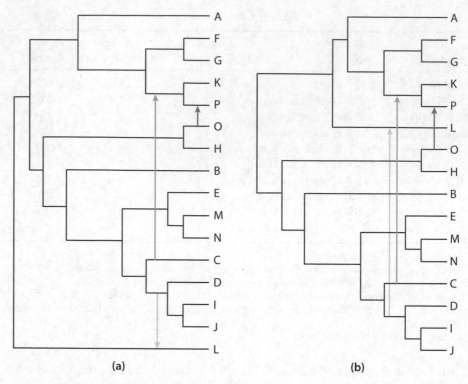

Figure 5.15. Analysis of a larger phylogenetic network. (a) A model phylogenetic network on 16 taxa with outgroup taxon Z. (b) The phylogenetic network inferred by the divide-and-conquer method of [859].

criteria resulted in the correct network when the number of reticulations was set at 3. The log pseudolikelihood and MDC score of the resulting network are −158498 and 173, respectively. However, when the number of reticulations was set at 5, both criteria resulted in two additional reticulations (the true network with three reticulations, plus two additional reticulations, for a total of five reticulations) with log pseudolikelihood and MDC score of −158483 and 160, respectively. Once again, these results illustrate that inference based on these criteria yields good results but that both of them suffer from the problem of determining the correct number of reticulations. Finally, using the divide-and-conquer method of [859], the inferred network is shown in figure 5.15b. As the result shows, the only error in the inferred network is the placement of taxon L. However, despite this wrong placement, the reticulation involving L and the clade (D,(I,J)) is inferred correctly. It is important to highlight here that the number of reticulations need not be specified as the trinets were inferred using the Bayesian method MCMC_SEQ, which determined the number of reticulations in each trinet individually. Furthermore, it is worth mentioning that when the true trinets were used in NetMerger, the correct network was inferred, which points to inaccuracy in some of the inferred trinets as the cause of the wrong placement of L.

With respect to the running time, maximum pseudolikelihood inference took 13.8293 CPU hours for 10 runs when the maximum number of reticulations was set to 3. Under the pseudolikelihood criterion, the tree-based augmentation approach took 7.2995 and 10.3108 CPU hours when the maximum number of reticulations was set to 3 and 5, respectively, for 10 runs. Under the MDC criterion, when the number of reticulations was set to 3 and 5, the tree-based inference took 3.9351 and 24.4727 CPU hours, respectively, for 10 runs. The divide-and-conquer inference took 1337.8253 CPU hours inferring the 680 three-taxon subnetworks by MCMC_SEQ (this can be trivially parallelized, but the time we report is the total time for inferring all 680 three-taxon subnetworks sequentially), while it took only 2.34 minutes for the merger procedure of obtaining the full network from the 680 three-taxon subnetworks.

5.6 Comparison and Summarization of Networks

While comparing and summarizing trees is easy, the same tasks are very hard for networks. For example, while testing if two trees are isomorphic can be done in polynomial time, the problem is NP-hard for phylogenetic networks in general. This is why various dissimilarity measures for restricted classes of phylogenetic networks have been devised [529, 113, 114, 110, 326, 115, 775]. While PhyloNet has utilities to compare phylogenetic networks based on their constituent trees, tripartitions, and clusters, it also has a function that computes the distance measure described in [529]. However, all of these measures are very sensitive and return "inflated" dissimilarity values even for what would, visually, be considered a small difference between the networks. We now discuss five features in PhyloNet for summarizing sets of phylogenetic networks with the goal of elucidating common substructures in sets of networks that are obtained by an inference method. The input to a summarization method is a set of networks obtained by an inference method (e.g., a set of equally optimal networks under the MDC criterion, or a set of the different network topologies collected from MCMC). The output is a summary that differs depending on the method used, as we now describe. The input to the summary task is a set of networks $\{\Psi_1, \Psi_2, \ldots, \Psi_n\}$, and the output is a set of pairs $\{(S_1, v_1), (S_2, v_2), \ldots, (S_m, v_m)\}$, where each S_i is a substructure obtained from the networks and v_i is its frequency in the input networks (normalized by the number of networks so that all frequencies add up to 1).

5.6.1 DISPLAYED TREES

A network with k reticulations displays up to 2^k trees, where each tree is obtained from the network by removing one of the two reticulation edges for each reticulation node. The summary based on displayed trees consists of the union of the displayed trees of all networks, and the frequency of each tree is derived based on the number of networks by which it is displayed.

5.6.2 BACKBONE NETWORKS

Backbone networks extend the notion of displayed trees. Given a network Ψ with k reticulations, a backbone of it is a network Ψ' that is obtained by removing one of the two reticulation edges for each node in a *subset* of the set of all reticulation nodes. If the subset contains all reticulation nodes, the resulting backbone is also a displayed tree. If the subset is the empty set, the resulting backbone network is the network Ψ itself. Thus, there are up to $\sum_{i=0}^{k} \binom{k}{i} 2^i$ backbone networks. The summary based on backbone networks consists of the union of the backbones of all networks, and the frequency of each backbone is derived based on the number of networks of which it is a backbone.

5.6.3 TREE DECOMPOSITIONS

A network can be decomposed into a forest of trees by applying the following operation to each reticulation node u whose parents are v_1 and v_2:

- Create two nodes x_1 and x_2 and label both of them by a unique name, say, Q;
- Delete the two edges (v_1, u) and (v_2, u); and,
- Add two edges (v_1, x_1) and (v_2, x_2).

This operation was illustrated in figure 3 of [749]. Given the tree decomposition of all networks in the input, a summary consists of all the unique trees in the union of resulting forests, along with the frequency of each tree.

5.6.4 TRIPARTITIONS

Each reticulation node u can be represented in the form of tripartition as $X : P_1 | P_2$, where X is the set of taxa that are under this reticulation node, and P_1 and P_2 are two sets of taxa that are under the two siblings of this reticulation node, respectively. When one of the parents of the reticulation node u is also a reticulation node, u has only one sibling, and the tripartition is of the form $X : P$. When both parents of u are reticulation nodes, u has no siblings, and the tripartition is of the form X. For example, for the reticulation node denoted by the medium grey circle in figure 5.1, the tripartition is

$$\{K, P\} : \{C, L\} | \{F\}.$$

The rationale behind a tripartition is that it illustrates that the clade containing the leaves in X is a hybrid and that the two clades given by the sets of taxa P_1 and P_2 are the parents of the hybrid. Once the tripartitions of all networks in the input are computed, they are displayed along with their frequencies across all networks (each unique tripartition appears once).

5.6.5 MAJOR TREES

For inference methods that produce networks with estimated inheritance probabilities, the major tree of a network is the one that is obtained from the network by removing for each reticulation node the reticulation edge with the lower inheritance probability (one is chosen arbitrarily if the inheritance probability is 0.5). The summary, in this case, consists of all major trees of the networks (each unique tree appears once) and for each the number of networks of which it is the major tree.

5.7 Reticulate Evolutionary Processes in PhyloNet

We believe the proliferation in the literature of specialized terms of *reticulate evolution* hampers, rather than helps, the development of general tools for inferring reticulate evolutionary histories. It is unclear, for example, what the difference between hybrid speciation and hybridization is from a modeling point of view. While the former results in a new species and the latter does not, from a modeling and inference perspective that distinction seems irrelevant. Gene flow has been defined as the "exchange of genes between two populations as a result of interbreeding" whereas admixture is "a sudden increase in

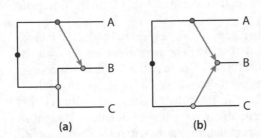

Figure 5.16. Visualizing the same network in different ways conveys different biological stories. (a) The phylogenetic network is drawn to convey the message that (A,(B,C)) is a species tree and that there is hybridization involving an ancestor of A and an ancestor of B. (b) The phylogenetic network is drawn to convey that hybrid speciation occurred between (ancestors of) species A and C and gave rise to new species B. Nodes with the same shading in the two networks correspond to each other.

gene flow between two differentiated populations" [580]. However, it is unclear if this distinction and naming have any implications on developing computational methods for inference. As far as PhyloNet application is concerned, all of these processes are collectively referred to as reticulation; elucidation of the actual biological process (e.g., whether it is bacterial conjugation or homoploid hybridization) that gave rise to the reticulation is not within the capabilities of phylogenetic method inference methods in PhyloNet.

Let us illustrate the issue with the phylogenetic networks in figure 5.16. What is shown is *one* phylogenetic network that is visualized in two different ways. That is, in graph-theoretic jargon, the two networks shown in figure 5.16 are *isomorphic*. Presented with multilocus data from the three taxa A, B, and C, the various inference utilities in PhyloNet yield a single phylogenetic network (let us assume here that the inferred network is unique). Calling the reticulation event hybridization and drawing the network as in figure 5.16a or calling the event hybrid speciation and drawing the network as in figure 5.16b is not a distinction made by PhyloNet nor is it an aspect that is inferable from data. Such a distinction can be made after the network is inferred and using some knowledge that is "external" to the inferred network topology.

Along the same lines, if A, B, and C were different species, the reticulation event illustrated in figure 5.16 would be referred to as hybridization or hybrid speciation. On the other hand, if the same exact network was inferred from data collected from different individuals collected from subpopulations of the same species, the reticulation event would be referred to as gene flow or admixture. And to hammer the point home even further, if the data consists of a sequence alignment from three individuals in one population and recombination was present, the phylogenetic network of figure 5.16 would be referred to as an *ancestral recombination graph*, and the reticulation node would denote a recombination event.

Therefore, it is important to distinguish between the networks that are inferred by methods such as those in PhyloNet, where the non-tree-like events should be generically referred to as reticulation, and the actual biological process that took place (e.g., hybridization vs. hybrid speciation vs. admixture) and whose elucidation is not within the realm of what a phylogenetic inference method is capable of. This very important point is a major theme in the excellent book on phylogenetic networks by David Morrison [520]. As Morrison wrote, "Unfortunately, there is little that a mathematical network can necessarily tell us about any underlying biology. That a particular taxon is apparently the descendant of a reticulation event does not actually tell us what that event was, nor even whether the correct taxon has been identified. For example, how to distinguish HGT and hybridization mathematically is not obvious." A method's inability to distinguish, say, horizontal gene transfer from hybridization, is not a weakness or limitation of the method. To the contrary, a claim that a method can make such a distinction from molecular data should raise questions.

Why, then, emphasize the process of hybridization with respect to PhyloNet, including in the titles of manuscripts as in [841, 840, 229]? The reason has to do with the model employed by PhyloNet and the detectability of reticulation. While a thorough study is currently lacking, we believe the signal in data obtained from a single individual within each subpopulation is not sufficient to infer the phylogenetic network that represents the network-like demographic structure of the entire population. This insufficiency is why the more popular methods in population genetics, namely STRUCTURE [587] and TreeMix [579], rely on allele frequency data that is obtained from many individuals within the population. Similarly, we believe all the complexities that arise in prokaryotic evolution owing to processes such as gene fusion and fission [487], coupled with the fact that in some cases only a single gene could be transferred, make it very challenging and potentially not useful to apply current functionalities in PhyloNet to analyze the evolutionary history of prokaryotic genomes.

5.7.1 ANALYSIS OF POLYPLOIDS

Despite the fact that polyploidization plays an important role in plant evolution, few studies attempt specifically to handle reticulations arisen from polyploidy, especially in the presence of ILS. Lott et al. [452] presented a method to construct a consensus tree by summarizing a collection of multilabeled gene trees. Jones et al. [351] developed a statistical framework that could accurately infer the evolutionary history of allopolyploids but restricted to networks of diploids and tetraploids. Gregg et al. [278] developed an algorithm to detect the polyploidy events by reconciling the known (multilabeled) species tree and a set of (multilabeled) gene trees but not accounting for ILS. Recently, Oberprieler et al. [547] designed a workflow for reconstructing the reticulate evolutionary history involving polyploid complexes from multilocus data using MDC-based inference in PhyloNet. The workflow uses a permutation approach to assign homoeologs in polyploids and then constructs the multilabeled (MUL) species tree under the MDC criterion. The algorithm proceeds in the following steps:

1. For each polyploid species and each gene tree, it generates all possible diploid parental inheritance histories of all possible combinations of allele pairs. To determine the polyploid-specific and locus-specific optimal allele pairs inheritance history, the MDC score is computed by running the Infer_ST_MDC command for each of the possible histories, and the one with minimum MDC score is kept.

2. Based on the polyploid-specific and locus-specific optimal allele combinations at all sequenced loci generated from step 1, it uses the Infer_ST_MDC command to compute the MDC score and find the polyploid-specific optimal allele combinations across loci.

3. After obtaining the polyploid-specific optimal allele combinations across loci for all polyploids, it runs the Infer_ST_MDC command to reconstruct the

overall MUL-species tree and finally converts the resulted MUL-species tree into species network.[5]

However, such a permutation approach can be executed directly using the Infer-Network_MP command. The InferNetwork_MP command would compute the optimal MDC score by converting the species network into the corresponding MUL-tree. Then all possible allele mappings to the leaves of the MUL-tree are produced, and the mapping that gives rise to the minimum MDC score is selected as the optimal coalescent history. Through this way, the optimal assignment of homoeologs to parental lineages in polyploids based on the MDC criterion is obtained. To illustrate this approach, we reanalyzed the *Leucanthemopsis* test data set of [547], which consists of 12 diploid, tetraploid, and hexaploid representatives of *Leucanthemopsis*, to reconstruct reticulate evolution involving hybridization and polyploidization and compared the result from the InferNetwork_MP command to that from the permutation approach through the Infer_ST_MDC command [547]. It is important to note that the given data set consists of only five inferred gene trees and that this small number of loci might not be sufficient to reconstruct a reliable history.

To construct the reticulate evolutionary history, we set the maximum number of allowed reticulations to 0, 1, 2, and 3. With increasing number of reticulations from 0 to 3, the sequence of MDC scores obtained was 139, 136, 135, and 135, respectively; the networks are shown in figure 5.17. It took less than 3.5 minutes to infer each of the resulting networks over 20 runs using one processor. Observe that the improvement in the MDC score from the zero reticulations to the one with one/two reticulation(s) is very small. And when setting the maximum number of allowed reticulations to 3, two of the three returned species networks with equivalent optimal MDC score have only two reticulations. This observation shows that it is very unlikely to have more than one reticulation for this data set (given the five gene trees used). One major difference between the species network with zero reticulations and with one reticulation is that the tetrapolyploid *L. alpina* is found to have autopolyploid origin in the former whereas it is assumed to be allopolyploid and formed from the hybridization between *L. pallida* subs. *virescent* var. *bilbilitanum*. The inferred species network from the permutation approach yields an MDC score of 155 (the DeepCoalCount_network command), which is greater than the one with 0/1/2/3 reticulations found by InferNetwork_MP.

If some species are known to be hybrid, InferNetwork_MP allows the user to restrict the search space of the species networks by forcing a set of species to be hybrid species using the -h option. We ran the analysis on the same data set, with one specified hybrid species, *L. alpina* subsp. *cuneata*, and with two specified hybrid species namely *L. alpina* subsp. *cuneata* and *L. longipectinata*, separately. The inferred species networks are shown in figure 5.18. For the former case, we ran the InferNetwork_MP with the maximum number of allowed reticulations set to 1, 2 and 3, while for the latter one, the maximum number of allowed reticulations was set to 2 and 3. When only *L. alpina* subsp. *cuneata* was forced to be a hybrid species, the inferred species networks are of equivalent MDC score 143 (figure 5.18a, b), and only one reticulation event was identified no matter what the maximum number of allowed reticulations was. When it comes to two specified hybrid species, with maximum number of allowed reticulations set to 2 and 3, unique species networks were returned (figure 5.18c with MDC score 138 and figure 5.18d with MDC score 136, respectively). Why would searching in a more restricted phylogenetic network space result in a network topology with better MDC score? It is worth repeating the message above that InferNetwork_MP uses the

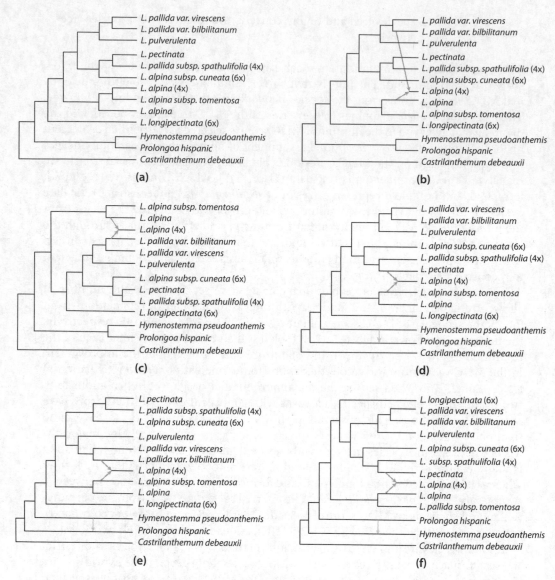

Figure 5.17. Inference results using the MDC criterion on the example data set of *Leucanthemopsis* representatives in [548]. (a)–(i) Species networks returned by InferNetwork_MP. (a), (b)–(c), (d)–(f), and (g)–(i) are optimal networks inferred when the maximum number of reticulations was set to 0, 1, 2, and 3, respectively, and the corresponding optimal MDC scores are 139, 136, 135, and 135, respectively. (j) The phylogenetic network inferred by the permutation approach of [548] and whose MDC score is 155 as computed by the DeepCoalCount_network command.

hill-climbing strategy to search for the optimal network. When only one hybrid species *L. alpina* subsp. *cuneata* was specified, in the limited number of runs, InferNetwork_MP had not yet searched over the more restricted space (i.e., the search space of two specified hybrid species *L. alpina* subsp. *cuneata* and *L. longipectinata*), and thus it did not find a network with the same or better MDC score.

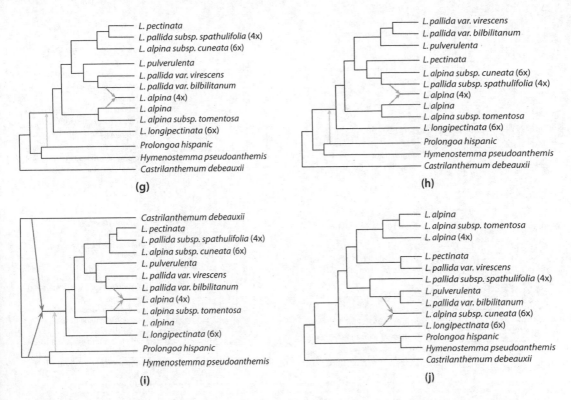

Figure 5.17. (*continued*)

In [841, 840], MUL-trees were used as the representation on which likelihood and parsimony computation were done. A thorough analysis of the connection between MUL-trees and phylogenetic networks, including cases of uniqueness (a MUL-tree uniquely defines a phylogenetic network) is given in [325]. As such, in addition to InferNetwork_MP, other methods in PhyloNet could be used to analyze data sets with polyploids.

Finally, while we demonstrated how methods in PhyloNet can be used to analyze data with polyploids, much work is still needed to model accurately the complexity that arises in the presence of polyploids, in particular when the mode of polyploidy—allo or auto— needs to be elucidated, as one is about reticulation and the other is effectively about whole-genome duplication. The recent review of Oxelman et al. [556] discusses existing methods and their applications to data with polyploid species, as well as challenges and future directions in the field.

5.8 Conclusions

Our group has been developing the PhyloNet software package for over 10 years. Our ultimate goal is to contribute to the development of a wide array of tools that will make phylogenetic network analysis and use in evolutionary biology as simple and natural as that of phylogenetic trees.

In this chapter, we discussed practical aspects that relate to data analyses using PhyloNet. This software package has been developed to enable evolutionary analyses of

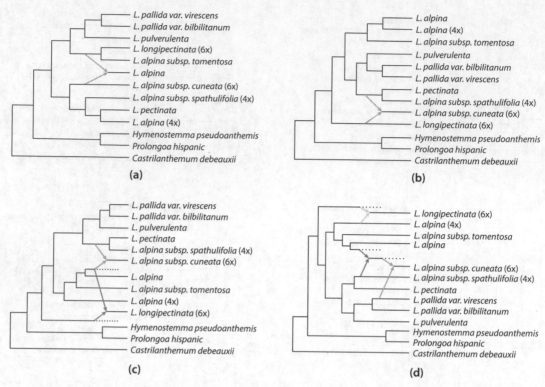

Figure 5.18. Inference results using the MDC criterion on the example data set of *Leucanthemopsis* representatives in [548] with specified hybrid species. When forcing the hexapolyploid *L. alpina* subsp. *cuneata* to be a hybrid species with the -h option, and setting the number of reticulations to 1, 2 and 3, InferNetwork_MP returned two distinct species networks (a, b) with identical MDC scores of 143. When specifying the hexapolyploid *L. alpina* subsp. *cuneata* together with the hexapolyploid *L. longipectinata* to be hybrid species, and setting the number of reticulations to 2 and 3, InferNetwork_MP inferred a network with a MDC score of 138 (c), and a network with a MDC score of 136 (d), respectively.

data sets where reticulation is suspected and consists of several inference functionalities that employ different criteria and inference algorithms. Analyzing and interpreting phylogenetic networks is much more challenging than analyzing and interpreting trees; therefore, it is important that the results obtained by network inference methods are inspected carefully. A major challenge in this area is determining the true number of reticulations that had occurred during the evolutionary history of the taxa under consideration. While the Bayesian approach allows for incorporating model complexity in a natural way via the prior distribution, extra caution must be taken when inferring networks using any of the inference methods in order to distinguish between real reticulations and false ones whose addition is simply a case of making von Neumann's elephant wiggle its trunk.[6]

While PhyloNet is capable of inferring phylogenies with reticulations, it cannot determine the biological processes behind those processes. One major direction for research and addition of features to PhyloNet relates to modeling gene duplication and loss—in addition to reticulation and ILS, which are currently handled by the software package. Furthermore, PhyloNet neither makes assumption about species nor delimits them. At an abstract level, whatever taxa the user specifies as different species, PhyloNet handles them as such.

The relationship between species networks and trees is intricate. While networks extend trees, inferring a network with no reticulations—that is, a tree—does not necessarily mean a correct tree (whatever that tree means) is inferred. The flip side of this coin is that network inference by first inferring a tree and then "completing" it into a network is not necessarily a good approach [108]. Methods implemented in PhyloNet infer evolutionary histories by directly searching the space of phylogenetic networks, and not the space of completions of a given tree into a network, though we provide such a feature for those interested in using it.

Last but not least, computational requirements of phylogenetic network inference remain the major bottleneck for almost all methods in PhyloNet. Direct applications of maximum parsimony, maximum likelihood, and Bayesian inference to large data sets, especially in terms of the number of taxa, could be infeasible. Inferences based on pseudolikelihood functions and divide-and-conquer approaches are very promising in such cases and are enabled by PhyloNet.

Notes

1. Mathematically, the definition of an inheritance probability is not this simple, as each locus could have its own inheritance probability, as defined and discussed in detail in [842]. Defining one inheritance probability for each locus and for each reticulation edge poses significant challenges in terms of inference, which is why all functionalities in PhyloNet assume a single inheritance probability across all loci per reticulation node. An alternative is to integrate out these probabilities during inference, which can be done analytically [754].

2. One could use methods in the second category in an optimization setting by returning from all samples collected the one with the highest posterior, which would be the *maximum a posteriorio*, or MAP, estimate.

3. For example, a drop in the MDC score from 4 to 3 is a 25% improvement, but that is just an artifact of the small denominator value of 4.

4. While MLE_BiMarkers utilizes pseudolikelihood computations, these can be computationally very expensive on large networks since full likelihood calculations are carried out on subnetworks.

5. A multilabeled tree, or MUL-tree, is similar to phylogenetic trees in terms of the topology, but differ only in that a MUL-tree can have more than one leaf labeled by the same taxon name [323].

6. According to Enrico Fermi, the famed mathematician John von Neumann said, "With four parameters you can fit an elephant to a curve; with five you can make him wiggle his trunk."

CHAPTER 6

Network Thinking: Novel Inference Tools and Scalability Challenges

Claudia Solís-Lemus

6.1 Introduction: The Impact of Gene Flow

Typically, phylogenetic analyses estimate a species tree to represent the evolutionary relationships among a group of organisms. A species tree implicitly assumes the absence of reticulate evolution, and this assumption is violated when there is some form of gene flow, such as horizontal gene transfer, hybridization, hybrid speciation, or introgression. In recent years, there has been an explosion of phylogenetic network methods that can represent reticulate evolution by estimating species networks, instead of species trees. However, there is still the belief that tree methods will be able to estimate the "major vertical signal" of the data, and network methods remain underutilized in the field. It turns out that this belief can be erroneous in some real-life data sets, which motivates the development of more accurate and more scalable network methods for phylogenetic analyses.

Just as incomplete lineage sorting (ILS) creates a pattern in the sample of gene trees that is poorly represented with concatenation methods [385, 624], gene flow also causes gene tree discordance that is not captured by coalescent-based tree methods alone. In the first case, concatenation methods assume that all genes have the same phylogeny. Given that ILS can cause gene tree discordance, especially for species trees with short internal branch lengths, it has been proven that concatenation methods are not robust to the presence of ILS, and thus can estimate the wrong phylogeny with high support [385, 624].

Coalescent-based tree methods such as ASTRAL [506, 850] and NJst [439] are a huge improvement over concatenation by explicitly modeling ILS via the multispecies coalescent model [372]. However, these methods are still limited under reticulate evolution as they do not account for gene flow. Indeed, it has been proven that when gene trees are simulated under some form of gene flow, tree methods can estimate the wrong species tree with high support [713]. Thus, under certain patterns of gene flow, tree methods cannot recover the main "vertical signal" of the data.

The rationale for the lack of robustness of tree methods to gene flow lies in the existence of anomalous gene trees (AGT): gene trees that do not have the same topology as the species tree and have a higher probability under the coalescent model than gene

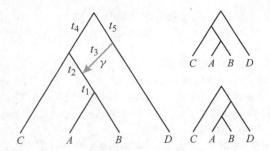

Frequencies among gene trees			
Quartet	$\gamma = 0.0$	$\gamma = 0.1$	$\gamma = 0.3$
$AB\,|\,CD$	**0.347**	**0.298**	**0.260**
$CA\,|\,BD$	0.327	0.351	0.370
$CB\,|\,AD$	0.327	0.351	0.370

$$t_1 = t_2 = 0.01, t_3 = t_4 = t_5 = 1$$

Figure 6.1. Anomalous gene trees under a network model. Left: Network model with one hybridization event (arrow). Center: By turning on/off the hybrid edge (arrow), we have two displayed trees, both containing clade (A,B). Right: The discordant gene trees with clade (A,C) or (A,D) have higher frequency than the gene tree that has the same clade as the species tree (A,B) for $\gamma > 0$ (gene flow) and short internal branches ($t_1 = t_2 = 0.01$ in coalescent units).

trees that have the same topology as the species tree. These anomalous gene trees can be unrooted (AUGT) or rooted (AGT). Degnan and others [178, 173] showed that there are no AUGT for the case of four taxa for gene trees generated from a species tree with no gene flow under the coalescent tree model. This result leads to the accuracy of quartet-based methods, like ASTRAL, as quartets with higher frequencies in the sample could not be anomalous and then would serve to reconstruct the correct species tree. Under the ILS + gene flow scenario, however, this is not the case anymore. Solis-Lemus and others [713] showed that there are AUGT for the case of four taxa under the network coalescent model, and thus species tree methods could reconstruct the wrong species tree by being misled by anomalous gene trees in the sample (see [445] for similar study under a model of continuous gene flow between sister species).

For example, in figure 6.1 we see a four-taxon network with one gene flow event shown with a grey arrow (more about networks description in section 6.3). This gene flow event has a γ parameter associated with it that represents the proportion of genes that were transferred through this reticulation arrow. If $\gamma = 0$, then there is no gene flow, and thus it would be as if the grey arrow did not exist, and the figure would represent a tree, not a network. In fact, by ignoring this grey arrow, we get the "major tree" in black (more about displayed trees on a network in section 6.3.3). It turns out that if we select very short lengths for the ancestral branches to (A,B) ($t_1 = t_2 = 0.01$ in coalescent units) and we simulate gene trees under the coalescent model on this network (ILS + gene flow), there will be some discordant gene trees (different from the major tree in black) with higher probability under the coalescent model than the gene trees that agree with the major tree. Intuitively, one might think that all gene trees should have the (A,B) clade because the gene flow is ancestral to the speciation event (see the two displayed trees in black in figure 6.1 next to the network; both trees have the (A,B) clade). However, as we know from the coalescent model (see chapter 1), there is random coalescence in the populations ancestral to the (A,B) clade that is more variable as the internal branch shortens. If, in addition, there is gene flow in this branch, the majority of the simulated gene trees no longer agree with the species tree topology. In fact, as γ increases (table in figure 6.1 right), the probability of observing the clade (A,B) in the unrooted gene trees is smaller than the probability of the other two possibilities: (A,C), and (A,D). Thus, the discordant trees (without (A,B) clade) appear in the sample with higher frequency than the gene tree in agreement with the species tree (with (A,B) clade). This results in a sample with a strong discordant signal that deceives tree methods into reconstructing a species tree without the (A,B) clade.

In conclusion, tree methods are not suited to handle gene flow. So, when there is the possibility of reticulate evolution among the organisms under study, network methods should be used instead. However, this raises the question: when is this the case? That is, when is a tree model good enough to describe the sample of gene trees at hand, and when do we need to resort to network models instead?

In the next section, we will describe a goodness of fit test that will allow us to identify if a tree model is a good fit for the data at hand. Later, we will also describe what we mean by a species network (section 6.3), several difficulties in network thinking, such as the concept of displayed trees (section 6.3.3), and the difficulty in comparing different networks (section 6.3.4). Finally, we will explain a scalable method to reconstruct phylogenetic networks (section 6.4) as well as examples on how to use the software tool PhyloNetworks (section 6.5).

6.2 Trees versus Networks

If we want to estimate the evolutionary relationships among a group of organisms, we can use a coalescent-based method that takes a sample of gene trees as input. A reasonable question is when to choose methods that estimate a species tree and when to choose methods that estimate a species network. Basically, we need a way to know if the data at hand follows a tree-like pattern of evolution (and thus, we can use a tree method) or if the data follows a non-tree-like pattern (and thus, we need a network method). There exist already multiple tests that allow us to identify whether there is hybridization within a given group of taxa. For example, the ABBA-BABA test for SNPs [277, 208] tries to identify if there is hybridization within a set of four taxa, HyDe [141, 77] tests if there is a potential hybridization among three taxa plus an outgroup by means of phylogenetic invariants, and MSCQuartets [508, 616] tests if the quartet frequencies of a four-taxon subset follow the expected probabilities under the multispecies coalescent model. Despite the accuracy of these tests, the disadvantage is that they test one subset of taxa at a time.

Unlike ABBA-BABA and HyDe, TICR (tree incongruence checking in R) [727] is a goodness of fit test that combines all four-taxon sets into a single test. TICR tests whether a specific species tree is a good enough fit to the data at hand by comparing the expected concordance factors under the coalescent tree model with the observed concordance factors from the sample of gene trees. The concordance factor (CF) of a given quartet (or split) is the proportion of genes whose true tree displays that quartet (or split) [53]. For example, if A is a hybrid intermediate between B and C, the CFs of $AB|CD$ and $AC|BD$ would be around 0.5 while the CF of $AD|BC$ would be near 0. On the contrary, if there is no hybridization among A, B, C, D, and the species tree has the split $AB|CD$, we would expect the two discordant splits: $AC|BD$ and $AD|BC$ to have equally minor CF, and the major split $AB|CD$ to have a higher CF than the other two (see figure 6.2 top). These CFs depend on the branch lengths in the species tree, represented in coalescent units. We use the term "concordance factor" as opposed to "probability" to emphasize that CFs measure genomic support, unlike probabilities (such as posterior probabilities or bootstrap values) that most often measure statistical uncertainty [27].

The TICR goodness of fit test requires a given species tree with branch lengths in coalescent units, which can be estimated with the average CF of all quartets that map onto

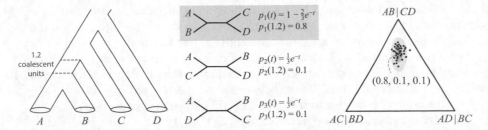

Figure 6.2. Left: four-taxon species tree with internal branch length of 1.2 coalescent units. Center: Under the coalescent model, the expected concordance factors (CFs) for the three quartets $(p_1(t), p_2(t), p_3(t))$ depend on the internal branch length. For $t = 1.2$, the expected CF for the major split $AB|CD$ is 0.8, and the expected CF for the minor splits $AC|BD$, $AD|BC$ are equal to 0.1. Right: Dirichlet distribution centered on the expected CFs $(0.8, 0.1, 0.1)$. Under the coalescent model, the two minor CFs are expected to be equal. This equality is disrupted by gene flow.

a given internal branch, converted to a branch length (in coalescent units) by inverting the equation in figure 6.2: $t = -\log((3/2)(1 - CF_{major}))$.

With this metric tree, we calculate the expected CFs under the multispecies coalescent model for every four-taxon set. For example, in figure 6.2, the expected CFs: $(p_{AB|CD}(t), p_{AC|BD}(t), p_{AD|BC}(t)) = (1 - (2/3)\exp(-t), (1/3)\exp(-t), (1/3)\exp(-t))$ for any $t > 0$. Next, for each four-taxon set A, B, C, D, the observed quartet CFs $(x_{AB|CD}, x_{AC|BD}, x_{AD|BC})$ are modeled by a Dirichlet distribution $D(\alpha_1, \alpha_2, \alpha_3)$, with parameters α_i that depend on the expected CFs under the multispecies coalescent model. To simplify notation, we will use the subscripts $1, 2, 3$ instead of $AB|CD, AC|BD, AD|BC$. For example, the multispecies coalescent model implies that the expected CF for the major quartet is $p_1(t) = \alpha_1/(\alpha_1 + \alpha_2 + \alpha_3) = 1 - (2/3)\exp(-t)$ if t is the internal branch length on the quartet tree. Similarly, the expected CF for the minor quartets is given by $p_2(t) = p_3(t) = (1 - p_1)/2 = (1/3)\exp(-t)$. The parameters α_i can thus be obtained from t.

All four-taxon sets are assumed to share the same concentration parameter $\alpha = \alpha_1 + \alpha_2 + \alpha_3$, and we can estimate this parameter by maximizing the pseudo-log-likelihood from the Dirichlet distribution assumed for the observed quartet CFs, which has the form:

$$\log PL(\alpha) = \sum_{j=1}^{M} \log \Gamma(\alpha) - \log \Gamma(\alpha p_1(t_j)) - 2 \log \Gamma(\alpha p_2(t_j)) + (\alpha p_1(t_j) - 1) \log x_{1j} +$$

$$(\alpha p_2(t_j) - 1)(\log x_{2j} + \log x_{3j})$$

$$= M \log \Gamma(\alpha) - \sum_{j=1}^{M} (\log \Gamma(\alpha p_1(t_j)) + 2 \log \Gamma(\alpha p_2(t_j))) +$$

$$\alpha M \widetilde{\log CF} - 3M \overline{\log CF},$$

where t_j is the internal branch lengths of the tree reduced to the jth four-taxon set, $\widetilde{\log CF}$ is the average of the observed log CFs, weighted by their $p_i(t_j)$ values, (x_{1j}, x_{2j}, x_{3j}) are the observed CFs for the jth four-taxon set, and $\overline{\log CF}$ denotes the average log CF. Note that this is a pseudolikelihood (not a true likelihood) because we are treating the M four-taxon sets as independent (by summing over the log probabilities) when they are not, as all the four-taxon set are linked through the tree topology.

Using the estimated value of α, we can calculate a p-value for each four-taxon set based on how much the observed CF for the major quartet, x_1, departed from its expectation, p_1. Let d denote the distance between the expected and the observed CF: $d = |x_1 - p_1|$; the p-value for this four-taxon set is then calculated as $p = P_\alpha(|X_1 - p_1| \geq d)$ from a Beta distribution with parameter $(\alpha p_1, \alpha(1 - p_1))$.

After calculating a p-value for each of the four-taxon sets, we can bin all these p-values into arbitrary categories to test if the bin frequencies depart from the expected proportions under the null hypothesis. We choose the categories as: 0–0:01; 0:01–0:05; 0:05–0:10; 0:10–1:0. The overall test is then defined using a χ^2 test with 3 degrees of freedom to determine if the bin frequencies depart from the expected proportions $(0.01, 0.04, 0.05, 0.90)$. The null hypothesis is whether the given species tree used to compute the expected CF is a good enough fit to the observed CF. If this hypothesis is rejected by the overall p-value, then we would prefer to use a network method as there is evidence of reticulate evolution, which does not follow the pattern of the coalescent model on a tree.

It is important to note that the TICR test has simplifying assumptions, but most of these assumptions have been explored in literature for their robustness in real applications. For example, while the pseudolikelihood function ignores the dependency of the quartets, this model has been explored to be robust in different studies [727, 440, 710, 843]. For more details about this and other assumptions, check out the TICR paper [727].

6.3 Species Networks

Just like phylogenetic trees, networks can be rooted or unrooted. A rooted phylogenetic network on taxon set X is a connected directed acyclic graph with vertices $V = \{r\} \cup V_L \cup V_H \cup V_T$, edges $E = E_H \cup E_T$, and a bijective leaf-labeling function $f : V_L \to X$ with the following characteristics. The root r has indegree 0 and outdegree 2. Any leaf $v \in V_L$ has indegree 1 and outdegree 0. Any tree node $v \in V_T$ has indegree 1 and outdegree 2. Any hybrid node $v \in V_H$ has indegree 2 and outdegree 1. A tree edge $e \in E_T$ is an edge whose child is a tree node (or a leaf node). A hybrid edge $e \in E_H$ is an edge whose child is a hybrid node. Unrooted phylogenetic networks are typically obtained by suppressing the root node and the direction of all edges. We also consider semidirected unrooted networks (figure 6.3 center), for which the root node is suppressed and we ignore the direction of all tree edges, but we maintain the direction of hybrid edges, thus keeping information on which nodes are hybrids. The placement of the root is then constrained because the direction of the two hybrid edges to a given hybrid node informs the direction of time at this node: the child edge must be a tree edge directed away from the hybrid node and leading to all the hybrid's descendants. Therefore the root cannot be placed on any descendant of any hybrid node, although it might be placed on some hybrid edges.

Throughout this chapter, we use the following notation

- $n =$ the number of taxa,
- $h =$ the number of hybridization events (in grey arrows in figure 6.3), and
- $k_i =$ the number of nodes in the undirected cycle created by the ith hybrid node.

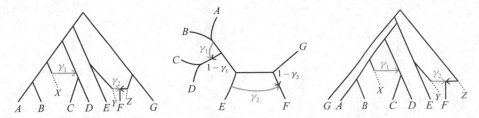

Figure 6.3. Example of rooted and semidirected phylogenetic networks with $h=2$ hybridization events and $n=7$ sampled taxa. Inheritance probabilities γ represent the proportion of genes contributed by each parental population to a given hybrid node. Left: Rooted network modeling several biological processes. Taxon F is a hybrid between two nonsampled taxa Y and Z with $\gamma_2 \approx 0.50$, and the lineage ancestral to taxa C and D has received genes introgressed from a nonsampled taxon X, for which $\gamma_1 \approx 0.10$. An alternative process at this event could be the horizontal transfer of only a handful of genes, corresponding to a very small fraction $\gamma_1 \approx 0.001$. Center: Semidirected network for the biological scenario just described. Although the root location is unknown, its position is constrained by the direction of hybrid edges (directed by arrows). For example, C, D, or F cannot be outgroups. Right: Rooted network obtained from the semidirected network (center) by placing the root on the hybrid edge that leads to taxon F (labeled by $1 - \gamma_2$).

For example, figure 6.3 shows a semidirected network (center) with two possible rootings (left and right). In this network, there are seven taxa ($n = 7$), and two hybridization events marked by the grey arrows ($h = 2$). One of the hybridization events (corresponding to γ_1) has three nodes in the cycle ($k_1 = 3$) and the second hybridization event (corresponding to γ_2) has four nodes in the cycle ($k_2 = 4$). The number of nodes inside each hybridization cycle will be important when we determine which hybridizations are identifiable and which are not (section 6.4.1.1).

The main parameter of interest is the topology $\mathcal{N}$ of the semidirected network, which can later be rooted by a known outgroup just as a tree can. The other parameters of interest are **t**, the vector of branch lengths in coalescent units, and a vector of inheritance probabilities γ, describing the proportion of genes inherited by a hybrid node from one of its hybrid parents (see figure 6.3). Each hybrid edge has a $\gamma < 1$ associated with it, and we distinguish between major hybrid edge (with $\gamma > 0.5$) and minor hybrid edge (with $\gamma < 0.5$). Only identifiable branch lengths are considered in **t**. For example, with only one sequenced individual per taxon, the lengths of external edges are not identifiable because branch lengths can be estimated only if there is the possibility of a coalescent event in the branch (not possible with only one taxon).

There are different classes of networks depending on their complexity, that is, how far they are from a tree. In this chapter, we will assume that the true network is of level-1 [339] (see figure 6.4), that is, any given edge can be part of at most one cycle. This means that there is no overlap between any two cycles. This is a huge limitation, and it is not entirely biologically reasonable. However, as we will see in the remainder of the chapter (in particular, section 6.4.1.1), level-1 networks are already quite complex and it is not straightforward to identify which reticulations can be detected and which cannot given the sample of gene trees. More research is needed in the area of identifiability of phylogenetic networks in order to relax the level-1 assumption and allow for more

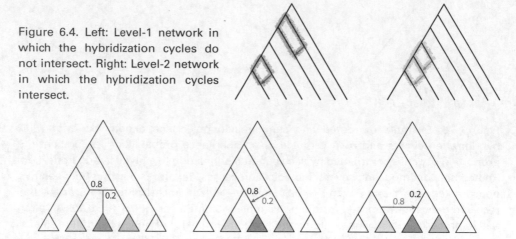

Figure 6.4. Left: Level-1 network in which the hybridization cycles do not intersect. Right: Level-2 network in which the hybridization cycles intersect.

Figure 6.5. Visual artifacts can mislead the interpretation (see text for explanation).

complex networks (refer to [339] for other types of evolutionary networks such as level-k or tree-child networks).

We also note that the network model uses one-time events to summarize episodes of continuous gene flow. Furthermore, the network model does not say anything about what biological process—introgression, hybridization, et cetera—is at work as all of these are modeled with the same network structure. However, in some cases, γ can provide some insight; for example, if $\gamma \approx 0.5$, we can suspect that a process of hybridization occurred.

Finally, visual artifacts can mislead the interpretation. For example, in figure 6.5, all three networks represent the same reticulation event, but they would appear as different processes by the way they are drawn. That is, the network on the left has two perfectly horizontal edges coming together into the hybrid node (ancestor of dark grey clade), which seem to illustrate a hybridization. This is not the same interpretation for the network in the center, which only displays one grey arrow with $\gamma = 0.2$ that flows into the black edge ancestral to the dark grey clade. This grey arrow seems to represent gene flow or horizontal gene transfer (HGT) from the ancestral population of the intermediate grey clade to the ancestral population of the dark grey clade. Finally, the network on the right displays the edge with $\gamma = 0.2$ as the major edge (in black), so it seems as if there is gene flow (e.g., HGT) from the ancestral population of the light grey clade into the ancestral population of the dark grey clade. Mathematically, all these three networks are the same, suggesting a reticulate clade (dark grey) with 80% contribution from the ancestral population of the light grey clade and 20% contribution from the ancestral population of the intermediate grey clade.

6.3.1 EXPLICIT VERSUS IMPLICIT NETWORKS

A huge variety of phylogenetic networks have been proposed (see [339]), but we can categorize them into two main classes: explicit networks and implicit networks (or split networks). In explicit networks, each internal node is associated to a specific biological process (like speciation or hybridization), whereas in implicit networks, internal nodes need not correspond to any biological mechanism or ancestral population.

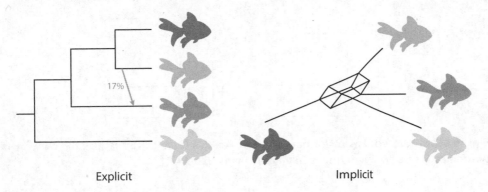

Explicit   Implicit

Figure 6.6. Explicit versus implicit networks: in explicit networks, internal nodes represents specific biological processes like speciation or hybridization. In implicit networks, internal nodes do not have any meaning.

Thus, explicit networks can be easily interpreted as phylogenetic trees are. For example, in figure 6.6 (left), there is a gene flow event represented by the grey arrow from the ancestral population of the lightest grey fish into the darkest grey fish. As explained already in previous sections, this gene flow event has an extra parameter associated with it (inheritance probability γ), which represents the proportion of genes that were transferred through this arrow (17% in this case). There is also a time progression from root to tips, with the root and every internal node representing ancestral populations. On the other hand, in the implicit network (figure 6.6 right), the internal nodes do not represent ancestral species anymore, and the edges are not directed, so we lose the time progression from past to present. Furthermore, the repeated parallel edges do not represent any particular form of gene flow. They simply represent gene tree discordance, which can well be ILS or gene flow, or even estimation error.

Many methods have been proposed to reconstruct implicit networks [265, 281, 825], which are a great way to summarize the data and visualize patterns of discordance. However, they are not as clear when we want to study the evolutionary relationships of organisms. Methods to reconstruct explicit networks are booming [842, 713, 21, 851, 10], but they still cannot handle the sizes of data sets that split networks can. Explicit network methods need more data and more computational time, but in return, they produce a species network that can be interpreted in biological terms.

6.3.2 EXTENDED PARENTHETICAL FORMAT

To represent network objects, we use the extended Newick format [517, 116, 749] This format uses the concept of minor hybrid edges (edges with $\gamma < 0.5$) and major hybrid edges ($\gamma > 0.5$). By default, we detach the minor hybrid edge at each hybrid node to write the extended Newick description of a network as we would for a tree, with a repeated label, that of the hybrid node ("#H1" in figure 6.7). This description can include edge information, formatted as :length:support:γ.

For example, the parenthetical format of the network in figure 6.7 can include γ values:

(((A,(B)#H1:::0.8),(C,#H1:::0.2)),D);,

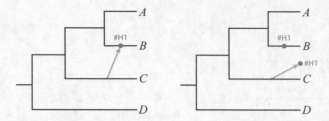

Figure 6.7. Extended Newick format for networks: (((A,(B)#H1),(C,#H1)),D); the network is written as a tree with two nodes having the same label #H1.

which are written after three colons because there is no information about branch length nor support for the hybrid edges. Other internal edges (tree edges) have information about branch lengths, which follow just one colon. These tree edges do not have information about γ because all tree edges have $\gamma = 1$.

6.3.3 DISPLAYED TREES AND SUBNETWORKS

As mentioned before, a $\gamma = 0$ in an explicit network means that the hybrid edge is not even present. Thus, by removing all the minor hybrid edges (the ones with $\gamma < 0.5$), we get a tree, denoted the "major tree" because it is the tree that the majority of the genes follow (figure 6.8), and hence, it could be used to study the evolutionary relationships of the group. By removing certain hybrid edges, we get a collection of trees denoted the "displayed trees." These trees are obtained by choosing one parent hybrid edge at each hybrid node and dropping the other parent hybrid edge.

For example, in figure 6.8 there is one network (left), with its major and minor trees in the center. The major tree (turning off the minor hybrid edge $\gamma < 0.5$) keeps the (A,B) clade, while the minor tree (dropping the major hybrid edge $\gamma > 0.5$) has the clade (E,A) now, as A is of hybrid origin between E and B.

If the network had more than one hybridization event, we could also get subnetworks by removing specific major/minor hybrid edges but leaving others. It is important to keep in mind the possibility of breaking down a network into displayed trees or subnetworks, as this will facilitate the comparison with other networks. That is, network estimation is difficult enough that recovering identical networks from different methods is unlikely. However, if we break down the networks into trees/subnetworks, we might be able to identify features that all the estimated networks have in common. As we will see in section 6.4.1.1, some reticulations are much harder to detect than others.

6.3.4 COMPARISON OF NETWORKS

One important ingredient in network inference is the ability to compare networks. For the case of trees, the most common approach involves the computation of the Robinson-Foulds distance [620]. However, there is not an equivalent distance for the case of networks, as the notion of splits or clades is not straightforward. Rooted networks can be compared by their hardwired cluster dissimilarity [339]: the number of hardwired clusters found in one network and not in the other. A hardwired cluster is associated with a given node in the rooted network, and it is defined as the set of all taxa that inherited at least some genetic material from that node. That is, the hardwired

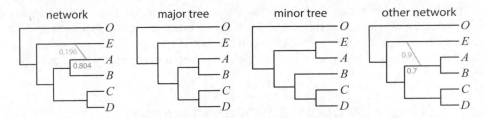

Figure 6.8. Network with its displayed trees and a second network to illustrate the hard-wired cluster distance of 4, which represents the number of hardwired clusters that are present in one network and not the other.

cluster of a node is the set of all tips descendant from that node. The hardwired cluster dissimilarity relies on the root position, so unrooted or semidirected networks need to be rooted with the same outgroup(s) first. Thus, this measure can be affected by rooting errors. Furthermore, the hardwired cluster dissimilarity is not a real distance for the space of all networks, but it is a distance for the class of level-1 networks. In fact, this dissimilarity is the Robinson–Foulds distance on trees. Recent work has produced distance measures on other classes of networks [529, 339, 117, 111].

To illustrate with an example, imagine that you want to compare two networks: one estimated with SNaQ (which will be described in section 6.4) and one estimated with a different network method (figure 6.8 left and right, respectively). Note that both networks need to be rooted in the same outgroup "O."

Visually, we can see that these two networks are different. The origin of gene flow is the same, E, but the target of gene flow is not: in the SNaQ network, it is the lineage ancestral to A, but in the other network, it is the lineage ancestral to (A,B). If we calculate the hardwired distance between these two networks, we get 4. This means that there are four hardwired clusters that appear in one network and not the other. We can list the hardwired clusters of any network by looking at the descendant taxa for every given internal node. For example, the hardwired clusters for the network on the right are {CD, AB, AB, ABE, ABCD, ABCDE, ABCDEO}, and the ones for the network on the left are {CD, A, AB, AE, ABCD, ABCDE, ABCDEO}. We can see that the hardwired clusters that do not appear in both networks are four: {AB, A, AE, ABE}. Finally, we can also note that if we compare the major trees (removing the minor hybrid edge), these trees are identical.

There is not a clear sense of whether 4 is small or big distance. Like the Robinson-Foulds distance, the hardwired cluster dissimilarity can be quite large for networks that are similar (i.e., having just one taxon in a different place). This is why the hardwired cluster dissimilarity is mostly used to determine whether two networks are equal (dissimilarity of 0), but see [112] for comparisons on network metrics.

6.4 Fast Reconstruction of Species Networks

Pseudolikelihood methods have been previously utilized to achieve computationally feasible inference when the full likelihood can be intractable. For example, Liu and others [440] used the pseudolikelihood approach to estimate phylogenetic trees under ILS in the software MP-Est. Here, we present a pseudolikelihood method to estimate phylogenetic networks denoted SNaQ (species networks applying quartets)

[713]. The rationale behind SNaQ is similar to MP-Est, except that SNaQ estimates phylogenetic networks, not trees, and SNaQ uses unrooted quartets in the computation of the pseudolikelihood, as opposed to rooted triplets. Using unrooted quartets is advantageous because it avoids potential rooting errors in the input gene trees.

6.4.1 MAXIMUM PSEUDOLIKELIHOOD ESTIMATION

Likelihood-based approaches to estimate phylogenetic networks rely on the computation of the likelihood of a network (with branch lengths and inheritance probabilities) given the data, which is a collection of gene trees estimated from DNA sequences. These methods can be very accurate as they can profit from all the statistical properties of maximum likelihood estimates. However, the computation of the likelihood can be very expensive, which, combined with the heuristic search in the space of networks (which is much bigger than the space of trees), results in an accurate method that is restricted to the case of small data sets (see [841, 842, 851] and [5]).

To overcome the scalability problems of likelihood-based methods, Solis-Lemus and Ané [713] devised a pseudolikelihood method, which first summarizes the sample of gene trees into CFs (thus being scalable in the number of genes) and then breaks down the computation of the likelihood of the whole network into the computation of the likelihood for four-taxon subsets. In this manner, only the likelihood of four-taxon networks is computed. Despite the increase in the number of four-taxon subsets as the number of taxa grows, this divide-and-conquer approach is still more scalable than computing the likelihood of the full network.

Species networks applying quartets implements the statistical inference method in [713]. The procedure involves a numerical optimization of branch lengths and inheritance probabilities and a heuristic search in the space of phylogenetic networks. The full inference scheme from multilocus sequences to phylogenetic network is shown in figure 6.9.

As mentioned already, the pseudolikelihood of a network is based on the likelihoods of its four-taxon subnetworks. That is, for a given network $\mathcal{N}$ with $n \geq 4$ taxa, we consider all four-taxon subsets $\mathcal{S} = \{s = \{a, b, c, d\} : a, b, c, d \in X\}$ and combine the likelihood of each four-taxon subnetwork to form the full network pseudolikelihood:

$$L(\mathcal{N}) = \prod_{s \in \mathcal{S}} L(s), \tag{6.1}$$

where $L(s)$ is the likelihood of the subnetwork of a given four-taxon subset s. These four-taxon likelihoods are not independent, which is why we get a pseudolikelihood when we multiply them, instead of a true likelihood.

To calculate the likelihood of a four-taxon network from gene trees $\mathcal{G} = (G_1, G_2, \cdots, G_g)$ at g loci, we note that for taxon set $s = \{a, b, c, d\}$, there are only three possible quartets, represented by the splits $q_1 = ab|cd$, $q_2 = ac|bd$, and $q_3 = ad|bc$. We then consider the number of gene trees $Y = (Y_{q_1}, Y_{q_2}, Y_{q_3})$ that display each of the three quartets. Assuming unlinked loci, Y follows a multinomial distribution with probabilities $(CF_{q_1}, CF_{q_2}, CF_{q_3})$, the theoretical CFs expected under the coalescent on the four-taxon network. These theoretical CFs expected under the coalescent model are already known if the network is a species tree [12]. In that case, these CFs do not depend on the position of the root and are given by $(1 - (2/3) \exp(-t), (1/3) \exp(-t), (1/3) \exp(-t))$

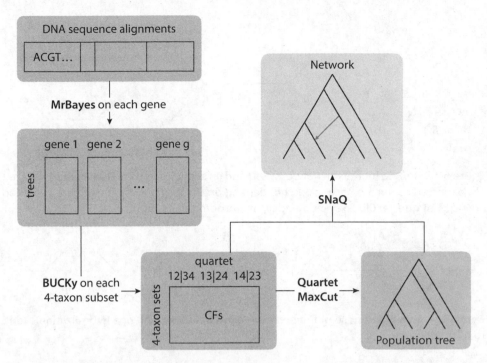

Figure 6.9. Flow chart of the procedure to estimate a network with species networks applying quartets (SNaQ) from multilocus sequences.

if the unrooted species tree is $q_1 = ab|cd$ with an internal edge of length t coalescent units. On a species network with reticulations, the probabilities of rooted gene trees were fully derived in [841], more efficiently in [842], and for unrooted four-taxon networks in [713]. We provide details on the computation of these probabilities in one example below.

By substituting the multinomial likelihood in equation 6.1, we get

$$L(\mathcal{N}) \propto \prod_{s \in \mathcal{S}} (CF_{q_1})^{Y_{q_1}} (CF_{q_2})^{Y_{q_2}} (CF_{q_3})^{Y_{q_3}}, \tag{6.2}$$

where $q_i = q_i(s)$ $(i = 1, 2, 3)$ are the three quartet resolutions on s. The data are summarized in the Y values, and the candidate network governs the CF values, which we explain below. The data, as mentioned already, could be either a collection of gene trees $\mathcal{G} = (G_1, G_2, \cdots, G_g)$ or the CFs estimated with BUCKy [27]. The advantage of estimating the CFs with BUCKy is that BUCKy tries to disentangle genomic discordance versus statistical discordance. Ideally, the CFs represent true genomic support: proportion of genes that support a given split. However, when we use estimated gene trees to calculate the CFs directly (by counting number of gene trees with a specific split), the resulting measure of support will have confounding effects of genomic support and statistical error, given that the gene trees can have some potential estimation error. When using BUCKy to estimate the CFs, BUCKy estimates the CFs as genomic support, accounting for statistical error in the reconstruction of the gene trees. Given that most coalescent-based network methods take a collection of estimated gene trees as perfectly known for

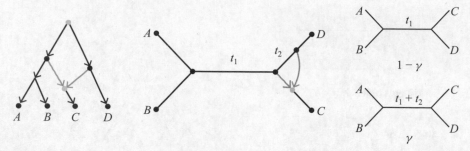

Figure 6.10. Rooted four-taxon network (left) and its semidirected version (center). Quartet CFs expected under the network do not depend on the root placement and are weighted averages of quartet CFs expected under the unrooted trees (right).

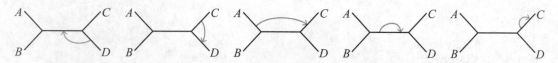

Figure 6.11. Five different semidirected four-taxon networks with one hybridization event, up to tip relabeling.

the method's input, methods like SNaQ that can take estimated CFs as input instead are more robust to the estimation error in the gene trees.

For a given four-taxon subnetwork, we want to calculate the expected CFs based on the multispecies coalescent model for networks [498, 841]. Take the subnetwork in figure 6.10 (left) with $h = 1$ hybridization event. Each gene from taxon C has probability γ of having descended from the hybridization edge sister to D and probability $1 - \gamma$ of having descended from the original tree branch, sister to (A,B). Therefore, the expected CFs are weighted averages of CFs obtained on two species trees with ILS. Because the quartet probabilities do not depend on the root placement in each species tree, they do not depend on the root placement in the original network either. Figure 6.10 (center) shows the corresponding semidirected network, and all rooted networks displaying it share the same quartet CFs, obtained from the coalescent models on the two unrooted species trees shown in figure 6.10 (right). These trees have the same topology but different branch lengths in this case. Therefore, we get that $CF_{ab|cd} = (1 - \gamma)(1 - (2/3) \exp(-t_1)) + \gamma(1 - (2/3) \exp(-t_1 - t_2))$, and the other two quartets occur with equal probabilities: $CF_{ac|bd} = CF_{ad|bc} = (1 - \gamma)(1/3) \exp(-t_1) + \gamma(1/3) \exp(-t_1 - t_2)$.

For four-taxon subnetworks with $h = 1$ hybridization, there are five different semidirected topologies up to tip relabeling (figure 6.11). Similar procedures to the one described for figure 6.10 can be used to compute the expected CFs for each of the five cases (see [713]).

With more than one hybridization event ($h > 1$), there are an infinite number of semidirected four-taxon networks, but we can still calculate the quartet CFs if we assume that the cycles created by different reticulations do not share edges. This is where the assumption of level-1 networks is used. By assuming that hybridization cycles do not intersect, we can reduce each network to an equivalent network with $h = 0$ or 1 with transformed branch lengths. For example, the network in figure 6.10 leads to equal CFs of the two minor quartets $ac|bd$ and $ad|bc$, so it is equivalent to the unrooted species tree $ab|cd$ with internal branch length $t_3 = -\log((1-\gamma)\exp(-t_1) + \gamma\exp(-t_1 - t_2))$ to ensure $(1/3)\exp(-t_3) = CF_{ac|bd}$ given above. This new quartet tree and the original four-taxon network have the same expected quartet CFs. Thus, the assumption of a level-1 network guarantees nonoverlapping reticulation cycles, such that we can find an equivalent four-taxon network with $h = 0$ or 1 and the same expected quartet CFs.

The maximum pseudolikelihood (MPL) estimate is the network $\mathcal{N}$, branch lengths t and heritabilities γ that maximize the pseudolikelihood (equation 6.2). This MPL optimization was fully implemented in SNaQ and is part of the open source package PhyloNetworks [711] in Julia [67]. The numerical optimization of branch lengths and γ parameters for a fixed topology is performed with a derivative-free methodology in the NLopt Julia package. The heuristic optimization of the network topology uses a strategy similar to that in [842]. Given a fixed maximum number of hybridizations (h_m), we search for the MPL network with at most h_m hybridizations. Since the pseudolikelihood can only improve when hybridizations are added, we expect the final network to have $h = h_m$ exactly.

For a given h_m, the search is initialized with a tree from a very fast quartet-based tree estimation method such as ASTRAL [506] or Quartet MaxCut [707, 38]. The length of each branch is initialized using the average observed CF of the quartets that span that branch exactly, $\overline{CF}$, transformed to coalescent units by $t = -\log(1 - (3/2)\overline{CF})$, just as in the TICR test. The search then navigates the network space by altering the current network using one of five proposals, chosen at random: (1) move the origin of an existing hybrid edge, (2) move the target of an existing hybrid edge, (3) change the direction of an existing hybrid edge, (4) perform a nearest-neighbor interchange (NNI) move on a tree edge, and (5) add a hybridization if the current topology has $h < h_m$. Any new proposed network is checked to verify that it is a semidirected level-1 network with $h \leq h_m$ and with at least one valid placement for the root (see section 6.4.2).

Although the deletion of a current hybridization is not proposed (because the MPL network should have $h = h_m$), this deletion is still performed when suggested by the data if the numerical optimization of parameters returns a $\hat{\gamma} = 0$. In this case, the corresponding hybrid edge is removed and the search attempts to add it back at random in the neighborhood of the original hybrid edge. If this attempt fails for all neighbors, the hybridization is deleted entirely and the search continues from a network with one fewer hybridization. Similarly, if the numerical optimization returns a branch of length $t = 0$, an NNI move is proposed immediately on that branch. The search continues until the pseudolikelihood converges or until the number of consecutive failed proposals reaches a limit.

In [324], Huber and others proved that the space of unrooted level-1 networks is connected by local subnetwork transfers, which generalize the NNI operations on trees and which are similar to our moves 1, 2, and 4. Although we do not have a formal proof that the MPL network can be reached from the starting tree using our proposals, the results in [324] suggest that it is the case.

After the pseudolikelihood estimation, we get the maximum pseudolikelihood estimated network (with estimated branch lengths in coalescent units and estimated inheritance probabilities) with $h = h_m$ hybridizations. A network can be estimated for various values of h_m, followed by a model selection procedure to select the appropriate number of hybridizations (see section 6.4.3). Furthermore, we are interested in measuring statistical uncertainty as well. We have mentioned already one way to account for

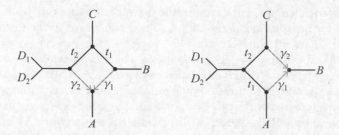

Figure 6.12. Networks with $k=4$ nodes in the reticulation cycle and identical unrooted topologies. They differ in their hybrid position. If D_2 is not sampled ($n=4$), the two networks are not distinguishable from each other. Furthermore, the set of numerical parameters is not identifiable, only $\gamma_i(1 - \exp(-t_i))$ for $i = 1, 2$ are identifiable.

statistical uncertainty by using CFs estimated with BUCKy as input instead of estimated gene trees. In addition, one can do a bootstrap analysis on the estimated network. This will be described in section 6.4.4.

6.4.1.1 Identifiability: What We Can and Cannot Learn from Data

Identifiability is a basic requirement if one seeks to learn about parameters from data. Here our parameters are the network topology $\mathcal{N}$, branch lengths $\mathbf{t}$ and inheritance values γ. We already know that quartet CFs do not depend on the root placement, so the rooted network is not identifiable and we only consider semidirected networks.

The pseudolikelihood model is identifiable if two different combinations of parameters $(\mathcal{N}, \mathbf{t}, \gamma)$ and $(\mathcal{N}', \mathbf{t}', \gamma')$ yield different sets of quartet CFs. It turns out that some reticulations and some parameters are impossible (or hard) to detect. By identifying the reticulations and parameters that can be recovered with the data at hand, SNaQ explores a reduced parameter space to avoid network and parameter combinations that are not identifiable.

On $n = 4$ taxa, we already showed that the network in figure 6.10 is equivalent to a tree with some appropriate internal branch length. In fact, the same holds true for all four-taxon networks with $k = 2$ or 3 nodes in their reticulation cycle: these reticulations cannot be detected. If $k = 4$, that is, if the reticulation involves more distantly related taxa, then the presence of the hybridization can be detected based on the quartet CFs. However, networks with the same unrooted topology are unidentifiable from each other from only four taxa, like the two networks in figure 6.12, if only D_1 is sampled ($n = 4$). They differ only in the placement of the hybrid node, which is therefore not identifiable with only $n = 4$, even if the presence of a reticulation is.

In general, for networks with $n \geq 4$ taxa, the presence of the hybridization event of interest can be detected if the quartet CFs from $(\mathcal{N}, \mathbf{t}, \gamma)$ cannot all be equal to the quartet CFs from $(\mathcal{N}', \mathbf{t}', \gamma')$ simultaneously, where $\mathcal{N}'$ is the network topology obtained from $\mathcal{N}$ by removing the hybrid edge of interest. Assuming that all $\binom{n}{4}$ four-taxon sets are used in the pseudolikelihood, the network $\mathcal{N}$ gives us $3\binom{n}{4}$ quartet CFs equations expected under the coalescent model. The assumption of level-1 networks allows us to consider one hybrid node at time in this identifiability study.

Intuitively, if both $\mathcal{N}$ (with h hybridization events) and $\mathcal{N}'$ (with $h - 1$ hybridization events) produce the same set of CFs, then there is no possibility to detect the extra hybridization event in $\mathcal{N}$ with the CFs as input data. Thus, we want to identify the region

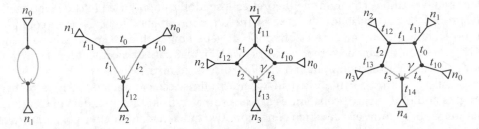

Figure 6.13. Networks with k nodes in a hybridization cycle: $k=2,3,4$, and 5 from left to right. When $k=2$, the presence of hybridization is not identifiable. When $k=3$, the hybridization is detectable if $n_i, n_j \geq 2$ for any i, j, but the set of parameters is not identifiable. In this case, a **good triangle** corresponds to $n_1, n_2, n_3 \geq 2$, in which case setting $t_{12}=0$ makes the other parameters identifiable. When $k=4$, the hybridization is detectable if $n_i \geq 2$ for any i, but parameters are not all identifiable for the **bad diamond I** ($n_0 = n_2 = n_3 = 1$ but $n_1 \geq 2$) and for the **bad diamond II** ($n_0 = n_1 = n_2 = 1$ but $n_3 \geq 2$). When $k \geq 5$, the hybridization is detectable and all the parameters are identifiable.

in parameter space where $\mathcal{N}$ and $\mathcal{N}'$ produce different CFs, that is, the region in which we are certain that we can distinguish $\mathcal{N}$ and $\mathcal{N}'$. To do this, we can match both systems of CF equations (one for $\mathcal{N}$ and one for $\mathcal{N}'$) by means of the algebraic geometry software Macaulay2 [275] and check the values of $(\mathbf{t}, \gamma)$ that produce the same CFs on both $\mathcal{N}$ and $\mathcal{N}'$. By avoiding these values, we can differentiate $\mathcal{N}$ and $\mathcal{N}'$ from the input CFs, and thus we can detect the presence of the hybridization of interest.

Apart from the obvious case $\gamma = 0$ for the hybrid edge absent in $\mathcal{N}'$, $\mathcal{N}$ and $\mathcal{N}'$ are also not distinguishable when $t_b = 0$ or $t_b = \infty$ for some tree branches b, implying either a hard polytomy or a branch with no ILS. We can ignore these cases with the following reasonable assumption:

A1: $t \in (0, \infty)$ for all tree branches and $\gamma \in (0, 1)$.

However, **A1** is not a sufficient condition to ensure that the presence of each hybridization in $\mathcal{N}$ can be detected. Increasing taxon sampling helps detect a hybridization only if the added taxa increase the size of the reticulation cycle. Namely, if the cycle only involves $k = 2$ nodes (see figure 6.13), then $\mathcal{N}$ is not distinguishable from $\mathcal{N}'$, regardless of n. For $k=3$, some hybridizations are detectable and some are not. If any two of the three subtrees defined by the hybridization cycle (figure 6.13) have only one taxon, then the hybridization is not detectable [713]. It is if instead at least two subtrees contain more than one taxon. In general, hybridizations with $k \geq 4$ can be detected if $n \geq 5$. Here and below, we use the terms detectable or identifiable in their generic sense [9, 16], which simply means that some conditions on $(\mathbf{t}, \gamma)$ are required, like **A1**, but that all these conditions are met except on a subset of measure zero.

Aside from the presence of the hybridization, we are also interested in knowing whether the direction of gene flow is identifiable. For example, figure 6.12 shows two networks that differ only in the placement of the hybrid node but otherwise have the same unrooted topology. It turns out that these two networks yield different sets of quartet probabilities and therefore are distinguishable from each other, showing that the direction of the hybridization becomes identifiable when $n \geq 5$ [713]. However, in practice, it is quite common that the pseudolikelihood function from small sample size data behaves differently than the theoretical pseudolikelihood used in the identifiability proofs, in particular for the case of the direction of gene flow. Thus, it could be impossible to detect the correct direction of gene flow with a given sample size, and it is important to consider other networks with an alternative placement for the hybrid node in the hybridization cycle. This will be discussed further in section 6.4.2.

Finally, we are also interested in whether numerical parameters such as branch lengths and inheritance probabilities can be estimated from the input data of CFs. Like before, we determine under which conditions two different combinations of parameters $(\mathbf{t}, \gamma)$ and $(\mathbf{t}', \gamma')$ yield different sets of quartet probabilities for a fixed network $\mathcal{N}$. Just as before, the identifiability depends on the type of network (figure 6.13). With only four taxa, there are more parameters than equations (three quartet CFs), so $\mathbf{t}$ and γ are not separately identifiable; to estimate these parameters, we need to have $n \geq 5$.

If $n \geq 5$, parameter identifiability is again easier if the reticulation involves more distantly related taxa. If $k \geq 5$, all the parameters are identifiable. If $k \leq 3$, parameters are not identifiable. If $k = 4$, parameters are identifiable if either $n_0 \geq 2$ (or n_2, symmetrically) or if both n_1 and $n_3 \geq 2$ (see figure 6.13). We call this a good diamond. Parameters are not all identifiable in the remaining two cases, which we call bad diamonds I and II (see figure 6.13). The bad diamond I already lacked identifiability under a different model in [579].

A naive search for the most likely network would get stuck alternating between nondistinguishable networks or parameter sets. Hence SNaQ reduces the searchable space to consider only networks whose reticulations involve enough nodes. Indeed, all reticulations with $k = 2$ and most with $k = 3$ are either not detectable at all or their parameters are not all identifiable. Thus, SNaQ will skip proposed networks with $k = 2$, and most with $k = 3$, and it reparametrizes the cases when parameters are not identifiable.

6.4.2 ROOTING OF SEMIDIRECTED NETWORKS

Species networks applying quartets estimates semidirected networks, where the root node is suppressed and we ignore the direction of all tree edges but maintain the direction of hybrid edges, thus keeping information on which nodes are hybrids. To interpret the estimated network, one might want to root it at a known outgroup. However, as mentioned before, unlike rooting trees, the placement of the root is constrained because the direction of the two hybrid edges to a given hybrid node inform the direction of time at this node: the third edge must be a tree edge directed away from the hybrid node and leading to all the hybrid's descendants. Therefore the root cannot be placed on any descendant of any hybrid node. For example, in figure 6.14 the network on the left was estimated with SNaQ, but it has a randomly chosen root. If taxon "O" is meant to be the outgroup, then this network is clearly not rooted properly. In this case, the root cannot be placed in any descendant edge of the hybridization event. That is, the root cannot be placed on edges 9, 10, or 11.

6.4.2.1 Conflicts between the Direction of a Reticulation and the Root

As mentioned already, the direction of hybrid edges constrain the position of the root. The root cannot be downstream of hybrid edges. Any hybrid node has to be younger than, or of the same age as, both of its parents. So time has to flow "downwards" of any hybrid node, and the root cannot be placed "below" a hybrid node.

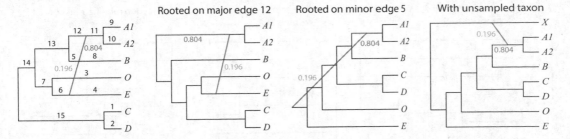

Figure 6.14. Left: Network estimated with SNaQ with a randomly chosen root. Black numbers denote edge labels. The root cannot be placed in edges 9, 10, or 11 as they are "below" the hybridization event. If the clade (A1,A2) is the outgroup, we need to root the network on the hybrid edges as following. Center left: Network rooted on the major hybrid edge (edge 12). Center right: Network rooted on minor hybrid edge (edge 5). This network requires an unsampled/extinct taxa. Right: Network rooted on the minor hybrid edge (edge 5) but with an added unsampled taxon X.

For example, see figure 6.14 (left) with edge numbers to illustrate where to root the network. Let's imagine that the A1 and A2 are outgroups, and this is the network estimated with SNaQ. According to this network, time must flow from the hybrid node toward A1 and A2. So any attempt to reroot the network with A1 as the outgroup, or with A2 as the outgroup, or with the A clade (on edge 11), will fail.

In this case, however, it is possible to root the network on either parent edge of the hybrid node (edges 5 and 12), resulting in the rooted versions in figure 6.14 (two networks in the center). The network rooted on major edge 12 represents gene flow between E and an ancestral population to the clade (A1,A2), while the network rooted on minor edge 5 requires an unsampled or extinct taxa to be included, so that gene flow would be between the unsampled taxa and the ancestral population to the clade (A1,A2) (see figure 6.14 right). The network rooted on the major edge 12 is more plausible if we think that the species tree is the major tree, meaning that any gene flow or introgression event replaced fewer than 50% of the genes in the recipient population.

In other cases, it may not be possible to reroot the network with a known outgroup: for example, if only A1 is the only outgroup, and if A2 was an ingroup taxon. In such a case, the outgroup knowledge tells us that our estimated network is wrong, as the placement of the reticulation contradicts the root position. Its placement might be correct, but its direction would be incorrect. In this situation, we can explore other candidate networks compatible with a known outgroup (see section 6.4.2.2).

6.4.2.2 Candidate Networks Compatible with a Known Outgroup

When estimating semidirected networks with SNaQ, there is always the possibility that the estimated network is impossible to root with a known outgroup. This is the case as SNaQ does not impose any rooting constraint on the network: the search for the highest pseudolikelihood score considers all level-1 networks, including those that are incompatible with a known outgroup. The monophyly of outgroups is not imposed either, as is the case for many other methods.

If the estimated network cannot be rooted with the known outgroup, we can recall the discussion in section 6.4.1.1 where we noted that sometimes it is impossible to detect

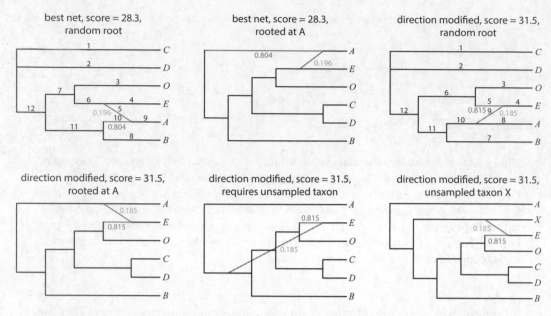

Figure 6.15. Top left: Best network estimated with SNaQ with random root placement. Black numbers represent edge labels. Top center: Same network but rooted with the outgroup A. Top right: Second best network with the only difference being that the direction of the hybrid edge is from A to E instead of from E to A. The second best network allows us to root in A (bottom left). Two options to root the second best network at A. Bottom left: The second best network rooted at A on edge 8. Bottom center: Option 2 of the second best network rooted at A on edge 10. This option requires the existence of an unsampled taxon X. Bottom right: Second best network rooted at A, with the inclusion of the unsampled taxa X.

the correct placement of the hybrid node in the hybridization cycle, depending on the specific network topology and the data at hand. Thus, we might want to compare the pseudolikelihood score of alternative placements of the hybrid node in the hybridization cycle on the estimated network (see figure 6.12). If we find in these alternative placements of the hybrid node a modified network that has a score close to that of the best network, and that can be rerooted with our known root position, then this modified network is a better candidate than the network with the best score. For example, in figure 6.15 we show several networks: the best network estimated by SNaQ with a pseudolikelihood score of 28.3 (the smaller, the better in this plot because the pseudolikelihood score denotes negative log-pseudolikelihood) and a randomly chosen root (top left), and the second best network with the direction of gene flow modified and score of 31.5 (not far from the best network), also with a random root (top right). Now imagine that our outgroup is taxon A. If we try to root the best network at "A" (edge 9), we will get an error. But we could root this network on the major parent edge to A (edge 10) in figure 6.15 (top center).

For the second best network (figure 6.15 top right), there are two ways to root it with A: on the external edge 8 to A (figure 6.15 bottom center) or on its parent edge 10 (figure 6.15 bottom left). These two options give quite different rooted versions of the network, one of which requires the existence of an unsampled taxon, sister to BOECD, that would have contributed to introgression into an ancestor of E (figure 6.15 bottom right). The rooted version on edge 8 (bottom left) says that an ancestor of A contributed to the introgression into the ancestor of E. Taxon A is the outgroup in both cases, but this last case is more parsimonious, in the sense that it does not require the existence of an unsampled taxon. If we compare this network (figure 6.15 bottom left) with the best network rooted at A (figure 6.15 top center), we can see that the only distinction is the direction of the minor hybrid edge: in one case going from A to E and in the other case going from E to A. Both networks have similar pseudolikelihood score, so additional biological knowledge could be use to choose the more appropriate evolutionary interpretation.

6.4.3 GOODNESS OF FIT TOOLS

6.4.3.1 Choice of the Number of Hybridizations

When using SNaQ (and other network methods), users need to specify the maximum number of hybridizations allowed because the pseudolikelihood will improve as h increases, in the same way that the likelihood [841, 842] and parsimony score [840] improve as hybridizations are added. Thus, model selection tools are necessary to estimate the number of hybridizations. Existing tools involve using cross validation to determine the best parameter h [842]. For the pseudolikelihood framework, the cross-validation error could be measured from the difference between the quartet CFs observed in the validation subset and the quartet CFs expected from the network estimated on the training set. Because K-fold cross validation requires partitioning the loci into K subsets and re-estimating a network K times at each h value, this approach can be computationally heavy.

Information criteria have already been used to select h (e.g., [387]), but these criteria are inappropriate if the full likelihood is replaced by a pseudolikelihood. Theory is missing to compare the pseudolikelihood scores of different networks because of the possible correlation between quartets from different four-taxon sets. It can be shown, however, that quartets from two four-taxon sets s_1 and s_2 are independent if s_1 and s_2 overlap by at most one taxon and if the true four-taxon subnetworks share no internal edges. Future work could exploit this partial independence to construct hypothesis tests. Global tests like TICR have recently been extended to the case of networks [105], but there is still room for more research on model selection tools for networks.

One alternative is to use data-driven tools; for example, slope heuristics can indeed be used with contrast functions (like pseudolikelihood) for model selection in regression frameworks [72, 51]. In figure 6.16 left, we show the log pseudolikelihood profile with h. A sharp improvement is expected until h reaches the best value and a slower, linear improvement thereafter. Based on this plot, we can estimate the best $h = 1$, as there is no longer an improvement in the pseudolikelihood for $h > 1$.

6.4.3.2 Visualization of Comparisons between Observed CF versus Expected CF

A good way to visualize the goodness of fit of a given estimated network to the data is to plot the observed CF from the input against the expected CF (figure 6.16 right). There are always many points overlapping on the bottom-left corner: concordance factors of 0.0 for quartet resolutions not observed and not expected. We could highlight quartets

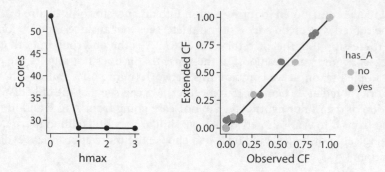

Figure 6.16. Left: Negative of log-pseudolikelihood used as score to choose the optimal number of hybridizations (hmax) to be included in the network: here it is 1. Right: Comparison of quartet CFs observed in gene trees and quartet CFs expected from the network as a measure of goodness of fit. Shade depends on whether the specific quartet (point) has taxon A (dark grey) or not (light grey).

that include taxon A, say, if we suspect that it is an unrecognized hybrid. With this plot, there is not a discernible pattern, but if A were indeed an unrecognized hybrid, then we would expect the light grey dots to be farther from the $x = y$ line than the dark grey dots.

6.4.4 BOOTSTRAP ANALYSIS

To measure uncertainty in the network, one may re-estimate the network on bootstrap data sets and compare the bootstrap networks. In order to create bootstrap data sets, either we can use the credibility intervals of the CFs obtained with BUCKy to sample CFs for each quartet or we can use bootstrap gene trees to sample input gene trees at random, one per gene.

As will be described in the next sections, to summarize the networks estimated from these bootstrap sets, we first calculate the support for edges being in the major tree: the tree obtained by suppressing the minor hybrid edge (with $\gamma < 0.5$) at each reticulation. We then summarize the support for the placement of each minor hybrid edge on that tree, considering two edges as equivalent if they are of the same type (hybrid or tree edges) and define the same clusters in the networks [842].

6.4.4.1 Summarization of Support for Tree Edges

After sampling 100 bootstrap data sets (CFs or gene trees), we estimate 100 bootstrap networks, and we need to summarize what they have in common (highly supported features) and what they do not (areas of uncertainty). The first thing to do is to summarize the bootstrap support of tree edges, which can be easily done by extracting the major tree in each of the 100 bootstrap networks. This results in a sample of 100 bootstrap major trees, and usual techniques to summarize bootstrap support on tree edges work.

In figure 6.17, we map the bootstrap support for internal tree edges (100 in every case) onto the estimated network (left). These bootstrap support values mean that all bootstrap major trees contain the same internal edges. So, the main vertical signal of the data is strong and highly supports the major tree in the estimated network.

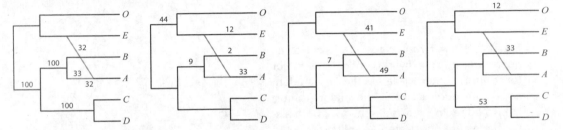

Figure 6.17. Left: Best network with bootstrap support on tree edges (all 100), on the two hybrid edges (32 and 33) and on the hybrid node (32). Black numbers represent bootstrap support. The bootstrap support value for the hybrid node represents the proportion of bootstrap networks that have the same two hybrid edges. Center left: Boostrap values on hybrid clades that appeared on the bootstrap networks. For example, the clade (O,E) was a hybrid clade in 44 out of the 100 bootstrap networks. Center right: Bootstrap values for the possible origin of gene flow (minor sister clade) that appeared in the bootstrap networks. For example, taxon E was the origin of gene flow in 41 out of 100 bootstrap networks. Recall that in 32 of those, the recipient of gene flow was taxon A, as shown by the bootstrap support of 32 on the minor hybrid edge (from E to A) in the network on the left. Right: Bootstrap values for the possible major sister clades in the bootstrap networks. For example, clade (C,D) was a major sister clade in 53 out of 100 bootstrap networks.

6.4.4.2 Summarization of Support for Hybrid Edges and Hybrid Nodes

Summarizing a set of networks is more complex than summarizing a set of trees because edges in a network do not uniquely correspond to bipartitions of the taxon set. In the previous subsection, we just illustrated that for a tree edge, we calculate its bootstrap support as a major edge, that is, the proportion of bootstrap networks whose major tree displays the edge of interest.

To summarize bootstrap support for reticulation events, we consider each hybrid node separately. We remove the minor hybrid edges at all the other reticulation events and keep only one reticulation event of interest. We then identify three clades: the hybrid clade, the major sister clade, and the minor sister clade of the hybrid node (figure 6.18). The descendants of a given hybrid node form the "recipient" or "hybrid" clade, which is obtained after removing all other reticulations. Because of the reticulation event, the hybrid clade has two sister clades, not one: the major sister (through the major hybrid edge with $\gamma > 0.5$) and the minor sister (through the minor hybrid edge with $\gamma < 0.5$). We can calculate the frequency that each clade is a hybrid clade, or a major or minor sister for some other hybrid, in the bootstrap networks.

For example, in figure 6.17 (left), we can plot the bootstrap values of the two hybrid edges in the best network: percentage of bootstrap networks with an edge from the same sister clade to the same hybrid clade. In this case, there is 32% bootstrap support for the minor hybrid edge (this hybrid edge appears in 32% of the bootstrap networks) and 33% for the major hybrid edge (this hybrid edge appears in 33% of the bootstrap networks). In addition, we could show the bootstrap support for the full reticulation relationship in the network, shown at the hybrid node in figure 6.17 (left). In this case, the support for same hybrid with same sister clades is 32%. This means that in 32% of the bootstrap networks we have the same reticulation event (exact same hybrid clade, major hybrid

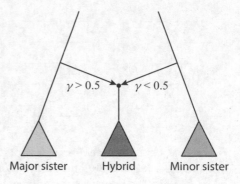

Figure 6.18. A hybrid clade is the set of all descendants of a hybrid node. The major (or minor) sister clade is the descendant set of the edge sister to the major (minor) hybrid edge. In unrooted networks, both sister clades are considered as splits (unrooted bipartitions) because the root can be placed anywhere outside the hybrid clade.

clade, and minor hybrid clade). Thus, this reticulation event (and the two hybrid edges) is not well supported by the data. So, we next explore what are the other reticulation events that appear in the bootstrap networks.

Hybrids in bootstrap networks. In figure 6.17 (center left), we show the bootstrap support for alternative hybrid clades mapped on the parent edge of these nodes. For example, taxon A is estimated as a hybrid in only 33% of our bootstrap networks. This percentage seems to contradict the 32% in the hybrid node in figure 6.17 (left); however, these two quantities represent different proportions. In figure 6.17 (center left), 33% represents the proportion of times that taxon A is a hybrid clade *regardless* of the sister clades, whereas in figure 6.17 (left), 32% represents the proportion of times taxon A is a hybrid clade, *with E and B as sister clades*. Continuing with other hybrid clades in figure 6.17 (center left), the lineage to (E,O) is estimated as being of hybrid origin 44% of the time;, in 12% it is taxon E, in 9% it is clade (A,B), and in 2% it is taxon B. So, there is quite a bit of variability in the hybrid clades among the bootstrap networks, with clade (O,E) as the one more supported, but with only 44%.

Origin of Gene Flow. Just as we explored the hybrid clades in the bootstrap networks (that is, the recipient clades of gene flow), we can ask which clades are the most supported as sister clades to the hybrid clade. That is, we are interested in identifying the clades that serve as the origin of gene flow in most of the bootstrap networks.

In figure 6.17 (center right), we plot the support for the various placements of the gene flow origin (minor sister clade), mapped along the parent edge of these nodes. We filtered clades to show only those with bootstrap support >5%.

In the best network, the lineage to E is estimated as the origin of gene flow (figure 6.17 left), but this is recovered in only 41% of our bootstrap networks. In another 49%, it is the lineage to A that is estimated as the origin of gene flow: so gene flow is estimated in the opposite direction. Finally, there is 7% support for gene flow originating in (A,B).

Mapping the support for major sister clades might be interesting too (figure 6.17). In the best network (figure 6.17 left), B is estimated as the major sister clade, with 33% bootstrap support. The clade (C,D) has the higher bootstrap support as major sister clade with 53%.

These examples illustrate some of the complications of summarizing samples of networks. These tools can also be used to summarize a posterior distribution of networks generated by other programs (e.g., [21, 851]). PhyloNet can summarize a set of networks by listing the networks with highest support (and averaging branch lengths and inheritance probabilities for each unique topology). The tools described here provide summaries about local relationships on individual edges and nodes. This lets us identify hybridization events that are highly supported, regardless of the other hybridizations in the network, which is crucial as we showed that some hybridizations are harder to be detected than others.

6.5 Appendix: Installation and Use of the PhyloNetworks Julia Package

Julia is a high-level and interactive programming language (like R or Matlab), but it is also high performance (like C). The instructions to install Julia are in https://julialang .org/downloads/. The instructions to install the PhyloNetworks package are in https:// crsl4.github.io/PhyloNetworks.jl/stable/.

SNaQ is a method implemented in the package to estimate a phylogenetic network from multiple molecular sequence alignments. There are two alternatives for the input data:

- A list of estimated gene trees for each locus, which can be obtained using MrBayes or RAxML, or
- A table of concordance factors (CF), that is, gene tree frequencies, for each four-taxon subset. This table can be obtained from BUCKy to account for gene tree uncertainty (see figure 6.9).

In the package documentation, we present a pipeline to obtain the table of quartet CF needed as input for SNaQ (see also https://github.com/crsl4/PhyloNetworks.jl /wiki). The tutorial starts from the sequence alignments, runs MrBayes and then BUCKy (both parallelized), producing the table of estimated CFs and their credibility intervals. Additional details on this TICR pipeline describe how to insert data at various stages (e.g., after running MrBayes on each locus).

Besides the scalability of SNaQ compared with other network methods, SNaQ has the advantage of not using branch lengths in the estimated gene trees. By using only gene tree topologies, SNaQ avoids the dangerous assumptions that all genes and/or lineages evolve at the same rate. For reconstructing species tree, methods that ignore branch lengths in gene trees tend to be more robust.

Also, SNaQ uses unrooted gene trees as input, which also prevents a potential layer of error if the outgroup is involved in ILS, or saturation or long branch attraction. For example, [261] shows that rooting errors explain incongruence in a yeast data set.

6.5.1 MAIN FUNCTIONS IN PHYLONETWORKS

First, we need to read the input data into Julia which can be a concordance factors table from BUCKy [27]:

```
buckyCF = readTableCF("bucky-tableCF.csv"),
```

or a list of estimated gene trees from MrBayes [332] or RAxML [720]:

```
raxmlCF = readTrees2CF("raxml-trees.tre").
```

If the input data is a list of estimated gene trees, then the readTrees2CF function will create the table of CFs automatically.

Finally, we need to read a starting tree or network for the optimization. Usually, users estimate a species tree with a coalescent-based method like ASTRAL [506], which we can read into Julia with the following command:

```
tre = readTopology("astral.tre").
```

Next, we describe how to estimate a network in Julia. The function snaq! in the PhyloNetworks package estimates a semidirected level-1 network from a collection of gene trees or a table of concordance factors. SNaQ also needs a starting topology for the optimization in the space of networks, and it needs to impose a constraint on the maximum number hybridizations allowed (hmax). Usually, one would start with hmax=0, and then use the resulting tree as starting point to estimate a network with one hybridization event (hmax=1), and so on.

We can run the snaq! function on a given starting tree (tree), input data (CF), and maximum number of hybridizations allowed (hmax=0 in the first example). The function will return a network object (net0), which can be used in subsequent analyses like comparative methods [50] or simply plotting or rerooting (see section 6.4.2).

```
net0 = snaq!(tree, CF, hmax=0, filename="net0", seed=1234);
net1 = snaq!(net0, CF, hmax=1, filename="net1", seed=2345);
net2 = snaq!(net1, CF, hmax=2, filename="net2", seed=3456);
net3 = snaq!(net2, CF, hmax=3, filename="net3", seed=4567);
```

Here, we are estimating networks up to three hybridization events in a sequential manner: first we estimate a tree (hmax=0), then using this estimated tree as starting topology, we estimate a network with one hybridization event (hmax=1), and so on.

For a full documentation and tutorial, see https://github.com/crsl4/PhyloNetworks.jl/.

PART II

Empirical Inference

Because many of the challenges of species tree inference, interpretation, and application to evolutionary questions are common across studies, we have included a section of empirical applications. These chapters demonstrate in practice the methodological developments and theoretical considerations presented in the first part. As such, these chapters represent some of the decisions made by empiricists when choosing among the many existing software packages that are available, as the authors of each chapter consider the features that are both taxon and data specific to each of their studies.

In chapter 7, Walker and Smith focus on the frequently observed phylogenomic conflict in the evolution of plants at broad taxonomic scales and present a detailed dissection of potential underlying mechanisms for the conflict. They focus on two examples within angiosperms: placement of the genus *Amborella* and relationships among the carnivorous plants in the order Caryophyllales. The phylogenetic placement of *Amborella* has been debated for decades and represents the difficulties with resolving evolutionary relationships that occurred during the early diversification history of groups. The relationships among the carnivorous plants in the order Caryophyllales represent an interesting and dramatic ecological shift in plants, where strong support for particular nodes in the phylogeny are key to testing evolutionary hypotheses about when and how many times these shifts might have occurred. In both examples, the empirical data show considerable discord among the gene trees of different loci used in the phylogenetic inference. They seek to reconcile conflict in gene trees by determining the different causes for conflict to obtain species tree estimates, focusing on some of the common causes of conflict (e.g., low levels of phylogenomic signal, incomplete lineage sorting, horizontal transfer, gene duplication and loss, hybridization, and errors in data processing, such as orthology identification), for which they provide a detailed review. Their work highlights, in particular, the effect of gene and genome duplication on species tree inference, as well as estimation of divergence times, and the importance of identifying sources of phylogenomic conflict (beyond the discord associated with the random coalescence of gene lineages) in order to resolve species relationships across the plant tree of life.

Chapter 8, contributed by Gates, Pilson, and Smith, focuses on detecting hybridization, using both experimental (i.e., generated in the lab) and natural hybrids in the plant subfamily Iochrominae, a diverse clade that contains several species that are known or suspected to hybridize. Because hybridization is central to addressing evolutionary questions about the processes involved in speciation, the authors focus on the performance of methods for detecting hybridization. Specifically, the authors apply several methods to the experimental and natural data, including internode and tree certainty (IC/TC) [647], tree incongruence checking in R (TICR) ([727]; chapter 6 in this volume), HyDe [77], D-statistics [208], and SNaQ ([710]; chapter 6 in this volume). The results obtained with these methods are mixed, with most generally detecting hybridization to some extent but few performing well in terms of accurately identifying parental species. In recognition of this uncertainty, the authors discuss different explanations for the mixed performance of the methods, including the possibility of recently formed hybrid populations whose genomes are also affected by the processes of incomplete lineage sorting and periodic gene flow. Given that incomplete lineage sorting is ubiquitous in recently diverged species, and that gene flow might be common as well, this chapter highlights the challenges with not only estimating species trees but also making inferences about the placement of the evolutionary process of interest (in this case, hybridization) when multiple processes may be associated with discord among gene trees and between gene trees and the species tree, making this chapter of broad interest beyond the focal taxa studied here.

The difficulties involved in inferring species-level relationships for taxa subject to hybridization is also the subject of chapter 9, by Blischak, Thompson, Waight, Kubatko, and Wolfe. However, the role of both hybridization and polyploidy in evolution is the focus here, as exemplified by the subsections *Humiles* and *Proceri* within the plant genus *Penstemon*. Applying many of the methods used in the previous chapter, this chapter makes for a nice comparison of the performance of the methods when the multiple processes contributing to discord also include polyploidy. In addition, this chapter presents a newly developed method based on the computation of quartet concordance factors (QCF) that can be used for species tree or network inference using TICR or PhyloNetworks (chapter 6), which is especially well suited to studies that include polyploid species. As with the previous chapter, the results here reinforce the challenges of inference in the presence of hybridization, especially for recent, rapid radiations. However, the authors identify some trends that emerge across the methods examined, which when coupled with the newly proposed QCF method, appear to be especially useful for analyzing species with differing ploidy levels.

The final chapter in the section, contributed by Cobb and Oaks, takes a different approach to some of the challenges empirical data pose for estimating species divergence histories. The focus in this case is on the interplay between sequencing error and assumptions about the extent of shared genealogical information between sites in an empirical analysis. Specifically, the authors consider both "linked" data (e.g., sites sampled from the same locus), for which there is some correlation across sites due to their shared genealogical history, and "unlinked" data, for which each individual site has its own underlying genealogical history (e.g., for sequences from different genes, or random SNPs across the genome as with RADseq data). The unlinked model for data presented in this chapter is similar in spirit to the "coalescent independent sites" data in chapter 4; however, in this chapter the authors present an analysis using a method implemented in the software package ecoevolity to compare analyses based on "linked" data with the "unlinked" data model based on computer simulations. The accuracy of estimates of the species history in the presence of sequencing errors is examined by introducing errors either by changing singleton patterns to constant patterns at random or by replacing heterozygous sequences with homozygotes at random. The general finding is that the common assumption of unlinked data results in more susceptibility to sequencing errors when they are present; however, unlinked models perform better in general when the majority of loci are short (e.g., for RADseq data). Together, the results of this study are pertinent to best practices for inferring divergence histories from empirical data. The authors also provide an in-depth discussion of the implications of all of their findings in the context of typical characteristics of empirical data, as well as a brief discussion of additional properties of empirical data not covered by the simulations that may be important to consider.

The chapters in this part collectively touch on many of the commonly encountered decisions to be made in applying species tree estimation methods, decisions that will vary among empirical settings depending on the data type and the evolutionary processes involved in the divergence of species. While the diversity of issues addressed in these empirical studies may not span the entire range of what is theoretically possible, they do provide useful perspectives on the analytical choices empiricists face, as well as the downstream impacts of such decisions. Specifically, the issues addressed in this section (e.g., whether to assume a tree or network structure; how to estimate horizontal edges; how to account for differing ploidy levels among species; how to reconcile and describe conflicting gene trees; how to estimate divergence times; and whether to process/filter data to match assumptions about correlations among sites) represent some common decisions that are essential to robust phylogenetic inferences and the evolutionary questions that might be addressed using the species tree framework.

CHAPTER 7

Phylogenomic Conflict in Plants

Joseph F. Walker and Stephen A. Smith

7.1 Introduction

The last few decades of phylogenetic and genomic research have highlighted that genomes are a composite of evolutionary histories. The relationships among species vary in their degree of concordance and conflict among gene evolutionary histories, creating challenges for the reliable inference of relationships among lineages (figure 7.1). Conflicting gene histories can originate from biological processes, including incomplete lineage sorting (ILS), gene duplication and loss, horizontal gene transfer (HGT), and hybridization [198, 455]. Lineage-specific factors potentially influence the origins of conflicting evolutionary histories because of the differing prevalence of the sources of biological conflict. For example, bacterial genomes have been shaped heavily by HGT [548]. High ILS characterizes insect genomes owing to periods of rapid diversification [797], while plant genomes, the focus of this particular chapter, have a notable propensity for gene and genome duplication and hybridization [769, 714, 74, 138]. Distinguishing these sources of conflict is often tricky [319], but with the rapid accumulation of genomic data from nonmodel species and the increase in large-scale phylogenetic analyses, inferring the prevalence of these processes and their influence on genomes is now possible.

Before the exponential increase in molecular data, morphology was the basis of plant classifications and phylogenetic analyses for decades [30, 626, 152]. Even without the insights offered by molecular data, researchers identified and discussed conflicting morphological patterns, assuming them to be a mixture of hybridization, convergence, and uncertainty. As molecular data became a dominant data source for species tree inference, beginning with allozymes and random amplified polymorphic DNA [3], so did the identification of conflict between data sources [291]. Conflicting phylogenetic relationships reconstructed from molecular and morphological sources were often compared [92, 189]. For example, analyses involving the placement of gymnosperm order Gnetales identified conflict among data types, including morphology and molecules [196, 189]. The anthophyte hypothesis that placed Gnetales sister to angiosperms was an intriguing possibility with enormous implications for morphological evolution, but support for this placement was heterogeneous across morphology and molecules. Larger data sets were assumed to be the solution for resolving this and other recalcitrant relationships across the plant tree of life.

The collection of genetic data grew dramatically with the use of polymerase chain reaction (PCR) and the availability of cost-efficient and effective protocols for extracting DNA from plants [199]. Plant systematists were early to organize broad sequencing initiatives that led to the development of widely used gene regions and protocols that would help resolve relationships from deep to shallow levels (e.g., ITS [43], ETS [42], *rbcL* [190, 130], and several additional chloroplast regions [671, 672]). These efforts resulted in significant advances in plant classification with novel implications for evolution across plants [159, 615, 715]. However, researchers were also quick to note conflicting results between chloroplast, mitochondrial, and ribosomal or nuclear regions [42, 670, 126]. The discussion of discordance became a prominent aspect of plant systematics (e.g., APG III). For example, the branching order of the earliest divergences in land plants, including the resolution of hornworts, liverworts, and mosses, has long been subject to debate, complicated by ancient divergence times. Mitochondrial data conflict strongly with other data sources, and the causes of this discordance have been an active area of discussion [592, 591]. Notably, these conflicts remain even with the addition of whole transcriptomes [798, 554].

As the magnitude of genetic data increased, the community struggled with the correct way to combine different data sources. Early studies combined data with supertree methods or concatenated data in a single matrix [593]. However, neither of these solutions managed to accommodate biological sources of heterogeneity adequately. Instead, researchers primarily assumed that the significant analytical problem was a lack of phylogenetic signal. If a lack of signal was the only issue, theoretically combining gene regions or genes and morphology could resolve low support or conflict [197, 594, 740].

Meanwhile, new methods that account for conflict originating from ILS became available [435, 388, 298, 405]. Unlike concatenation, these methods use the patterns of conflict among gene regions to model ILS and infer the species tree [456]. Further developments brought about methods that model other sources of conflict, such as gene duplication and loss [88]. Methods that explicitly model gene tree conflict overcome several important limitations of concatenation: concatenation in the presence of ILS is not statistically consistent [385, 624], and the anomaly zone, where the majority of genes will not contain the species relationship [176, 442, 495], can mislead majority-rule consensus trees from gene tree topologies [53]. Recent studies have also demonstrated that concatenation methods are particularly susceptible to small amounts of highly influential data (single genes or sites), helping to explain data set sensitivity in contentious plant relationships [675, 771]. As data sets have grown, theoretically alleviating sampling issues [867] and increasing accuracy through the use of more genes [629], the problems of multiple sources of conflict, low information, and heterogeneous evolutionary processes have become exacerbated.

Despite the challenges of phylogenomic analyses, large-scale sequencing efforts have exploded over the last few years. These studies have allowed researchers to collect enormous data sets, changing the face of phylogenetics to study plants [735, 348, 828, 488]. Several large collaborative projects such as 1kp [554], the Plant and Fungal Tree of Life Project (PAFTOL), and 10kp (Ten Thousand Genomes Project) [137] have contributed enormous resources. Individual labs have complemented these efforts by generating large data sets focused on specific clades. Next-generation sequencing (NGS) has provided unique evolutionary insights into historically recalcitrant relationships and provided evidence for genomic events such as ancient gene and genome duplication [107, 826, 554, 454]. Nevertheless, the addition of more data has not alleviated conflict and instead has revealed it to be a ubiquitous phenomenon [361, 376, 630, 778, 554].

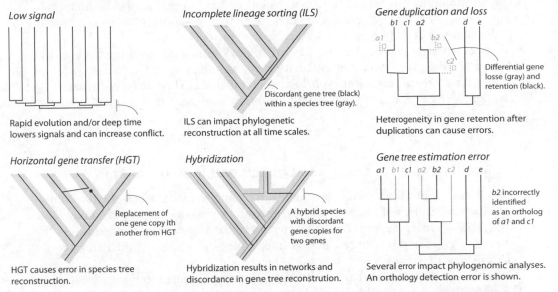

Some causes of gene tree conflict

Low signal

Rapid evolution and/or deep time lowers signals and can increase conflict.

Incomplete lineage sorting (ILS)

Discordant gene tree (black) within a species tree (gray).

ILS can impact phylogenetic reconstruction at all time scales.

Gene duplication and loss

b1 c1 a2   d e

a1   b2

c2

Differential gene losse (gray) and retention (black).

Heterogeneity in gene retention after duplications can cause errors.

Horizontal gene transfer (HGT)

Replacement of one gene copy ith another from HGT

HGT causes error in species tree reconstruction.

Hybridization

A hybrid species with discordant gene copies for two genes

Hybridization results in networks and discordance in gene tree reconstrution.

Gene tree estimation error

a1 b1 c1 a2 b2 c2 d e

b2 incorrectly identified as an ortholog of a1 and c1

Several error impact phylogenomic analyses. An orthology detection error is shown.

Figure 7.1. Several causes of gene tree conflict include low phylogenetic signal, incomplete lineage sorting (ILS), gene duplication and loss, horizontal gene transfer (HGT), hybridization, and analytical error. Low signal as a result of rapid evolution and/or deep divergences can cause gene tree conflict. Incomplete lineage sorting is a common and pervasive source of conflict throughout the tree of life. Gene duplication and loss is common in plants and can be a significant source of conflict in homologs. Horizontal gene transfer, while less common, is still present in some lineages of plants. Hybridization is widespread for closely related taxa in several plant lineages. Several sources of analytical error (orthology is shown here) can present problems for phylogenomic analyses and increase the appearance of conflict across gene trees.

While NGS data and methods have allowed researchers to identify and analyze the biological processes that cause conflict, the automated nature of high-throughput techniques has exacerbated errors from nonbiological sources (outlined below) that arise during the processing of large data sets.

New phylogenomic data sets have helped document the ubiquity of conflict, and they have also facilitated the development of a diverse set of tools for examining patterns of gene tree conflict. For example, there are methods to visualize and quantify the amount of conflict underlying the data [310, 311, 27]. Methods have also been developed to assess patterns of conflict in manners agnostic to the underlying processes [646, 700], to analyze the influence of taxon sampling and identify regions of the tree that would benefit from taxon sampling [578, 774, 857], and to explore overlapping patterns of ILS and gene and genome duplications [369]. While species tree reconstruction is still a significant challenge for these data, analyses of NGS data have led to critical biological discoveries in plants, including the hypothesis of hard polytomies within several major clades [658, 374, 407]. Pigmentation genes in tomatoes show signs of conflict among sites [568]. Gene tree conflict induced by gene duplication has helped explain the disjunct phylogenetic distribution of betalain pigmentation in Caryophyllales [673]. Network analyses of gene tree patterns have shown widespread hybridization and instances of reticulate evolution that can explain the distribution of adaptations [749, 355]. As a result of lateral gene transfer, conflict has helped explain the distribution of C4 photosynthesis among the grasses [142, 207].

While gene tree conflict has been a part of phylogenetic analyses for years, recent advances in data collection methods have made the analysis and accommodation of conflict an integral part of any phylogenomic analysis. This chapter discusses the consequences of gene tree conflict and examines how gene tree conflict influences phylogenetic analyses.

7.2 Two Examples of Gene Tree Conflict within Angiosperms

For plants, contentious relationships add uncertainty to fundamental questions about the early paleoecology of angiosperms and ancestral floral structure and genomics [238, 6, 655, 855]. While gene tree conflict is often prevalent at contentious relationships, the patterns of conflict can help researchers infer biological processes underlying species relationships.

Within angiosperms, perhaps no lineage has captured more significant debate than the monotypic genus *Amborella*. This small, unassuming tropical shrub from New Caledonia is hypothesized to be the sister lineage to all other extant flowering plants [476, 561, 848] but has also been supported as sister to the water lily order, Nymphaeales [820], or inferred as nested within the flowering plant phylogeny [271]. The position of *Amborella* helps determine when in earth's history the first flowers appeared and, in turn, identify what species codiversified with flowering plants. Genomes of both *Amborella* and the species *Nymphaea colorata* of the order Nymphaeales have been sequenced [6, 855]; their position in the angiosperm phylogeny greatly influences comparative methods. For example, angiosperms share a whole-genome duplication, and *Amborella* did not experience an additional whole-genome duplication event [6].

In contrast, Nymphaeales experienced an additional duplication event that may be as old as the split from the rest of the angiosperm lineages, more than 140 million years old [855]. Changes in the resolution of these taxa would determine the number and timing of genome duplications within angiosperms. Also, the Nymphaeales and many other flowering plants have scented flowers, and *Amborella* lacks scented flowers. The traditional placement of these taxa, along with analyses of the genes involved, suggests that scented flowers evolved twice, within Nymphaeales and mesangiosperms [855]. However, a significant change in the resolution of these lineages could suggest a different model of evolution for scented flowers in flowering plants. Finally, the female gametophyte development within angiosperms, once thought to be relatively simple, is more complex, with several different types of female gametophytes. *Amborella* has a four-celled egg apparatus and two developmental modules. Nymphaeales have a three-celled egg apparatus and one developmental module. Eudicots, monocots, and other flowering plants have a three-celled egg apparatus and two developmental modules [245, 254]. The divergence of *Amborella* with respect to Nymphaeales has a considerable impact on the interpretation of plesiomorphic characters, the early evolution of the female gametophyte in flowering plants, and even how the flower may have evolved along with insect pollinators.

While these recalcitrant relationships across the tree of life may occur more than 150 MYA, more recent edges also experience uncertainty due to extensive conflict. For

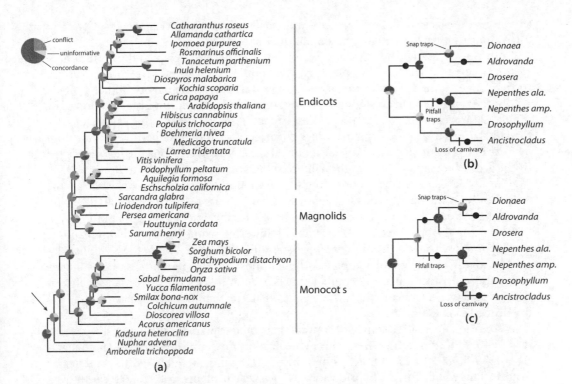

Figure 7.2. Patterns of conflict surrounding the contentious relationships of *Amborella* and the DA clade (a clade composed of the monotypic genus *Drosophyllum* and the post-carnivorous genus *Ancistrocladus*). (a) The inferred phylogeny of flowering plants from [798] with gene tree conflict and concordance mapped to each node. The conflict surrounding the debated position of *Amborella* is highlighted. (b and c) The contentious position of the DA clade from [774]. (b) Conflict from the amino acid data mapped to the tree depicts the group's affinity for being sister to the woody vines in the genus *Nepenthes*. (c) The CDS (transcriptome-coding DNA sequence) data relationships and depiction of the group's position inferred to be sister to all other carnivorous plants in the Caryophyllales.

example, nested within the eudicots, the carnivorous plants of the order Caryophyllales contain an equally recalcitrant lineage [774]. The placement of a clade (hereafter called the DA clade) composed of the monotypic genus *Drosophyllum* and the postcarnivorous genus *Ancistrocladus* has proven a challenge to infer. Morphologically, *Drosophyllum* is almost indistinguishable from *Drosera*, as both lineages are small succulent plants with sticky traps. Similarly, *Ancistrocladus* and *Nepenthes* are woody vines and live predominantly in tropical regions. Inference from morphology almost unanimously places *Drosophyllum* sister to *Drosera* and *Ancistrocladus* sister to *Nepenthes*. However, molecular data strongly supports the DA clade (figure 7.2). The divergence occurring either sister to all other carnivorous plants or sister to the *Nepenthes* alters the inferred ancestral condition of the first carnivorous lineages within Caryophyllales. It adds uncertainty as to whether ancestrally these plants were woody vines or small succulent sticky traps. Furthermore, *Nepenthes* and *Ancistrocladus* grow natively in tropical regions, whereas *Drosophyllum* is one of the only carnivorous plants to grow on dry soil;

therefore, the placement of the DA clade influences inferred historical biogeography and conditions that led to carnivory arising in the Caryophyllales.

Both *Amborella* and the DA clade exhibit a shared pattern in which the evolutionary history inferred from the chloroplast genome conflicts with species trees inferred from the nuclear genome. Hybridization in plants has been inferred using the conflict between chloroplast and the nuclear genomes. Hybridization and genome duplication followed by diploidization are sources of conflict. These processes provide a tempting explanation for the DA clade as this would explain the significant conflict and the hyperdiverse morphology exhibited by the carnivorous Caryophyllales. This explanation is further supported by the group having one of the highest inferred rates of whole-genome duplication [774]. The family Drosophyllaceae, consisting of the monotypic genus *Drosophyllum*, is the only family in the carnivorous Caryophyllales without its own paleopolyploidy event, making gene duplication and loss less likely to be responsible for the contention.

Regarding the placement of *Amborella*, all flowering plants possibly share a genome duplication [6]. Thus duplication or hybridization with the ancestor of the Nymphaeales could provide possible explanations. Although these are two potential sources of conflict, almost all other sources of conflict may be responsible for the discordance between the chloroplast genome and the nuclear genes. Thus further investigation to disentangle the processes is warranted.

A common alternative source of conflict that could explain these relationships is ILS. It is possible to distinguish ILS from hybridization by examining branch lengths and rates [349, 498, 387, 77, 389]. Conflict due to ILS will generally lead to coalescence deeper than conflict from hybridization. However, in plants, life history can have a massive impact on molecular branch lengths, with lineages with shorter life histories having increased evolutionary rates [698]. *Amborella* is a tropical tree lineage, and the Nymphaeales are herbaceous water lilies. Several life history shifts are associated with the transition to carnivory in the DA clade [300]. Therefore, molecular rates are likely to be obfuscated by life history and are challenging to interpret in both systems. An alternative to examining the molecular rate differences is to examine the patterns of conflict exhibited by the genes. In the case of *Amborella,* there are roughly equal numbers of conflicting genes supporting Nymphaeales sister to *Amborella* and Nymphaeales sister to all other flowering plants. This pattern indicates that the conflict likely originated, at least in part, from ILS (the primary alternative relationships appear in equal quantities). There is only one significant alternative in the DA case, making it difficult to determine if ILS would be the source of conflict.

7.3 The Consequences of Gene Tree Conflict in Phylogenomics

Gene tree conflict originates from many sources, and critical questions that continue into the genomic era is how we accommodate, understand the effects of, and incorporate conflict into downstream analyses. Below, we discuss the consequences of gene tree conflict and examine ways gene tree conflict affects species tree inference, gene and genome duplication analyses, and divergence time estimation.

7.3.1 INFERENCE OF SPECIES TREES

One of the most actively discussed problems regarding phylogenetic conflict involves the inference of species trees. Gene tree conflict causes different methods, data sets, filtering techniques, and parameters to infer alternative species tree resolutions despite hundreds of taxa and thousands of genes [798, 773, 407]. The complexity this presents has made the resolution of many important relationships in plants complicated and has made many early divergences almost impossible to infer with confidence. Many phylogenomic studies contrast concatenation methods that combine genetic information to increase weak signal with methods that accommodate ILS and incorporate patterns of conflict to resolve relationships. However, results may differ with slight changes to data sets even when using the same species tree reconstruction methods [675, 771, 95, 407].

In the face of these challenges, researchers do not have simple solutions for resolving important plant relationships. Low information content drives the use of concatenation, which is not statistically consistent in the presence of conflicting signals and might converge on the incorrect answer with the increase of data [385, 624]. Recent analyses have also demonstrated that the inferred topology of the concatenation method also may be driven by a small number of loci [95, 675, 771]. The common alternatives to concatenation methods—coalescent methods—also present several challenges. First, errors in gene tree estimation are common, especially as taxon sampling outside the relationship of interest increases and the problems of low information content emerge (discussed above). Using complete data methods [140] may account for errors in gene tree estimation; however, this comes at the cost of losing information on either molecular branch lengths or historical population sizes. Other coalescent methods (i.e., nonsummary-based methods) can adequately model heterogeneity in population processes and parameters; however, they are typically computationally prohibitive for phylogenomic data and model only one source of conflict (ILS, ignoring introgression). While ILS is prevalent in plants, as noted earlier, several other sources of conflict also affect gene tree conflict (but see PHYLDOG [88]).

Perhaps most importantly, for phylogenomics, summary coalescent methods are typically used instead of fully specified ILS methods because of the computational complexity of the latter [440, 506]. Summary methods provide an efficient approach that is computationally tractable and consistent under the coalescent. However, this consistency is only relevant for situations in which the gene trees are known. Summary methods also come with additional considerations. First, regardless of the disparity in likelihood, summary methods treat all genes equally, thereby losing influence that may have arisen from increased information content. Whether this is a reasonable assumption is not clear. Summary coalescent methods are also statistically inconsistent when data are not infinite [622]. Incomplete lineage sorting as a source of conflict involves several assumptions (discussed above). These assumptions may be positively misleading when incorrectly assuming ILS as the source of conflict (e.g., analyses assume neutrality, although selection likely plays a prominent role in shaping genomes; [306]). Furthermore, removing the most influential site from each gene can alter the topology a summary method infers [675].

Presented with these analytical challenges, researchers may have difficulty choosing the best path(s) forward. One possibility is to analyze specific phylogenetic questions instead of constructing the entire phylogenetic tree. For example, suppose our goal is to examine how much information we may have within a data set to reconstruct a particular relationship. In that case, our best route may be the thorough investigation of the specific relationship(s) of interest while allowing conflict to exist outside the relationship. To answer whether there is support in a data set for *Amborella* to be sister to the rest of the angiosperms, we may not need to resolve the relationships within the Nymphaealles. Furthermore, it may not be necessary to model hybridization in a nested part of the tree if hybridization is not a part of the story surrounding *Amborella*. Instead, we may focus on what information lies within our data set to address the specific relationships surrounding *Amborella*.

Researchers have recently developed methods that identify genes that exhibit disproportionate likelihood densities [95, 675, 771], perform a series of analyses specific to the relationship in question [590], and/or perform analyses specific to each gene as a means of allowing conflict to exist outside the relationship(s) of interest [31, 771, 772, 702]. Each of these methods helps characterize the support for relationships within a data set. For example, one set of analyses described in Walker et al. [771] and Smith et al. [702] consists of calculating the maximum likelihood, or posterior probability in the case of Bayesian analyses, for each gene while constraining one relationship, allowing the rest of the tree to vary. This calculation then repeats for every alternative resolution. As a result, the magnitude of support for different relationships for each gene, measured by differences in likelihood and how many genes support each relationship, can be summarized. This method has advantages over fixed tree comparisons, as relationships outside those of interest are allowed to vary. This approach allows for a detailed accounting of the support, or lack thereof, for each alternative relationship. Unfortunately, and similar to coalescent-based summary methods, branch lengths are not calculated as part of this analysis and can be affected by conflict (section 7.3.3). Nevertheless, and with these limitations, these and other focused investigations may help researchers more adequately address the question of whether our data support particular hypotheses.

Instead of, or in addition to, the resolution of species relationships, researchers may be interested in the processes that underlie the biological conflict. For questions regarding the processes leading to conflict, it may not be necessary to resolve species relationships. Attempts to accommodate conflict by explicitly modeling the biological processes that generate conflict [435, 388, 298, 405, 88] can also provide insight into species relationships. Even without confident resolution, we may be able to determine whether ILS, hybridization, HGT, or some other process is responsible for the conflict. Incorporating gene functional data and matching that with the phenotypes it codes for, the patterns of gene tree conflict may help provide a biological context to this conflict. For example, some conflicting genes may code for plesiomorphic traits. Species in the carnivorous DA clade discussed above are notable for their apparent three origins of a sticky trap and potential plesiomorphic distribution of being a woody vine. Assuming the trait is ancestral, then analyzing the genes that place the species with a sticky trap together could guide functional analyses of genes associated with sticky trap development. Functional work is not always possible but may become more accessible as molecular genomic sequencing continues to increase. Additional work into estimating the underlying sources of conflict between gene regions will continue to be a promising research program moving forward.

Finally, nonbiological sources of conflict (i.e., error) will continue to be a significant challenge for phylogenomic reconstruction. Unlike biological processes, errors do not provide insight into the tree of life. However, automation makes sources of error challenging to detect as they often occur because of data-set-specific stochasticity. Data-set-specific errors may be difficult to generalize and incorporate into a general use analysis pipeline. Therefore, thorough data set interrogation is crucial for robust phylogenetic inferences.

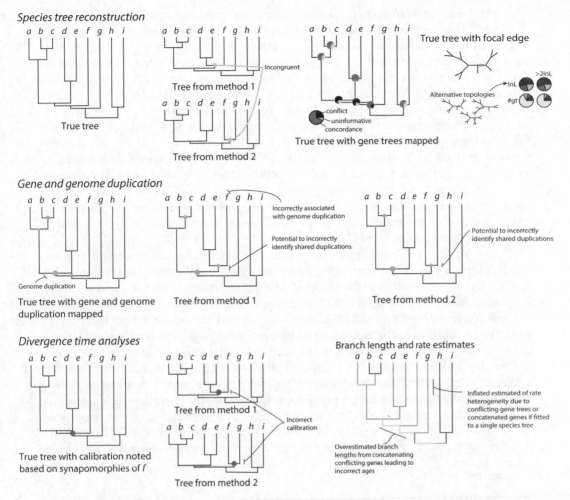

Figure 7.3. An illustration of several consequences of gene tree conflict, including complications in species tree reconstruction, determination of the placement of gene and genome duplication, and bias in divergence time estimates. Species tree reconstruction in the presence of conflict can result in incongruent species trees given different reconstruction methods. We can reconstruct a complete picture by mapping gene trees and focal edge analyses. While gene and genome duplications are influential in plant evolutionary history, conflict can influence the placement of these events. Divergence time analyses are essential for many evolutionary studies, but gene tree conflict can bias the inferred divergence times. Calibrations may be incorrect due to misplaced lineages in species trees, and conflict among the gene tree can inflate the number of inferred substitutions.

7.3.2 GENE DUPLICATION AND GENOME DUPLICATION

Gene tree conflict is not only generated by gene duplication and loss but the inferred location and frequency of these events may be altered because of conflict (figure 7.3). Any conflict that results in discrepancies in species tree inference or inaccurate gene tree reconstruction will cause an error in gene and genome duplication analyses. For example, taxa may be missing from genes because of orthology error [223] or sequencing effort. If so, a single misplaced taxon can lead to a different position for genome duplication. This sensitivity is essential to consider when examining unbalanced trees, as species-poor lineages are often challenging to place. The simplest way to alleviate this issue is to increase phylogenomic sampling and sequencing depth. However, often this is not practical. Rapid evolution and low signal also significantly affect gene and genome duplication analyses. In the flowering plant order Ericales, gene tree conflict, and putative genome duplication, are found at early diverging branches [407]. These relationships have low phylogenetic signal, possibly contributing to the error in the genome duplication(s) placement on the tree. The lack of signal may result from a rapid radiation and/or some form of life history shift as these are both shown to influence the molecular rate. Several alternative topologies are inferred for these early diverging branches, all of which affect the placement of gene and genome duplications (e.g., [407, 554]). This pattern of high conflict and genome duplications may be common across the plant tree of life [554].

Undoubtedly, as sequencing efforts continue to increase, more gene and genome duplications will be discovered. Still, the challenges of conflicting genes and accurate species tree reconstruction in the presence of extensive conflict remain. However, there are several promising ways forward. For example, the development of approaches that incorporate uncertainty into duplication analyses will allow researchers to integrate multiple possibilities in the placement of duplications in large gene families [865]. Considering that many of these events occur deep in the plant phylogeny where resolution is complex and gene trees have low support, incorporating support should significantly improve our accuracy, if not precision. The development of approaches and collection of data that eliminate the need for precise species trees should also allow for the analysis of genome duplications in the presence of conflict and unresolved species trees. Syntenic information, for example, enables researchers to verify results from phylogenetic analyses of duplications and can help place genome duplications in the presence of gene tree conflict.

7.3.3 DIVERGENCE TIME AND COMPARATIVE ANALYSES

There are several ways conflict can bias evolutionary inferences aside from species tree reconstruction and gene duplication analyses. Here, we discuss the potential biases in divergence time estimation. Transforming branch lengths to be proportional to time is a fundamental analysis for biologists and is required for many comparative studies. The fragmentary fossil record of plants poses many challenges for divergence time estimation, including the limited availability of informative fossil calibrations for many lineages. As a result, many studies rely on secondary dates in addition to fossil calibrations. Furthermore, the heterogeneous life histories and molecular evolutionary changes across the plant tree of life also present challenges for divergence time analyses [700, 57].

Branch lengths are required for divergence time analysis and may be estimated beforehand [648, 701] or as part of a joint analysis of divergence time and phylogenetic reconstruction (e.g., [203]). Whether methods rely only on a species tree with calibrations (R8S, PL) or they coestimate the species tree with divergence times (BEAST), with few exceptions (*BEAST2) [549], divergence time estimation methods assume no underlying conflict in the data. Nevertheless, conflict is common, and branch lengths estimated across a set of genes will be affected by conflicting signals, as genes are required to conform to branch patterns with which they may conflict (figure 7.3). When

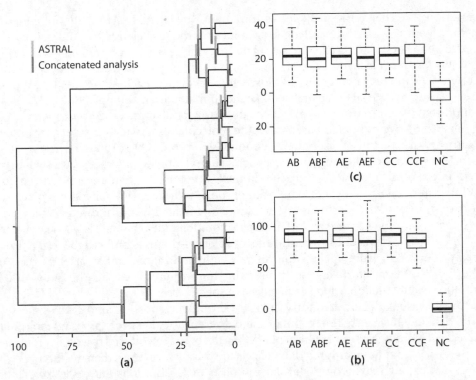

Figure 7.4. (a) Divergence time estimates for one simulated data set with no conflict. Light and dark grey bars represent the estimated divergence times based on ASTRAL and concatenated analyses, respectively, using a simulated data set with conflict due to ILS. Parameters for this analysis reflect a low level of conflict (e.g., smaller effective population size). (b) Distribution of the summed error in divergence time estimates for 50 simulations with small population size and conflict due to ILS. AB, ABF, AE, and AEF are results from ASTRAL analyses using original gene trees (AB, ABF) or re-estimated from underlying data (AE, AEF). Then final branch lengths are estimated from concatenating all genes (AB, AE) or only genes that reflect species tree relationships (ABF, AEF). CC and CCF are based on concatenated analyses with branch length estimates from the entire gene set (CC) or filtered based on gene tree concordance with species tree (CCF). NC are the results based on simulations of genes with no conflict. (c) Same analyses as (b) but with large population size and therefore more conflict.

branch lengths are estimated across conflicting genes, the molecular model is violated, leading to deeper branching and, consequently, overestimating dates [495, 644, 470]. One way to alleviate this issue is to filter the genes for concordance [696]. However, filtering may not be appropriate for conflict due to ILS and can result in very few genes with which to estimate branch lengths, especially as sampling increases.

We examined a simple case in which not accommodating conflict due to ILS in branch length estimates can bias divergence time estimates (figure 7.4). We simulated species trees with 30 taxa and 30 genes in the presence of ILS with effective population sizes of 100 and 500, low enough that resulting gene trees were not completely conflicting with species trees but high enough to generate some conflict and branch length variation. We then estimated species trees by means of ASTRAL [850] and concatenation, with the final branch length estimates on the species tree based on either the entire gene set or a filtered gene set, including only genes that were concordant with the species tree. Finally, we also simulated genes on the species tree without conflict to compare estimates with conflict to those without conflict. Each of the resulting species trees was then dated using penalized likelihood methods and compared. Generally, we found that as conflict increases, so does the error for divergence time estimates. Ages tended to be overestimated based upon the molecular data, which is especially important in plants that historically have a poor fossil record. Overestimation is expected, considering that conflicting gene trees modeled under the wrong species tree will overestimate the substitutions on adjacent branches (figure 7.3), noted in [495, 644, 470]. While filtering may alleviate some of the problems, it does not solve the problem of bias in divergence time estimates. The example presented here is meant only as a simple demonstration and not an exhaustive exploration of this issue. Also, while this has been explicitly discussed for ILS, in this chapter we discuss conflict that results from many sources, including gene duplication/loss, hybridization, and error. Regardless of the source of conflict, divergence time estimation requires accurate estimation of the number of substitutions along each edge of a species tree.

In the vascular plant literature, there are several hot spots of conflict that intersect with debated divergence time estimates, and discrepancies between fossil and molecular divergence times may, in part, be explained by the biases introduced by conflict [61, 600, 518]. Fossils may be discovered that help clarify these issues. Until then, researchers will continue to rely heavily on molecular estimates of substitutions on species trees, and conflict will continue to be a challenge.

7.4 Resolution of the Tree of Plant Life

A fully resolved tree of life is a fundamental goal for evolutionary biology. At all stages of the process, scientists have found it to be more complex than previously imagined. While a single tree has traditionally been the goal, a complete view of the tree of life includes documenting the patterns of conflict and complexity that shape the tree. Conflicting patterns among genes in systematics has been and continues to be a driving force behind methods development and data acquisition efforts. Studying the processes that generate conflict brings us closer to understanding the structure of the elusive tree of plant life.

CHAPTER 8

Hybridization in *Iochroma*

Daniel J. Gates, Diana Pilson, and Stacey D. Smith

8.1 Introduction

The impact of hybridization on evolution and speciation has been well established in evolutionary biology [185, 480, 481, 24, 723, 274, 618, 49, 462, 1]. Despite the historical notion that hybridization between distant relatives is rare and typically detrimental [480, 481], numerous studies have indicated that hybridization occurs frequently between both closely and distantly related species [461, 463]. Moreover, the introgression of genes via hybridization can be selectively advantageous [272, 751, 366]. The recognition that hybridization is a widespread and important mechanism of adaptation and speciation has placed a premium on the development of methods to detect signatures of reticulate evolution and identify the lineages and genomic regions involved [530, 297].

Inferring a history of hybridization, whether among close or distant relatives, has typically relied upon molecular markers. The earliest approaches were developed and applied in a population genetic framework, focusing on pairs or "swarms" of putatively hybridizing species [516, 36, 34]. Although population genetic methods are powerful at shallow timescales [208], they do not easily scale to cases involving dozens of taxa. The need to test for hybridization across entire clades has driven the development of phylogenetic methods that build on population genetic principles (e.g., [387, 841]). These methods are often computationally intensive and require large amounts of data to distinguish the signal of introgression from the background noise generated by the stochasticity of the coalescent process [845, 209]. Nevertheless, advances in sequencing and bioinformatics have made it increasingly feasible to build genome-wide data sets outside of model taxa (e.g., [232, 420, 640, 343]). Thus, future improvements in estimating evolutionary histories involving hybridization are likely to come from theoretical and computational advances (e.g., [389, 710]).

Current approaches for studying hybridization in a phylogenetic framework can be divided into two broad categories, global and local. Global methods aim to detect the signal of hybridization across the phylogeny without explicitly determining the specific taxa involved (table 8.1). These approaches often combine coalescent simulations with distance metrics or other indices to exclude the hypothesis that gene tree discordance can be explained by incomplete lineage sorting alone. In comparison to the global tests, local methods attempt to resolve specific hybridization relationships (i.e., identify the lineages involved in hybridization events) (table 8.1). The more recent methods

Table 8.1. Various approaches that have been used to detect hybridization.

Pattern	Type	Reference
Tree distance	Global	Buckley et al., 2006 [101]
Count deep coalescences	Global	Reid, Demboski, and Sullivan, 2012 [613]
Tree distance	Global	Blanco-Pastor, Vargas, and Pfeil, 2012 [76]
Network likelihood index	Global	Konowalik et al., 2015[a]
Genealogical sorting index	Global	de Villiers et al., 2013[b]
Shallow interspecific coalescences	Local	Joly, McLenachan, and Lockhart, 2009 [349]
Species network likelihood	Local	Kubatko, 2009 [387]
Minimize deep coalescences	Local	Yu et al., 2011 [845]
Species network likelihood	Local	Yu, Degnan, and Nakhleh, 2012 [841]
Four taxa nucleotide distances	Local	Durand et al., 2011 [208]
Five taxa nucleotide distances	Local	Pease and Hahn, 2015 [569]
Minimum pairwise sequence distance	Local	Rosenzweig et al., 2016 [636]
Species network pseudolikelihood	Local	Solis-Lemus and Ané, 2016 [710]
Sequence invariant patterns	Local	Kubatko and Chifman 2019 [389]; Blischak et al., 2018 [77]

Note: Global analyses are those that test for hybridization involving two or more branches of the phylogeny but do not attempt to localize the events. Local methods aim to determine the number and location of the reticulation events and therefore which lineages have been involved.

[a] Konowalik, K., Wagner, F., Tomasello, S., Vogt, R., and Oberprieler, C. (2015) Detecting reticulate relationships among diploid *Leucanthemum Mill.* (Compositae, Anthemideae) taxa using multilocus species tree reconstruction methods and AFLP fingerprinting. *Molecular Phylogenetics and Evolution* 92:308–328.

[b] de Villiers, M. J., Pirie, M. D., Hughes, M., Moller, M., Edwards, T. J., and Bellstedt, D. U. (2013) An approach to identify putative hybrids in the 'coalescent stochasticity zone', as exemplified in the African plant genus *Streptocarpus* (Gesneriaceae). *New Phytologist* 198:284–300.

in this category estimate the proportion of genes inherited by a hybrid lineage from each parent, in addition to inferring the topology of the species network [389, 710, 77]. After finding a global signal across the tree, local approaches can be used to pinpoint hybridization events. Still, the inference of the exact hybrid relationships remains challenging, and in some scenarios, it may be impossible to recover the full evolutionary history, regardless of the amount of available data [560].

Here we explore the ability of local and global methods to detect historical hybridization in a natural and experimental context, using the genus *Iochroma* as a model clade. In this group of roughly 25 Andean species in the tomato family, many of the taxa remain crossable and interfertile, even after millions of years of divergence [692]. Previous phylogenetic studies with nuclear markers have demonstrated substantial gene tree conflict and suggested at least three reticulation events [691, 693]. In this study, we apply phylogenomic approaches to quantify the signal of hybridization and test specific hybrid relationships based on morphology and biogeography. Given the interfertility of many *Iochroma* species, we also created an artificial hybrid to serve as a "positive control." Our approach is inspired by the pioneering work of Lucinda McDade [485, 486], who used natural hybrids to quantify the impact of hybridization on phylogenetic inference. Following her lead, we similarly manipulate the number of hybrids in our data set to test the effect on the inference of species trees and hybridization events. We predict that increasing the number of hybrids in the data set will increase discordance among gene trees and reduce the confidence in estimated species relationships. We also expect that artificial hybrids will be easier to detect as they have not had the opportunity to back-cross or accumulate new mutations as in natural hybrids. In addition to improving our understanding of the role of hybridization in the radiation of *Iochroma*, these results will shed light on the strengths and limitations of current approaches and provide direction for future development of computational methods.

8.2 Methods

8.2.1 STUDY SYSTEM

The subtribe Iochrominae and its largest genus, *Iochroma*, have been well sampled in previous systematic studies (e.g., [691, 526, 171, 259]), providing a strong framework for focused analyses of hybridization. This clade is sister to the large Physalinae (tomatillos and their relatives) [170] and is largely restricted to the Andes of South America [691]. Species of *Iochromina* often grow in sympatry [693] and, in some cases, form hybrid zones [698]. Phylogenetic studies have sampled several putative hybrids and recovered strongly conflicting relationships in nuclear gene trees [691]. The strongest support for hybridization from genetic, morphological, and biogeographic data corresponds to two taxa (*Iochroma ayabacense* S. Leiva and *I. stenanthum* S. Leiva, Quip. & N. W. Sawyer) (Figure 8.1). Native to northern Peru, *I. ayabacense* grows in a zone of sympatry with several species (*I. cyaneum* (Lindl.) M. L. Green, *I. confertiflorum* (Miers) Hunz., *I. arborescens* (Schltdl.) J. M. H. Shaw, and *I. lehmannii* Bitter (syn. *I. squamosum* S. Leiva & Quip.) [698]). Alleles from this species fall either in a clade with *I. lehmannii* or with *I. cyaneum* [691]. Such a biphyletic pattern (with alleles from the same locus in distant clades on the phylogeny) is a signature consistent with recent hybridization [47]. The chromosome number of *I. ayabacense* is 2n=24, like other *Iochroma* species, suggesting it is a homoploid hybrid [698]. Also, its morphology is intermediate, with sturdy funnel-shaped flowers like *I. lehmannii* but flushes of purple floral pigmentation and a long corolla tube like *I. cyaneum* [698].

The other putative hybrid taxon, *I. stenanthum*, has similarly intermediate coloration and morphology compared with its putative parental lineages. It occurs in another zone of sympatry in northern Peru and has been proposed to be a hybrid of the lineage comprising the purple-flowered *I. cornifolium* (Kunth.) Miers and *I. cyaneum* and a white-flowered lineage from the clade containing *I. confertiflorum* and *I. arborescens* [691, 694]. *Iochroma stenanthum* presents an intermediate morphology, being purple at the mouth of the corolla and fading to cream toward the base. It has long tubular flowers as in *I. cornifolium* and *I. cyaneum* but more triangular corolla lobes as in *I. confertiflorum* and *I. arborescens*. In individual gene tree analyses, alleles from *I. stenanthum* show affinity toward both the white- and purple-flowered lineages (Gates et al., unpublished data). In combined analyses, the phylogenetic position of *I. stenanthum* is unstable, but it often appears as the sister lineage to the clade containing the putative parental taxa [691, 166]. The latter phylogenetic pattern was proposed by McDade [485] as a common outcome for hybrid taxa.

In addition to these two putative natural hybrids, we included one artificial hybrid, generated in the greenhouse. This accession is a cross between *I. arborescens* (Bohs 2428 (UT), grown from seed collected from Las Cruces, Costa Rica) and *I. cyaneum* (Smith 265 (WIS), grown from seed collected from plants cultivated by W. G. D'Arcy at the Missouri Botanical Garden; likely provenance, Ecuador). These parental species are thought

Figure 8.1. Known and hypothesized hybrid relationships. (a) *Iochroma cyaneum* and *I. arborescens* were crossed in the greenhouse to create the F$_1$ hybrid on the right. (b) *I. cyaneum* and *I. lehmannii* are the putative parental taxa of *I. ayabacense*, right. (c) Members of the clade containing *I. cyaneum* and *I. cornifolium* (shown, left) and the clade containing *I. arborescens* and *I. confertiflorum* (middle) are the putative parental taxa of *I. stenanthum*, right.

to be 2 to 4 million years diverged [166], but crosses between the two produce vigorous offspring. The hybrid flower has an intermediate phenotype (funnel-shaped flowers, white with purple markings). Its fertility has not been thoroughly examined, although it seems to be capable of backcrossing to the parental taxa (S. D. Smith, unpublished data).

Table 8.2. Accessions used for this study of *Iochroma*.

Species	Relationship	Voucher	Locality	DNA ID #
I. arborescens	Putative parent	Smith 312 (MO)	Peru: Contumaza. 7.42409°S 78.90111°W	98
I. arborescens	Putative parent	Smith 209 (WIS)	Ecuador: Alluriquin. 0.32145°S 78.99764°W	250
I. arborescens × *I. cyaneum*	F1 Hybrid	Smith 687 (COLO)	University of Colorado-Boulder Greenhouses	-
I. ayabacense	Putative hybrid	Smith 337 (MO)	Peru: Ayabaca. 4.61462°W 79.71975°S	126
I. cyaneum	Putative parent	Smith 265 (WIS)	Univ. of Wisconsin-Madison Greenhouses	156
I. lehmannii	Putative parent	Smith 487 (MO)	Ecuador: Cañar 2.37168°W 78.96976°S	228
I. lehmannii	Putative parent	Smith 330 (MO)	Peru: Ayabaca. 4.64422°W 79.71975°S	176
I. stenanthum	Putative hybrid	Smith 313 (MO)	Peru: Contumaza. 7.40116°S 78.89658°W	99
P. peruviana	Outgroup	Smith 217 (MO)	Ecuador: Quito. 0.16761°S 78.48133°W	91

8.2.2 EXPERIMENTAL DESIGN

In order to assess our ability to detect hybridization with various phylogenomic methods, we created a series of data sets with different combinations of the two putative natural hybrids (*I. stenanthum* and *I. ayabacense*), their putative parental lineages (*I. arborescens*, *I. lehmannii*, *I. cyaneum*), the artificial F_1 hybrid (*I. arborescens* × *I. cyaneum*), plus an outgroup (*Physalis peruviana* L.). For two of the three putative parents (*I. arborescens* and *I. lehmannii*), we have two sampled individuals (table 8.2). All of the experimental data sets contain the same number of tips (six), and all have at least one individual of each putative parent plus the outgroup (figure 8.2). Data set 1 has no putative hybrids, while data set 2 has the F_1 hybrid substituting for one of the two *I. arborescens* individuals. Data set 3 has the F_1 and *I. ayabacense* replacing one of each of the *I. arborescens* and *I. lehmannii* individuals. As *I. ayabacense* tends to fall with *I. lehmannii* in species trees (Gates, unpublished) estimated with STAR [442], the topology for data set 3 is expected to mirror that for data set 2 (figure 8.2), albeit with less support owing to the expected hybrid ancestry. The fourth data set has the two putative natural hybrids (*I. stenanthum* and *I. ayabacense*) in the place of one of each of the *I. arborescens* and *I. lehmannii* individuals. The putative parents for *I. stenanthum* are the same as for the F_1 (*I. arborescens* × *I. cyaneum*), although STAR species tree analyses tend to infer a closer relationship with *I. arborescens* (Gates, unpublished). Thus, we expect a similar species tree topology to the previous data sets, but a weaker signal of hybridization since *I. stenanthum* is likely a late generation hybrid compared with the F_1 (figure 8.2).

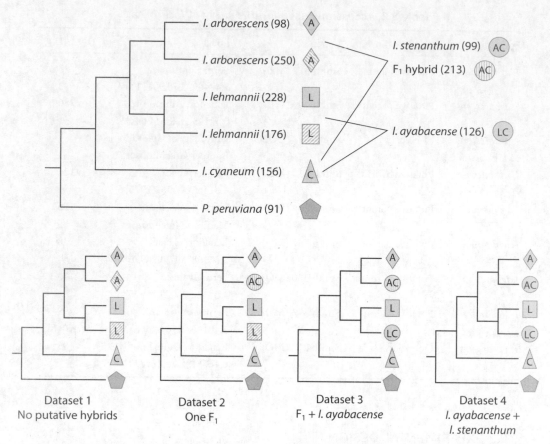

Figure 8.2. Hypothesized relationships among sampled tips and experimental design. The top tree shows the relationships among *I. arborescens*, *I. lehmannii*, and *I. cyaneum* based on previous studies (see text). Numbers indicate DNA accession number for reference (table 8.2). The lines connect putative or known hybrid taxa (figure 8.1) to their putative parental lineages. The letter scheme indicates the putative parentals (AC denotes an A × C hybrid). Tests for hybridization were carried out on the four six-taxon data sets at the bottom, which range from no putative hybrids to two putative hybrids. See text for complete description.

8.2.3 TARGET CAPTURE AND ASSEMBLY

We conducted targeted sequence capture to build a molecular data set for phylogenomic analyses following [259]. Briefly, we used genomic resources for tomato [654] and *I. cyaneum* ([260]; Powell et al., https://ecoevorxiv.org/dcuht/) to design probes for 242 single copy genes, most of which were pulled from the conserved orthologous set [814]. These loci totaled approximately 1 megabase and were used as the template for custom bait synthesis of 80 nucleotide targets at 2x tiling from MycroArray (MycroArray, Ann Arbor, MI). Genomic DNA was extracted from silica-dried leaf tissue using a 2XCTAB protocol [195] and sheared to approximately 500 bases with sonication. For all libraries except for the F_1 hybrid, we used the TruSeq library preparation kits for sequence adapter ligation and Mycroarray capture protocol V1.

These libraries were sequenced on a single lane of an Illumina HiSeq2500 on 100 bp pair-end rapid mode. For the F_1 hybrid, we used the NEBNext Ultra II kit (New England Biolabs, Ipswich, MA) for library preparation and Mycroarray capture protocol V4. This library was sequenced on an Illumina HiSeq 4000, also in 100 bp pair-end mode. We filtered raw reads by using Trimmomatic [81] and assembled filtered raw reads with the iterative read-mapping program YASRA [605]. We created a master alignment for all individuals by orienting assemblies to homologous bases of the reference using custom Blat [362] and samtools [429] scripts. We then used this master alignment to subset individuals into each of the four experimental data sets. We conducted gene tree searches on each gene in each of the four data sets using RAxML 8 [720] with 100 bootstrap replicates to assess support. These alignments and gene tree estimates were used for all downstream analyses (available at https://github.com/danjgates/HybrData).

8.2.4 DETECTION OF PATTERNS OF HYBRIDIZATION FROM GENE TREE DISTRIBUTIONS

To quantify the level of gene tree conflict associated with varying numbers of hybrids in our four data sets, we first applied the internode and tree certainty (IC/TC) measures of Salichos, Stamatakis, and Rokas [647] as implemented in RAxML 8 [720]. These information theory-based metrics can be applied to trees generated from any method and optimality criterion (e.g., parsimony, likelihood) and are computationally tractable with large, genome-scale data matrices. We computed IC for each of the three internodes in our experimental data sets (figure 8.2). These values represent the degree of conflict between relationships inferred with STAR [442] and the most common conflicting partition. When IC approaches 1, there is little conflict at the node, and when it is near 0, there is strong conflict, with similar numbers of gene trees showing the bipartition in the species tree and the next most common bipartition. For each of our four data sets, we also computed TC, which is the sum of the IC scores over all ingroup bipartitions [647]. We predicted that the addition of tips with known or inferred hybrid ancestry would increase gene tree conflict and thus decrease the certainty in relationships across the phylogeny.

We further characterized the distribution of the gene trees in each data set by using tree incongruence checking in R, or TICR ([727], chapter 6). This program uses pseudo-log-likelihoods (PLL) calculated from rooted quartet distributions to test the fit of gene tree distributions to different hypotheses of panmixia or species tree structure. While rejection of a population tree may be consistent with a history of hybridization and may be used as evidence of hybridization [727], the program does not directly test a model of hybridization against the other models of bifurcation or panmixia. We used TICR to compute the PLL and associated χ^2 statistic for both the panmixia (star-like ingroup tree) and fully bifurcating species tree in each of the four data sets. In comparing data sets with hybridization to those without (e.g., one F_1 data set vs. the no putative hybrid data set), we expect that there should be less support for a bifurcation model as indicated by a greater χ^2 value and more support for a panmixia model as indicated by a lowered χ^2 value in the data set with greater amounts of hybridization. The TICR analyses rely on concordance factors (CFs) for each quartet [27], estimated with BUCKY [405]. We extracted these CFs for each clade as an additional window into patterns of gene tree discord associated with adding hybrid taxa.

8.2.5 TESTING OF HYBRIDIZATION IN EMPIRICAL DATA SETS

In addition to examining uncertainty and discordance across the tree, we also implemented three specific tests for hybridization. While the two approaches above (IC/TC and TICR) may offer support for hybridization, these methods do not explicitly test whether a hybridization model is a more appropriate model than a model without hybridization. In addition to reticulation, conflict among gene trees and instability in species tree inference can arise because of other factors, such as incomplete lineage sorting [176], recombination [661], model misspecification [100], biased taxon sampling [303], missing data [800], sequencing errors [391], and alignment uncertainty [315]. Thus, without explicit modeling of processes of hybridization, metrics like IC and CF offer only weak support for hybridization.

In order to test for hybridization directly, we first applied HyDe (hybrid detection; [77]). This program localizes hybridization events (i.e., identifies hybrid taxa and their putative parental taxa) based on patterns in sequence alignments. Detecting these events relies on invariants, that is, phylogenetically informative site patterns in quartet subtrees, such as those used in the ABBA/BABA patterns used to compute D-statistics [564]. Specifically, HyDe uses the ratio of the differences between specific site patterns as this approach was found to provide the highest statistical power [141]. We also implemented a similar alignment-based method that computes D-statistics across all quartet subtrees to test for introgression [209]. While HyDe reports statistics for every combination of two parental tips and one hybrid tip, the D-statistic approach reports all possible triplet combinations that are consistent with the species tree. For the purpose of this analysis, we report only the maximum statistic for each method in each data set as this should be the strongest proposed hybrid relationship.

Along with these alignment-based approaches, we used an additional method, SNaQ (species networks applying quartets; [710], chapter 6), to fit reticulation events based upon gene tree distributions. Like the TICR method described above, SNaQ calculates a quartet-based PLL for models with and without hybridization. Unlike TICR, however, SNaQ does directly test support for hybridization. For each data set, we report the maximum PLL score for the strictly bifurcating species tree as well as the maximum score from the best network produced by a full search with up to six optimal reticulation events. For SNaQ as well as HyDe, we expect that inferred reticulation events will follow the predictions based on morphology and biogeography and that more recent hybridization events (the F_1 hybrid and the biphyletic *I. ayabacense*) will be more consistently and accurately recovered than the older events (those that formed *I. stenanthum*).

8.3 Results

8.3.1 ADDITION OF HYBRID TAXA INCREASES DISCORDANCE AND DECREASES TREE-LIKE SIGNAL

Comparisons between the no hybrid and one F_1 data set offer strong evidence that hybridization has an impact on a number of tree metrics. Values for internode certainty (IC) ranged from 0.19 to 0.60 in the no hybrid tree and dropped to 0 to 0.32 with the substitution of the F_1 hybrid for one of the putative parental individuals (figure 8.3a, b). Overall tree certainty (TC) fell from 1.04 to 0.32 (figure 8.3a, b). We also saw decreases in concordance factors (CF), from 0.47 to 0.71 without the F_1 to 0.26 to 0.53 with the F_1. Given that the F_1 is an artificial hybrid between *I. arborescens* and *I. cyaneum*

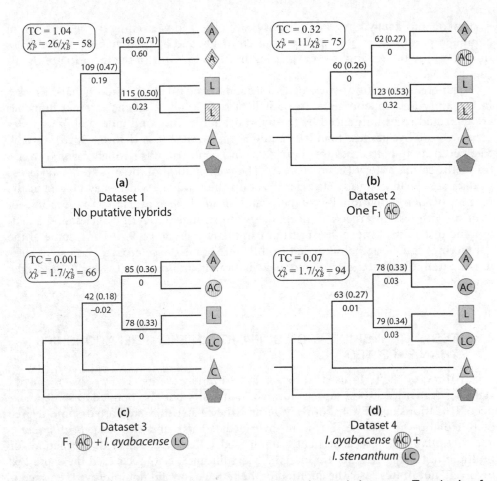

Figure 8.3. Gene tree conflict and signal across experimental data sets. Topologies for each data set follow the species tree inferred by STAR. The symbols for tips follow figure 8.2, with known or putative hybrids indicated with two letters (AC or LC) corresponding to their putative parental lineages. Numbers above the branches are the numbers of gene trees estimated to have this relationship from concordance analysis. Proportions of gene trees (concordance factors) are in parentheses. Numbers below the branches denote internode certainty values, rounded to two decimal places. Tree certainty (TC) metrics are given in the boxes for each tree along with the χ^2 value for two models (panmixia, P, and fully bifurcating, B) tested by tree incongruence checking in R. Values above 8 are significant at P = 0.05 (with three degrees of freedom, [727]), suggesting that the data reject the model. For example, data set 4 strongly rejects a bifurcating tree (χ^2_B of 94) but does not reject the panmictic model (χ^2_P of 1.7).

(figure 8.1), we would expect the branches connecting these taxa to be most strongly affected by adding the hybrid and, indeed, these were the only internodes with decreases in IC and CF values. In fact, the unaffected branch, uniting the two *I. lehmannii* individuals, saw a slight increase in IC and CF (figure 8.3a, b). These increases in gene tree conflict with the addition of the F_1 hybrid shifted the χ^2 values from the TICR test in line with decreasing tree-like signal. Although both data sets reject the two extremes (panmixia and a fully bifurcating history), we observed weakening of the rejection of complete panmixia (χ^2_P dropping from 26 to 11) and a strengthening of the rejection of a fully bifurcating tree (χ^2_B rising from 58 to 75) with the inclusion of the F_1 (figure 8.3a, b).

The increase from one hybrid tip to a second hybrid tip brought additional decreases in certainty and concordance. In both data sets 3 and 4, all the internal branches are expected to be influenced by the hybridization events, and indeed all internodes dropped to IC values less than 0.03 and CFs of less than 0.36 (figure 8.3c, d). Overall tree certainty in both cases was near zero, and panmixia was strongly favored over a fully bifurcating history (figure 8.3c, d). The combination of the F_1 and *I. ayabacense* in data set 3 had a slightly stronger effect on these values, which may be due to the recency of their formation. By comparison, *I. stenanthum* appears to be a more ancient hybrid (with less intermediate morphology, no strongly biphyletic alleles, and a geographic distribution more distant from its putative parents) [691]. Thus, some of the conflicting signal originally carried in this lineage may have eroded because of sorting of parental alleles, the fixation of new mutations, and/or backcrossing to one of the parents.

8.3.2 TESTS OF HYBRIDIZATION SUPPORT DIFFERENT RELATIONSHIPS THAN EXPECTED

All three of the tests used to detect hybridization events (HyDe, SNaQ, and D-statistics) provided strong support for hybrid ancestry but also pointed to some unexpected relationships. First, both HyDe and the D-statistics supported some degree of reticulation in data set 1 (the no-hybrid data sets), and two taxa, *I. arborescens* and *I. lehmannii*, were consistently implicated (table 8.3a, b). With correction for multiple tests, these results are marginally significant. SNaQ recovered the same tips involved, but in this case, the addition of the reticulation did not improve the psuedo-log-likelihood of the tree (table 8.3c). Although the implication of some reticulation involving *I. arborescens* and *I. lehmannii* was surprising, it is notable that *I. lehmannii* does show instability in traditional phylogenetic analyses and variously appears as more closely related to *I. arborescens* [171] or *I. cyaneum* [170], always with weak support (bootstrap < 75%). Species tree analyses favor the former relationship [259], but these have not previously included tests for hybridization. Still, the strength of support for hybridization was weak compared with the other three data sets (table 8.3). Given these mixed results, we consider that *I. lehmannii* does not have a large degree of hybrid ancestry and that the patterns may reflect a small rate of gene flow between *I. lehmannii* and *I. arborescens* in southern Ecuador and northern Peru, where the two grow in proximity (S. D. Smith, pers. obs.).

The data sets with one or more hybrid taxa consistently supported the presence of hybridization and, in a few cases, recovered the predicted relationship. We expected the best case scenario to be the one F_1 data set, in which the parents of the artificial hybrid (*I. arborescens* and *I. cyaneum*) are known and included in the tree. HyDe identified the F_1 as being involved but placed it as a parent, with *I. lehmannii* as the other parent and *I. arborescens* as the hybrid (table 8.3a). The D-statistics pointed to the same three taxa as involved in hybridization (table 8.3b). By contrast, SNaQ recovered the correct relationship among the parental taxa and the F_1 and with strong support for data set 3 (table 8.3c).

Table 8.3. Identification of hybrid relationships with HyDe, D-statistics, and SNaQ.

(a) HyDe

Data set	Hybrid	Parent 1	Parent 2	Test statistic	P
1. No hybrids	*I. lehmannii* (176)	*I. arborescens* (250)	*I. lehmannii* (228)	3.18	0.002
2. One F_1	*I. arborescens* (98)	F_1	*I. lehmannii* (176)	5.08	0*
3. F_1 + *I. ayabacense*	*I. arborescens* (98)	*I. ayabacense*	*I. cyaneum* (156)	8.99	0*
4. *I. ayabacense* + *I. stenanthum*	*I. cyaneum* (156)	*I. lehmannii* (228)	*I. stenanthum*	8.64	0*

(b) D-statistic analysis

Data set	A	B	C	Test statistic	P
1. No hybrids	*I. lehmannii* (176)	*I. arborescens* (250)	*I. lehmannii* (228)	0.13	0.002*
2. One F_1	F_1	*I. arborescens* (98)	*I. lehmannii* (228)	0.42	0*
3. F_1 + *I. ayabacense*	F_1	*I. ayabacense*	*I. arborescens* (98)	0.25	0*
4. *I. ayabacense* + *I. stenanthum*	*I. ayabacense*	*I. lehmannii* (228)	*I. cyaneum* (156)	0.29	0*

(c) SNaQ analysis

Data set	Hybrid	Parent 1	Parent 2	Reticulation PLL	Tree PLL
1. No hybrids	*I. lehmannii* (228)	*I. arborescens* (98)	*I. lehmannii* (176)	−1.86	−3.26
2. One F_1	F_1	*I. arborescens* (250)	*I. cyaneum* (156)	−2.11*	−26.80
3. F_1 + *I. ayabacense*	*I. lehmannii* (228)	*I. arborescens* (250)	*I. lehmannii* (228)	−31.99*	−91.32
4. *I. ayabacense* + *I. stenanthum*	*I. lehmannii* (228)	*I. arborescens* (250)	*I. lehmannii* (228)	−133.28*	−257.97

Note: In (b), columns A, B, and C represent the phylogenetic position in a three tip relationship ((A,B),C) and the p-value indicates hybridization between A and C. For (a) and (b), the asterisks indicate significance after correction for multiple comparisons ($\alpha - 0.05/30 = 0.0017$ for HyDe and $\alpha = 0.05/10 = 0.005$ for D statistics). For (c), the asterisks indicate significance in a likelihood ratio test with one degree of freedom (corresponding to the additional parameter introduced by the reticulation). Instances in which the method inferred a set of relationships consistent with the hypothesized hybrid ancestry are highlighted in gray.

The results for data sets with two of the six taxa being hybrids were similarly mixed. For data set 3 (F_1 + *I. ayabacense*), HyDe correctly grouped *I. ayabacense* with *I. arborescens* and *I. cyaneum* but inferred *I. arborescens* instead of *I. ayabacense* as the hybrid among the three. Neither D-statistics nor SNaQ estimated a predicted grouping of parents with hybrids for data set 3. For data set 4 with two natural putative hybrids

(*I. ayabacense* + *I. stenanthum*), only the D-statistics recovered a predicted grouping of tips involved in hybridization (*I. ayabacense* and its putative parents, *I. lehmannii* and *I. cyaneum*) (table 8.3).

8.4 Discussion

8.4.1 EFFECTS OF HYBRIDIZATION ON PATTERNS OF GENE TREE DISCORDANCE

Our manipulations of the taxon sampling show that adding known or putative hybrids increases discordance between gene trees and decreases the signal of tree-like relationships. All of the data sets in which putative or known hybrid taxa were swapped with putative parental taxa showed marked decreases in concordance factors and internode certainty and a strong preference for panmixia over tree-like relationships (figure 8.3). The comparison of the one F_1 data set compared with the no putative hybrid data set offers the strongest evidence on this front because the F_1 is an artificial hybrid generated in the greenhouse. In this data set, the concordance factors for the nodes connecting the two putative parental species were reduced by roughly half and the certainty for those relationships dropped to zero. Doubling the number of hybrid taxa (data sets 3 and 4) further increased gene tree conflict, and with hybrids distributed across the tree, complete panmixia could not be rejected by TICR (figure 8.3). Altogether, we see patterns consistent with prevalent, detectable hybridization in all trees that we expect to exhibit a signal of reticulate ancestry.

The methods designed to identify hybrid tips and their parental taxa also showed strong evidence of hybridization in data sets with more putative or known hybrids. For example, SNaQ and HyDe did not find significant support for hybridization in data set 1, with no known hybrids (table 8.3). In the remaining three data sets, all of the methods inferred significant reticulation, with the strength of that inference increasing with the number of hybrid taxa. These results indicate that the sharp discordance generated by the addition of hybrid tips, as indicated by decreasing concordance factors and tree certainty (figure 8.3), is generally correctly interpreted by HyDe, D-statistics, and SNaQ as evidence of hybridization during the evolutionary history of *Iochroma*.

8.4.2 CHALLENGES IN DETERMINING THE EXACT HYBRID RELATIONSHIPS

Our results also indicate that exact hybrid relationships may be difficult to assign. Looking across the three methods and the three data sets with augmented hybrid taxa, only two of these nine combinations estimated a predicted set of three taxa (two hypothesized parental tips along with the hypothesized hybrid tip). These differed across methods, with D-statistics inferring a predicted trio of tips in data set 4 and SNaQ in data set 2 (shaded in table 8.3). Although both HyDe and SNaQ directly estimate which tip in a trio is the hybrid tip, only SNaQ correctly inferred a predicted relationship, identifying the F_1 as the product of a cross between *I. arborescens* and *I. cyaneum*. This relationship between the artificial hybrid and its parents was intended to serve as our "positive control" for detecting hybrid ancestry, and it is worrisome that only one of the three methods recovered the correct topology.

The wide range of inferred and seemingly spurious hybrid relationships across methods and data sets may relate to both our experimental design and our study system. We intentionally built a small data set with two putative natural hybrids and their putative parents with the goal of maximizing the potential for methods to converge upon the correct sets of relationships. While we are not entirely certain of these relationships, the combination of morphological, biogeographic, and genetic data provides the strongest possible a priori predictions, beyond generating these hybrids ourselves. Still, a larger data set, with more taxa to break up the branches between the putative parents, might have provided more power to discriminate among possible placements for reticulation events. That hybridization appears to be common in *Iochroma* may also contribute to the challenge of estimating reticulation events. Controlled greenhouse crosses suggest that all *Iochroma* are able to interbreed to some degree, and they can also cross with closely related genera [692]. Apparent hybrid zones are well documented in herbaria as well as in the literature [691, 694]. Accordingly, all of the *Iochroma* lineages may contain a signal of hybrid ancestry, consistent with rejection of a fully bifurcating tree by TICR for the no hybrid data set (figure 8.3a). This episodic gene flow, combined with incomplete lineage sorting (ILS) along short branches of the backbone of the species tree [170], may result in relatively little information for making robust inferences about which tips have a significantly reticulate history.

It is important to note that the difficulty of making inferences about hybridization is already well documented in empirical and theoretical studies. Simulations attempting to reconstruct networks where there are few tips (<5) and multiple hybridization events have often failed to find the accurate network [845, 710]. Presumably, this is because with few taxa and multiple reticulations, the set of gene trees may be equally well explained by different network topologies [560]. That multiple hybridizations can confound inference could explain why SNaQ identified the correct relationships in the one F_1 data set but failed in datasets 3 and 4 with two hybrids (table 8.3). Moreover, SNaQ assumes that the edges can be involved in only reticulation event [710], and in data sets 3 and 4, *I. cyaneum* likely contributed to two events. Although making inferences about multiple reticulations from small trees is challenging, a similar empirical study was strongly able to support up to four reticulations in a data set with six species [788]. Given these conflicting notions in the literature about when hybridization can be confidently detected, we suggest that future theoretical studies explore a broad array of scenarios to determine which factors have the strongest influence on the power to infer reticulate relationships correctly. We expect that population size and divergence time would have major impacts because these parameters directly govern the amount of signal and noise (ILS) in multilocus data sets [176]. Violations of mutation rate assumptions may also have a large effect upon methods like HyDe and D-statistics because increases or decreases in mutation rates will make some lineages more similar to distant relatives without hybridization ever taking place. As empirical studies like ours continue to apply these relatively new methods, we expect that additional biological factors that potentially influence statistical power are likely to emerge.

8.4.3 HYBRIDIZATION IN *IOCHROMINAE*

One goal of this study was to assess support for the hybrid origins of *I. ayabacense* and *I. stenanthum* with genome-wide markers. While previous studies had assembled evidence for these hypotheses from a handful of nuclear markers along with morphological and biogeographic information [691], this is the first study to test directly for hybridization with phylogenomic methods. The analyses of gene tree discordance provide some support for hybrid ancestry in that the presence of these taxa in the data sets decreased tree-like signal and certainty in relationships among tips. Also, the analyses with the two natural hybrids (data set 4) showed similarly low concordance to those with one natural hybrid and one artificial hybrid (data set 3), suggesting that a natural hybrid (in this case, *I. stenanthum*) disrupted tree-like relationships to a similar degree as the F_1 (figure 8.3). Nevertheless, our results are far from conclusive in terms of which lineages gave rise to these two hybrid species. The local methods each returned different estimates of relationships among putative parental taxa and putative natural hybrids (table 8.3). Moreover, none of these estimated relationships followed those predicted based on previous studies (figure 8.1) despite the ingroup being composed of putative parental taxa. Accurate inference of the hybrid ancestry for *I. stenanthum* was expected to be challenging because its phylogenetic position is more suggestive of an ancient hybrid, for example, between stem lineages that gave rise to *I. arborescens* and *I. cyaneum* [691]. By contrast, we expected the relationships between *I. ayabacense* and its putative parents would be easily detected given that the strong biphyletic patterns (e.g. one *LFY* allele sister to *I. lehmannii* and the other sister to *I. cyaneum*) in gene trees [691]. The inability of multiple approaches to recover the predicted topology for *I. ayabacense,* not to mention the F_1 positive control, suggests that renewed attempts at testing these hybrid origins will require more statistical power (more loci, more individuals, and/or more tips). Ultimately, with a larger data set, we expect that the majority of local methods should converge on the same set of relationships, especially since they all build on quartet-based patterns.

8.5 Conclusions

The growth of phylogenomic data sets along with the well-established prevalence of hybridization in nature has driven the development of an array of statistical tools for detecting and localizing reticulation in phylogenies. The application of these methods has led to the inference of hybridization at multiple phylogenetic scales, from sister species (e.g., Turissini and Matute 2017; https://doi.org/10.1371/journal.pgen.1006971) to more distantly related taxa (e.g., [101, 209]). Our exploratory analyses, which included both known and hypothesized hybrid tips, suggest that while the broad signature of reticulate evolution is relatively easily detected [613, 76, 766], the accurate localization of hybridization on the phylogeny is significantly more difficult. While these inferences may be challenging in *Iochroma* because of periodic gene flow across the entire tree, we do not consider this clade to be an outlier. Hybridization has accompanied the diversification of many taxa (e.g., butterflies, birds, fishes) and is certainly well documented in plant clades. Groups in which past and ongoing hybridization is suspected are likely to be the primary targets for empirical applications of methods, such as SNaQ and HyDe. Thus, we encourage further empirical or experimental studies manipulating the degree of hybrid ancestry in addition to theoretical work to explore the power to make inferences across a spectrum of biologically realistic scenarios (including recently formed hybrid species in a backdrop of ILS and periodic gene flow). With more robust inferences, we may begin to arrive at a broader understanding of how reticulation affects patterns of genomic variation and how these impacts relate to the timing, duration, and scale of hybridization events.

CHAPTER 9

Hybridization and Polyploidy
in *Penstemon*

Paul D. Blischak, Coleen E. Thompson, Emiko M. Waight, Laura S. Kubatko, and Andrea D. Wolfe

9.1 Introduction

Phylogenetic inference with multiple gene sequences has emerged as a dominant paradigm in systematics, with multilocus data sets ranging in size from just a handful of genes to thousands of loci pulled from whole genomes. Discordant signals from these different gene regions can often be present, however, raising the issue of how to model the incongruence among the sampled gene trees from the underlying species tree. The multispecies coalescent (MSC) model is one approach for species tree estimation from multilocus data that can accommodate gene tree discordance caused by incomplete lineage sorting (ILS) (reviewed in [177]). The appeal of the MSC model stems from its connection with concepts in population genetics (Wright–Fisher model; [364]) and its explicit predictions regarding the amount of gene tree discordance that should be present for a given species tree [746, 559, 744]. Nevertheless, despite the popularity of the coalescent model, it has been shown that it can be a poor fit to empirical data sets [614, 280]. A potential reason for the poor performance of the MSC in empirical data is that it models only ILS, leaving other processes that generate genealogical discordance, such as gene flow and hybridization, unaccounted for [455].

An alternative to using the coalescent to model gene tree discordance is to use the concept of a "gene-to-tree map," wherein gene tree topologies are mapped to possible species tree topologies without assuming an underlying process. This was the approach taken by [27], who used a Bayesian framework to estimate a species tree by maximizing gene tree concordance. Implemented in the software BUCKy [405], this method relies on the concept of concordance factors, or the proportion of gene trees for which a given bipartition is true [53]. The resulting phylogenetic estimate is referred to as the *primary concordance tree* (PCT) and can be estimated even if ILS is not the only process affecting gene tree incongruence. Larget et al. [405] also introduced the concept of a population tree, which uses the average concordance factors for all quartets on an internal branch of the PCT to calculate branch lengths in coalescent units. Because concordance factors can contain information about both ILS and admixture/gene flow, [710] developed a method for estimating species networks (species trees with reticulate edges) from concordance factors estimated for quartets of species. Their method, called SNaQ (species networks applying quartets), uses these quartet concordance factors (QCFs) to maximize a pseudolikelihood function that matches the expected QCF values under the coalescent model with hybridization [498] and the observed QCFs.

Despite the availability of these concordance factor-based approaches, there remain several areas in which we believe the estimation of QCFs can be improved. First, estimating QCFs for multiple individuals or haplotypes per species is not easily accomplished using BUCKy or the methods available in the PhyloNetworks package that implements the SNaQ method [27, 711]. This is problematic not only because having multiple alleles sampled from a population can potentially help to increase phylogenetic resolution [22] but also because many hybrid plant lineages are polyploids, which means that not all of their homoeologs can be modeled simultaneously. Second, BUCKy requires the estimation of posterior distributions of gene trees, which can be computationally demanding for large numbers of loci and/or large numbers of sampled alleles. Using gene tree posteriors is a common way to deal with gene tree uncertainty (utilized by several methods in PhyloNet; [790]), but other methods, such as those using site pattern frequencies, would allow for faster computation of the gene tree quartet topologies that are used to estimate QCFs and could include measures of uncertainty from statistical resampling.

To address the issues listed above, we first developed a method for estimating concordance factors directly from sequence data for quartets of species. Our method accommodates multiple haplotypes sampled per species and can conduct bootstrapping to account for gene tree uncertainty. To validate the method, we simulated multilocus sequence data on both tree and network topologies to assess how accurately it could estimate QCFs. We then collected nuclear amplicon data for two subsections in the plant genus *Penstemon* (Plantaginaceae; subsections *Humiles* and *Proceri*). Subsections *Humiles* and *Proceri* are known to hybridize and have the additional complication of containing multiple polyploid species. Using phased haplotypes from the nuclear amplicon sequences, we estimated species trees and networks by means of four different approaches to evaluate if the current circumscription of the subsections are in agreement with our phylogenetic estimates. Overall, we found strong evidence for hybridization in subsections *Humiles* and *Proceri*, but phylogenetic support was generally lacking for many of the species relationships owing to large amounts of genealogical discordance. Given the pace with which *Penstemon* has recently radiated, these types of patterns are not unexpected [810]. Nevertheless, our phylogenetic estimates showed some stable relationships among the different methods used and suggest that the current taxonomy of the two subsections needs revising.

9.2 Approach

We begin with a brief description of our method for QCF estimation, which uses site patterns to estimate the gene tree quartet relationships that are used to calculate species-level concordance factors. The basis for this method stems from ideas regarding the use of phylogenetic invariants for the inference of phylogenetic trees [9, 141]. For a given quartet of species, the QCF values represent the proportion of gene trees that agree with each of the three possible unrooted topologies relating the four species: ((1,2)(3,4)), ((1,3)(2,4)), and ((1,4)(2,3)). Because they represent a single bipartition among four species, these topologies are referred to as "splits" and are denoted 12|34, 13|24, and 14|23 [141]. Estimating the species-level concordance factors then amounts to estimating quartet topologies for each gene, followed by tabulating which species-level split is supported by each gene. When there are multiple haplotypes at a locus, we consider all of their possible sampling combinations and calculate the gene tree quartet topologies that support each species-level relationship. Using this approach, we are able to quickly estimate QCF values for samples with different ploidy levels. Below we detail our method for scoring gene tree quartet topologies and combining them across sampled haplotypes to get species-level QCFs.

9.2.1 CALCULATION OF QUARTET CONCORDANCE FACTORS

Consider four species (1, 2, 3, and 4) with DNA sequence data collected at G independent loci and with haplotypes phased and aligned for each locus. For example, a diploid individual from species 1 might have two haplotypes at locus one, which we denote $1_{(1,1)}$ and $1_{(1,2)}$. For any species, S, we denote its haplotypes at each gene by indexing across the gene number ($g = 1, \ldots, G$) and the haplotype number ($h = 1, \ldots, s_g$; where s_g is the number of haplotypes present in species S at gene g). Then, for each gene, we score the three possible splits of all haplotype combinations using the frequency of matching site patterns. To get these scores, we first calculate the number of times a pair of species, A and B, have the same nucleotide at each site in the alignment:

$$m_{AB} = \sum_i \mathbb{I}(A(i) = B(i)). \tag{9.1}$$

Here $\mathbb{I}()$ is the indicator function and is equal to 1 if the two bases are the same and 0 otherwise.

We then use the m_{ab}'s to calculate scores for each of the three unrooted quartet topologies:

$$\mathcal{G}(1,2|3,4) = m_{12} + m_{34} - (m_{13} + m_{14} + m_{23} + m_{24}),$$

$$\mathcal{G}(1,3|2,4) = m_{13} + m_{24} - (m_{12} + m_{14} + m_{23} + m_{34}),$$

$$\mathcal{G}(1,4|2,3) = m_{14} + m_{23} - (m_{12} + m_{13} + m_{24} + m_{34}). \tag{9.2}$$

These scores can be considered raw dissimilarity scores, assigning patterns of nucleotide substitution that support a given split positive weight and those that support alternative topologies negative weight. Other scores such as the neighbor joining distance [645] or LogDet [444] distance could also be substituted here (see section 9.5). At the species level, we tabulate the number of gene trees supporting these same splits and add 1 to the species topology that corresponds to the gene split with the highest score. If there is a tie for the highest score, we add 0.5 to the two species-level splits for the highest scoring gene trees. If all three gene tree splits have the same score, then one-third is added to each species-level split.

The calculation of concordance factors for each species-level quartet is then done by summing over all genes and tabulating how often each possible split is supported by each gene. This sum is also taken over all possible combinations of haplotypes, giving the following equation for calculating QCFs:

$$\mathcal{CF}_{12|34} \propto \sum_{g=1}^{G} \left(\sum_{a=1}^{1_g} \sum_{b=1}^{2_g} \sum_{c=1}^{3_g} \sum_{d=1}^{4_g} \mathbb{I}_{max} \left[\mathcal{G}\left(1_{(g,a)}, 2_{(g,b)} | 3_{(g,c)}, 4_{(g,d)}\right) \right] \right). \tag{9.3}$$

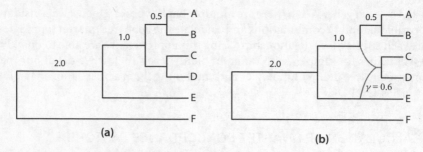

Figure 9.1. Simulation setup for (a) tree and (b) network topologies. Internal branches are annotated with their lengths in coalescent units (CUs). The total tree height is 4.0 CUs.

Here, $\mathbb{I}_{max}()$ is shorthand for an indicator function that is 1 if $\mathcal{G}(1, 2|3, 4)$ is greater than both $\mathcal{G}(1, 3|2, 4)$ and $\mathcal{G}(1, 4|2, 3)$ but is 0 otherwise. Ties are handled as described above. The calculation of $\mathcal{CF}_{13|24}$ and $\mathcal{CF}_{14|23}$ is the same as above but with the species and sums switched into the correct order for the split under consideration. All three species-level concordance factors are then normalized by their sum.

9.2.2 BOOTSTRAPPING AND GENE TREE UNCERTAINTY

To deal with uncertainty in gene tree quartet estimation, we can also conduct bootstrap resampling of sites within genes when calculating the gene tree split scores. If we conduct B rounds of resampling, the gene tree contributions to the species-level splits can then be calculated across bootstrap replicates, with each gene tree split getting a weight proportional to the number of times it was the best scoring topology across all replicates:

$$\tilde{\mathcal{G}}\left(1_{(g,a)}, 2_{(g,b)}|3_{(g,c)}, 4_{(g,d)}\right) = \frac{1}{B}\sum_{b=1}^{B}\mathbb{I}_{max}\left[\mathcal{G}\left(1_{(g,a)}, 2_{(g,b)}|3_{(g,c)}, 4_{(g,d)}\right)\right]. \tag{9.4}$$

The species-level QCF value is then taken as the sum over these bootstrap weighted gene tree splits:

$$\mathcal{CF}_{12|34} \propto \sum_{g=1}^{G}\left(\sum_{a=1}^{a_g}\sum_{b=1}^{b_g}\sum_{c=1}^{c_g}\sum_{d=1}^{d_g}\tilde{\mathcal{G}}\left(1_{(g,a)}, 2_{(g,b)}|3_{(g,c)}, 4_{(g,d)}\right)\right). \tag{9.5}$$

As before, $\mathcal{CF}_{13|24}$ and $\mathcal{CF}_{14|23}$ are calculated in a similar way, such that their indices are in the correct order. These $\mathcal{CF}$ values are then normalized by their sum.

9.2.3 VALIDATION OF QCF ESTIMATION

We validated our approach for QCF estimation by means of simulations on both tree and network topologies (figure 9.1). The details of these simulations can be found in the online supplemental materials available through https://press.princeton.edu/ISBN/9780691207605. In general, our method for calculating QCF values produced accurate estimates when compared with the true simulated data, with root mean squared deviation (RMSD) values ranging from 0.019 to 0.042 (0.023–0.059 for bootstrapped) and 0.019 to 0.036 (0.021–0.043 for bootstrapped) for the tree and network topologies, respectively (tables S4 and S5). Figures S1 and S2 show plots of fitted linear regression models for each quartet of species and the corresponding QCF estimate (x-axis) for each of the three unrooted topologies compared with the true simulated value (y-axis). Simulated QCFs were tabulated using PhyloNetworks [711].

9.2.4 IMPLEMENTATION

We have implemented our new method in the open-source software package qcf, which is wrtten in C++ and is available under the GNU GPL v3 on GitHub (https://github.com /pblischak/QCF). Documentation and tutorials for using the software can be found on ReadTheDocs (https://qcf.readthedocs.io).

9.3 Materials and Methods

9.3.1 STUDY SYSTEM

Penstemon Schmidel (Plantaginaceae) is the largest group of flowering plants endemic to North America, with circa 300 species distributed from Alaska to Guatemala and from the Pacific to Atlantic coasts [810]. The center of diversity for *Penstemon* is the Intermountain West of the United States, with the biogeographic origin of the genus hypothesized to be in the Columbia Plateau [730, 809, 810]. *Penstemon* has undergone a recent and rapid radiation, which is thought to be driven by Pleistocene glaciation cycles as well as adaptation to different ecological habitats and pollinators [122, 810, 804]. The most comprehensive molecular phylogeny of *Penstemon* was published by [810], with 193 species sampled for their analyses of the nuclear ribosomal ITS region and two chloroplast genes (*trnCD* and *trnTL*). While support for many species-level relationships was lacking, [810] were able to make several inferences regarding higher-level relationships within *Penstemon*. A more recent study conducted high-throughput sequencing for 70 species of *Penstemon* [791] and recovered high support across the entire tree. However, the limited taxon sampling of their phylogeny did not contain many members of the subgenus *Penstemon*, limiting the interpretation of their results in the context of the whole genus.

Two groups within *Penstemon* subg. *Penstemon* that are of particular interest are the subsections *Humiles* and *Proceri*. Species in these subsections are primarily distributed in the Pacific Northwest of the United States and occur at subalpine to alpine elevations in a variety of habitats, including some that are atypical for species of *Penstemon* in the western United States (e.g., mesic meadows; [358, 541]). These subsections are also morphologically distinct from other members of the genus, with their inflorescences organized into verticillasters. Many of the species also have glandular hairs on the inflorescence, a character present in all species of subsect. *Humiles* but only in some members of subsect. *Proceri* [358]. The traditional taxonomic division between these groups is based on a single leaf character: members of subsect. *Humiles* have serrate leaf margins and members of subsect. *Proceri* have entire margins [358]. However, there have been observations of hybridization between the two subsections, such that members of subsect. *Proceri* can sometimes have toothed leaf margins [732]. Cases of hybridization at the diploid level are well documented in *Penstemon* [729, 155, 812, 811, 161],

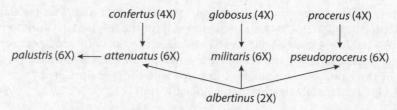

Figure 9.2. Hypotheses of allopolyploid formation in *Penstemon attenuatus* according to [358]. Varieties of *P. attenuatus* are placed in the center, with their putative diploid parent below (*P. albertinus* for all) and their putative tetraploid parents above. *Penstemon attenuatus* var. *palustris* is marked with a dashed arrow owing to the uncertainty of its placement.

and numerous instances of polyploidy, mostly within sect. *Penstemon* and subg. *Saccanthera*, have been studied as well [358, 93]. Most of these polyploids are thought to be allopolyploids (formed through hybridization) [810], but the majority of these hypotheses remain untested (but see [408]).

Given the lability of the leaf character dividing these two subsections, as well as their similarities in geographic ranges and morphology, we sought to evaluate the monophyly of subsections *Humiles* and *Proceri* by means of nuclear amplicon data. We also aimed to investigate the extent to which hybridization has occurred in these groups, as well as gain an understanding of the origin of any polyploid taxa (auto- vs. allopolyploid). In the case of subsections *Humiles* and *Proceri*, the *P. attenuatus* species complex presents a compelling test for understanding polyploidy in these groups. According to [358], the four varieties of *P. attenuatus* are all hypothesized to be allopolyploids, forming through hybridization between *P. albertinus* in subsect. *Humiles* and three different species in subsect. *Proceri* (see figure 9.2). Earlier molecular phylogenetic analyses recovered the subsections as polyphyletic [810]. However, these patterns are based on two uncombined gene trees, which would not allow for processes that cause gene tree incongruence to be modeled. Here we use species tree and network approaches to account for genealogical discordance caused by both ILS and hybridization to estimate a phylogeny for subsections *Humiles* and *Proceri*.

9.3.2 SAMPLE COLLECTION, DNA EXTRACTION, AND AMPLICON SEQUENCING

We extracted DNA from field-collected leaf tissue that was dried on silica gel. We used a modified CTAB protocol for DNA isolation [808], quantified all samples using a Qubit fluorometer (Invitrogen, Carlsbad, CA), and normalized all samples to 20 ng/μL. Normalized DNA samples for 38 accessions representing 17/22 and 20/27 currently circumscribed taxa from subsections *Humiles* and *Proceri*, respectively, plus an outgroup taxon from *Penstemon* subg. *Dasanthera* (*P. davidsonii* var. *davidsonii*), were sent to the IBEST Genomics Resources Core at the University of Idaho (Moscow, ID) for sample preparation and sequencing (listed in table S1). Amplification of targeted amplicons and the addition of sample barcodes and Illumina adapters were done using microfluidic PCR on the Fluidigm 48 × 48 Access Array (Fluidigm, South San Francisco, CA), followed by 300 bp paired-end sequencing on an Illumina MiSeq (Illumina, San Diego,

CA) [759]. Primers for the 48 loci used in this study were designed and tested as described in [79] and are given in tables S2 and S3.

Raw, paired-end sequencing reads were returned from IBEST and processed using Fluidigm2PURC v0.1.2 (https://github.com/pblischak/fluidigm2purc; [78]). Fluidigm2PURC trims reads using Sickle [352], joins paired reads using FLASH2 [459], and prepares the input file for clustering and chimera detection using the program PURC [637]. After clustering with PURC, haplotypes are inferred based on cluster sizes and user-specified ploidy levels. Three rounds of chimera detection and clustering were performed using default settings in the script purc_recluster2.py, a modified version of the original script distributed with PURC (https://bitbucket.org/crothfels/purc). To get haplotypes, clusters were first cleaned of excessive gaps using Phyutility (threshold = 33%; [699]) and then realigned using MAFFT [356]. Haplotypes were then inferred for all sampled taxa assuming known ploidy levels reported in [358], [732], and [93]. Information regarding haplotype dosage (i.e., number of haplotype copies) was ignored, resulting in only unique haplotypes being returned for each taxon at each gene.

9.3.3 SPECIES TREE INFERENCE

Haplotype-level gene trees for each locus were inferred with RAxML v8.2.11 using the GTRGAMMA model of nucleotide substitution and 500 rapid bootstrap replicates [721, 720]. We then inferred a taxon-level species tree with ASTRAL v5.5.9 (ASTRAL-III) using a mapping file to link haplotypes with their respective taxa [507]. To increase the thoroughness of the ASTRAL-III search algorithm, we added the following command line options: –polylimit 20 (maximum size of polytomy), –samplingrounds 100 (number of rounds of subsampling haplotypes from taxa), and –extraLevel 2 (increase the number of bipartitions added to the search space).

A species tree was also inferred using methods from the TICR pipeline (https://github.com/nstenz/TICR; [727]). The TICR (tree incongruence checking in R) pipeline estimates QCFs using BUCKy [405] and then uses the concordance values of these quartets to infer a species tree using the QuartetMaxCut algorithm [707]. Here, we instead used our new method to estimate QCFs and inferred a species tree using the script get-pop-tree.pl. Average concordance factors and branch lengths (in coalescent units) were then estimated for this tree using the getTreeBranchLengths.R script.

As a final estimate of phylogeny, we used only the majority haplotype (haplotype inferred from the largest cluster) returned by Fluidigm2PURC to estimate a species tree with RAxML using a supermatrix as input. It has been shown previously that concatenating multilocus data can result in incorrect inferences of phylogeny when gene tree discordance is present [385]. However, congruence among different methods can also be a good indicator of stable species relationships. Inference with RAxML was conducted using a partition file to estimate separate model parameters for each gene, and 1000 rapid bootstrap replicates were used to assess statistical support.

9.3.4 CANDIDATE HYBRIDIZATION EVENTS FROM ROOTED TRIPLES

To generate a list of candidate hybridization events, we used the program HyDe v0.4.2 to test for hybridization on all possible triples of taxa from subsections *Humiles* and *Proceri* [77]. HyDe tests for hybridization using site pattern frequencies [389] and estimates the amount of admixture occurring between two parental taxa to form a third hybrid taxon. Using *P. davidsonii* var. *davidsonii* as an outgroup and a mapping file to assign haplotypes to taxa, we tested all triples in all directions with the run_hyde_mp.py script. Statistical significance was assessed at the $\alpha = 0.05$ level with a Bonferroni correction for the number of hypothesis tests conducted.

9.3.5 SPECIES NETWORK INFERENCE

Our species tree analyses with ASTRAL-III, QuartetMaxCut, and RAxML (supermatrix) recovered two clades with corresponding taxon membership (see section 9.4), which we refer to as clades A and B (figures 9.3–9.5). To reduce the computational burden of estimating a large network (>30 taxa), we chose to analyze these clades independently. Haplotypes from taxa belonging to each clade were extracted from the original sequence alignments and written to new files. *Penstemon davidsonii* var. *davidsonii* was included in the data set for both clades as an outgroup. We then estimated haplotype-level gene trees as before and inferred a taxon-level species in ASTRAL-III using a mapping file and default search settings. We also estimated QCFs for each clade by using the qcf software with 500 bootstrap replicates. The resulting species trees and QCF estimates for clades A and B were then used as input for network estimation using the SNaQ method implemented in the software package PhyloNetworks v0.7.0 [710, 711]. We varied the maximum number of hybridization events from h = 1 to h = 5 and used the resulting log-pseudolikelihood values to determine the most likely number of hybridization events. The log-pseudolikelihood for the case of no hybridization (h = 0) was calculated by maximizing the fit of the observed QCF values on the fixed tree topology estimated by ASTRAL-III. All network analyses were conducted on the Oakley cluster at the Ohio Supercomputer Center (https://www.osc.edu).

9.4 Results

9.4.1 NUCLEAR AMPLICON DATA

Of the 48 loci that were amplified using the Fluidigm Access Array, 43 (all nuclear) recovered sufficient data for processing and downstream phylogenetic analyses. Data processing with Fluidigm2PURC on these 43 loci produced phased haplotypes for all 38 taxa, with many of the polyploid taxa containing three or more unique haplotypes. The supermatrix of majority haplotypes used for analysis with RAxML had a total alignment length of 18,207 bp.

9.4.2 SPECIES TREE INFERENCE

Species trees were inferred with three different methods, all of which produced mostly similar phylogenetic estimates for subsections *Humiles* and *Proceri* (figures 9.3–9.5). For each method, four species were consistently recovered as a grade outside the rest of the ingroup: *P. anguineus*, *P. rattanii*, *P. watsonii*, and *P. whippleanus*. *Penstemon ovatus* was also recovered outside the two subsections in the ASTRAL-III analysis. For ASTRAL-III and QuartetMaxCut, two species from subsect. *Proceri*, *P. attenuatus* var. *attenuatus* and *P. attenuatus* var. *pseudoprocerus*, were recovered in a clade consisting primarily of species from subsect. *Humiles*. These two methods also recovered a clade consisting almost entirely of species from subsect. *Proceri*, with the exception of *P. radicosus* being present in the ASTRAL-III tree. The supermatrix analysis inferred a tree with a number of relationships that differed from the other two approaches. Three of the

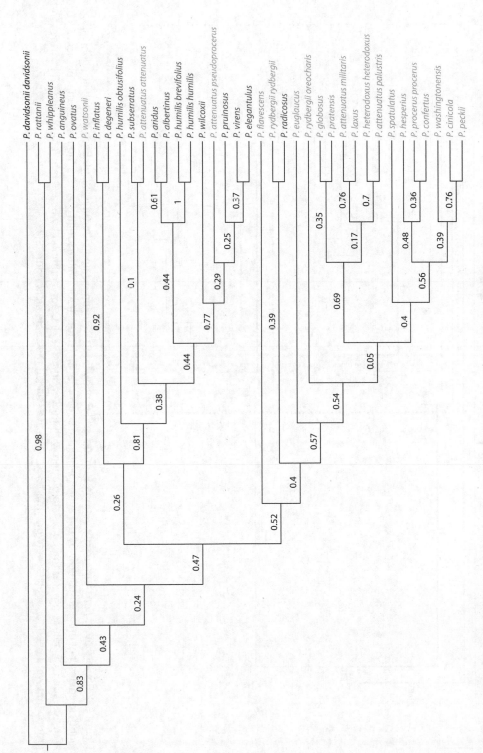

Figure 9.3. Phylogeny of *Penstemon* subsections *Humiles* and *Proceri* inferred by ASTRAL-III. Labels on branches are local posterior probabilities [657]. Taxa are shaded based on their current taxonomic classification: light grey = subsect. *Proceri*, dark grey = subsect. *Humiles*.

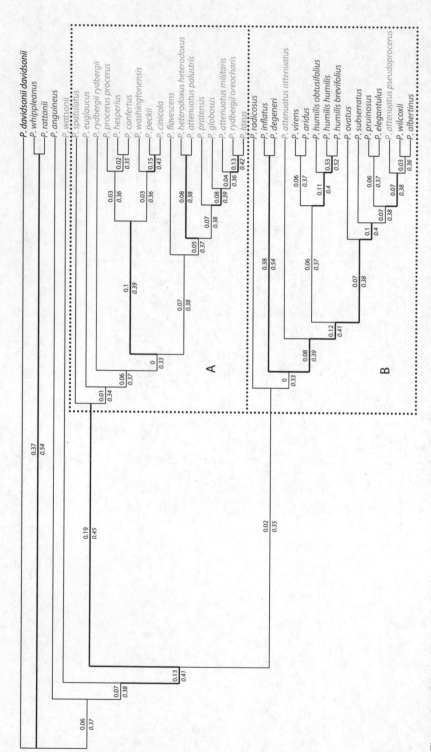

Figure 9.4. Phylogeny of *Penstemon* subsections *Humiles* and *Proceri* inferred using qcf and QuartetMaxCut. Each branch is labeled above by its length in coalescent units and below by the average quartet concordance factors (QCF) value for all quartets induced by that branch. All branches with average QCF values greater than 0.38 are plotted with thicker lines [727]. Clades outlined by boxes labeled "A" and "B" correspond to the groups that were used for species network inference. Taxa are shaded the same as in figure 9.3.

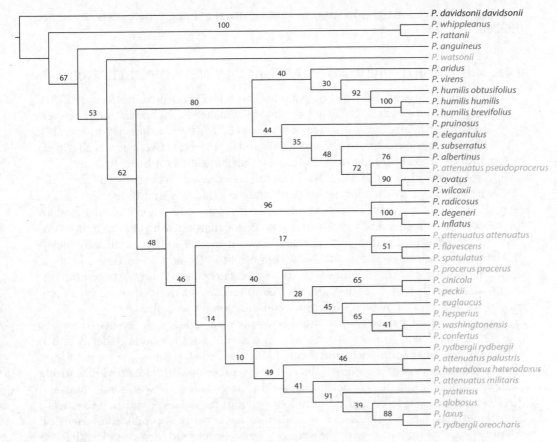

Figure 9.5. Phylogeny of *Penstemon* subsections *Humiles* and *Proceri* inferred with RAxML using a supermatrix of 43 loci. Labels on branches are support values from 1000 bootstrap replicates. Taxa are shaded the same as in figure 9.3.

four varieties of *P. attenuatus* were inferred to belong to the same clade, but *P. attenuatus* var. *pseudoprocerus* was still recovered in a clade of *Humiles* taxa. This analysis also shifted a clade of three species, *P. radicosus*, *P. degeneri*, and *P. inflatus*, to be sister to the clade consisting of *Proceri* species with high support (bootstrap support = 96%). Another notable difference among the methods was that ASTRAL-III did not recover the three varieties of *P. humilis* as monophyletic, but the QuartetMaxCut and supermatrix approaches did.

Estimated branch lengths in coalescent units from ASTRAL-III and QuartetMaxCut were extremely short for all internal branches, indicating rampant genealogical discordance (figures 9.4, S3, and S4). Branch lengths from the RAxML supermatrix analysis were also very short for the branches along the backbone of the tree, demonstrating that few substitutions were present to inform relationships for these deeper bipartitions. Support values were generally low across the different trees, with only a few relationships showing high levels of support. This is likely a result of the short branches observed in the different trees and support the hypothesis that speciation has occurred rapidly, with little time for informative substitutions to occur. Another possible reason for these low support values is the occurrence of hybridization. As we show below, there is strong evidence for hybridization in these groups, making phylogenetic inference difficult.

9.4.3 TESTS FOR HYBRIDIZATION AND SPECIES NETWORK INFERENCE

Analyzing all possible triples of ingroup taxa with HyDe resulted in a total of 23,310 hypothesis tests, of which 282 showed significant evidence for hybridization. The average value for the hybridization parameter (γ) was 0.513 (standard deviation = 0.114), with a minimum and maximum value of 0.205 and 0.843, respectively, across all triplets tested. Out of 37 total ingroup taxa, 24 had a significant signal for hybridization.

Species network inference with SNaQ was then conducted on the two primary clades that were recovered in the QuartetMaxCut analyses (clades A and B; figure 9.4). The reason for using these clades was that this analysis recovered a pattern of relationships for subsections *Humiles* and *Proceri* that was most consistent with the current taxonomy. Although hybridization was detected between species in these clades by means of HyDe, network inference with SNaQ cannot currently handle 38 taxa. However, since the members of these clades were recovered fairly consistently between the different methods that we used for species tree inference, we decided to analyze them independently to make network inference computationally feasible.

Using a range of values on the number of possible reticulation events (h = 0 to h = 5), we were able to infer species networks in all cases for both clades A and B within the amount of compute time allotted (10 cores, 80–100 hours). For both clades, adding reticulation events greatly reduced the log-pseudolikelihood, providing strong evidence that hybridization is occurring within these clades. Networks with four and three reticulations had the highest pseudolikelihoods for clades A and B, respectively, but the network topology with four reticulations for network A produced nonsensical relationships (hybridization with the outgroup), so we preferred the network with h = 3 (figures S5 and S6). In addition to having the highest (or second highest for clade A) log-pseduolikelihood, the networks estimated with three reticulations were among the only estimates that had sensible branch lengths. For most other networks in both clades A and B, one of the reticulate edges was always inferred to have a branch length of ~9.5 coalescent units. Given the amount of gene tree discordance present in the data set, and the short branch lengths estimated by ASTRAL-III and QuartetMaxCut, these estimates are most likely incorrect.

The best networks for clades A and B showed different patterns for the timing of hybridization (figure 9.6). For clade A, all of the reticulation events occurred closer to the present and only involved pairs of species hybridizing. The hybridization events inferred include (1) *P. spatulatus* × *P. attenuatus* var. *palustris* → *P. flavescens*, (2) *P. rydbergii* var. *oreocharis* × *P. globosus* → *P. laxus*, and (3) *P. confertus* × *P. cinicola* → *P. washingtonensis*. Clade B, on the other hand, was estimated to have a deep reticulation event involving two ancestral populations, with the resulting hybrid lineage then diversifying into 12 different taxa. The other hybridization event in this clade was *P. ovatus* × *P. humilis* var. *humilis* → *P. humilis* var. *brevifolius*.

9.5 Discussion

In this chapter, we investigated the phylogenetic relationships and taxonomic affinities among the taxa within *Penstemon* subsections *Humiles* and *Proceri* using nuclear amplicon sequencing. We found strong evidence for hybridization in these groups, but

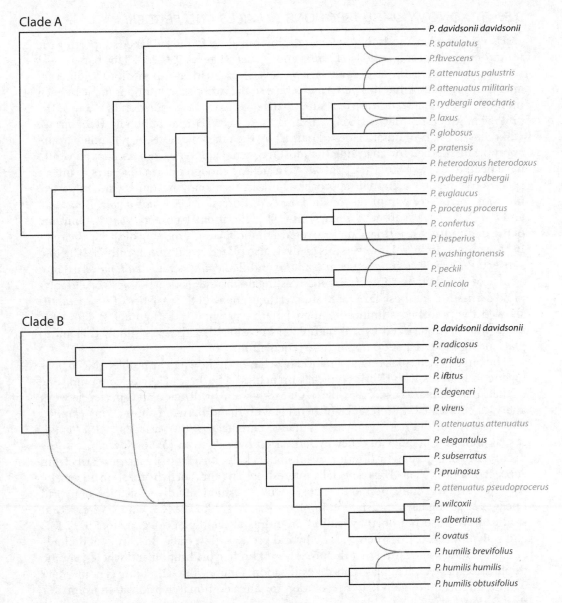

Figure 9.6. Best maximum pseudolikelihood (ML) networks for clades A and B estimated by SNaQ. The maximum number of hybridization events for these ML networks is h = 3. Taxa are shaded the same as in figure 9.3.

the rapid diversification in these two subsections made the exact inference, localization, and interpretation of reticulation events extremely difficult. Despite these shortcomings, there are some clear trends regarding the taxonomic implications of our phylogenetic estimates, as well as what they may mean for character evolution and biogeography in the group. Our work also highlights the difficulties of estimating phylogeny for recently radiating groups with variable ploidy levels and high amounts of hybridization, an issue that is common for many groups of angiosperms.

9.5.1 TAXONOMY OF SUBSECTIONS *HUMILES* AND *PROCERI*

Using several methods for phylogenetic inference, we found evidence supporting the nonmonophyly of subsections *Humiles* and *Proceri* (figures 9.3–9.5). This pattern was recovered for all methods despite variable levels of statistical support for the different analyses, suggesting that this pattern is robust to the various assumptions made by each method. For example, our use of a supermatrix approach is in violation of many of the biological processes that are likely affecting these taxa (hybridization and ILS), but we still recover the two major clades identified by the other methods. A potentially more appropriate method for analyzing the most frequent haplotype from each taxon would have been to use SVDQuartets, but we were lacking enough informative sites to infer a species tree reliably using this approach [140]. In each analysis, four species in particular were recovered completely outside the two subsections: *P. whippleanus*, *P. rattanii*, and *P. anguineus* (all subsect. *Humiles*), as well as *P. watsonii* (subsect. *Proceri*). Only one of the methods, QuartetMaxCut, recovered a monophyletic grouping of species belonging to subsect. *Proceri*. However, this analysis also placed two taxa currently classified in subsect. *Proceri* (*P. attenuatus* var. *attenuatus* and *P. attenuatus* var. *pseudoprocerus*) into a clade of species from subsect. *Humiles*, casting doubt on Keck's hypotheses of allopolyploid formation for these taxa [358]. Interestingly, none of the varieties of *P. attenuatus* showed the predicted affinities for their putative parental taxa (figure 9.2). However, given their hypothesized hybrid nature, it is possible that they are simply difficult to place using models that do not include reticulation.

The phylogenetic placement of hybrid taxa has been shown to be problematic in phylogenetic analyses, with hybrids typically branching at the base of a clade containing one of their parental taxa (e.g., [485, 486]). Our tests for hybridization and species network analyses confirmed the presence of hybrids within and between subsections *Humiles* and *Proceri* but did not support the putative parentage of the varieties of *P. attenuatus*, as well as a number of other hypothesized hybrids from [358]. Nevertheless, the occurrence of a deep hybridization event in the clade consisting of taxa primarily from subsect. *Humiles* (clade B) is especially interesting. The effect of hybridization on genetic variation and its connection to the subsequent speciation and diversification of hybrid lineages has long been understood in plants [23, 724, 26, 273, 462], and several recent studies have observed deep hybridization events at the diploid (e.g., [246, 257, 247]) and polyploid (e.g., [515]) levels. The hybrid group within clade B is a mix of diploids and polyploids, suggesting that the processes of both hybridization and whole-genome duplication have been at play in its diversification. Future work with more genomic data will be important to gain better resolution for any additional hybridization events for this clade.

A particular strength of our method of QCF estimation is the ability to analyze taxa with different ploidy levels, allowing us to analyze all diploid and polyploid taxa in subsections *Humiles* and *Proceri* simultaneously. From our network analyses with SNaQ, the only polyploid that was inferred to be a hybrid was *P. flavescens* (hexaploid; clade A). However, the simplicity of our raw dissimilarity metric could also be leading to some inaccuracies in our calculated QCF scores. Future work into the effectiveness and accuracy of alternative distance measures for calculating QCFs will be an important next step for better understanding and improving our QCF method. Our tests for hybridization with HyDe found far more evidence for hybridization, likely because it tests only three species at a time rather than trying to infer an entire network. Nevertheless, out of the 24 taxa inferred to be hybrids using HyDe, only 4 were polyploids. This casts doubt on the hypothesis that most of the polyploids in subsect. *Proceri* are of hybrid origin. However, a lack of phylogenetically informative variation could also be preventing us from detecting the full extent of hybridization potentially occurring in these polyploids, as well as the fact that so much hybridization is likely violating SNaQ's assumption of a level-1 network topology [711].

9.5.2 CHARACTER EVOLUTION AND BIOGEOGRAPHY

Given the reticulate history of subsections *Humiles* and *Proceri*, there are several patterns of morphological character evolution that can be interpreted in the context of their past genetic exchanges. Of particular interest is the presence of glandular hairs on the inflorescence, a trait with potentially adaptive importance [424] that is present in all species of subsect. *Humiles* but is absent in the majority of species in subsect. *Proceri*. For the species in subsect. *Proceri* for which it does occur, it is hard to determine if it is simply a labile trait that has arisen several times or if there is a single clade of species that all have the trait. A perhaps more interesting scenario could be that this trait was gained through hybridization or introgression; however, testing this hypothesis is currently not feasible due to a lack of methods for discrete character reconstruction on phylogenetic networks (but see [346, 50] for examples of continuous character evolution). A possible workaround would be to construct all possible resolutions of the underlying trees displayed by the networks and to reconstruct the character history on each tree. Nevertheless, future model development on discrete character evolution on networks will help to address this type of question.

The biogeographic context of these hybridization events is also of interest, with reconstructions of species' geographic ranges potentially helping to shed light on the plausibility of hypotheses about the occurrence of reticulation. The current geographic distribution of the taxa in subsections *Humiles* and *Proceri* is concentrated in the Pacific Northwest of the United States, an area with several well-established biogeographic and phylogeographic hypotheses regarding the occurrence of species in the Cascade Range, the Northern Rocky Mountains, and the Sierra Nevada in northern California [97, 120, 96]. Two recent studies that have investigated the biogeography of hybridization events include [103] and [247], who take different approaches to reconstructing ancestral contact zones where hybridization could potentially have occurred. Burbink and Gehara [103] found a single deep reticulation event in the phylogeny of New World kingsnakes and used the resulting parental trees (trees where a hybrid clade is sister to either parent) to infer ancestral areas for the hybrid clade. Folk et al. [247] used climatic niche reconstructions to find likely regions where ancestral lineages of *Huechera* and *Mitella* may have occurred in sympatry and hybridized. These approaches could be used in concert to illuminate the dynamics of vicariance and dispersal for lineages of subsections *Humiles* and *Proceri*, as well as helping to locate geographic regions where hybridization, as well as whole-genome duplication, could have occurred in the past.

9.5.3 PHYLOGENETICS OF HYBRIDS AND POLYPLOIDS

As was seen from our phylogenetic analyses, the internal branch lengths of our species trees and networks were very short (most were less than 0.5 coalescent units), highlighting the prevalence of incomplete lineage sorting and genealogical discordance in our data. Previous research has shown that *Penstemon* is a young genus (crown age

2.5–4.0 MYA) that has radiated extremely rapidly, with hybridization and polyploidy occurring frequently [810, 791]. These types of processes are likely not uncommon for other groups of angiosperms, and having methods to deal with them will be especially important for making future inferences about the evolutionary history of these groups. To resolve hybridization events, especially when they involve polyploids, there are several methods that have already been developed. Some of these are not coalescent based but instead try to reconstruct a network from gene trees that have all of the haplotypes from a polyploid sampled (a so-called multilabeled tree; [451, 467]). Other studies have relied on coalescent-based assignment of homoeologous haplotypes into putative, diploid subgenomes, but these approaches can be computationally limited due to the cost of exploring all permutations of haplotypes assignments [65, 547]. The only approach to simultaneously infer a network topology and homoeolog assignment in a coalescent framework is the method of [351]. However, this method uses a hierarchical Bayesian framework that does not scale well to large numbers of loci or taxa.

If homoeolog assignment is the goal, then it may be beneficial to first identify parental taxa so that the number of comparisons for determining haplotype origin is reduced. Kamneva et al. [353] used such an approach in strawberries to identify potential parents for several different polyploid species. They used a two-step approach to generate and test hypotheses, first constructing networks using consensus methods, followed by evaluating the likelihood of candidate networks using PhyloNet [790]. Their analyses were limited to no more than five haplotypes per taxon and also did not include an actual search over network space. Our method for QCF estimation was able to analyze all inferred haplotypes for all 38 taxa sampled in this study, and our network analyses with SNaQ were used to conduct an actual search over network topologies. The appeal of these types of approaches is that they do not require a priori knowledge about parental taxa when inferring a network. For nonmodel taxa for which cases of hybridization and allopolyploidy are being investigated for the first time, the ability to model these processes with little input from the user regarding putative hybridization events should help to facilitate the discovery of reticulate evolutionary events in virtually any group of taxa where they may be occurring.

9.6 Conclusions

Hybridization and polyploidy are processes that obscure phylogenetic inference for many groups of taxa and are a particular problem for lineages of angiosperms, where they are especially common. Using the concept of quartet concordance factors (the proportion of gene tree quartets supporting a species-level quartet), we developed a method for estimating these concordance factors that can accommodate taxa with variable ploidy levels. Using this approach, and several others, we then inferred species trees for *Penstemon* subsections *Humiles* and *Proceri*, finding that the subsections were not reciprocally monophyletic. Tests for hybridization and species network inference also revealed that reticulation has been a common occurrence in these groups. In general, this study highlights the difficulties of inferring phylogeny in a rapid species radiation where hybridization and whole-genome duplication are common. However, our approach for QCF estimation, in combination with the network inference method SNaQ, helped to disentangle the complex patterns of hybridization in these subsections and should provide a useful tool for other researchers interested in reticulate evolution as well.

CHAPTER 10

Comparison of Linked versus Unlinked Character Models for Species Tree Inference

Kerry Cobb and Jamie R. Oaks

10.1 Introduction

Current model-based methods of species tree inference require biologists to make difficult decisions about their genomic data. They must decide whether to assume (1) sites in their alignments are each inherited independently ("unlinked"), or (2) groups of sites are inherited together ("linked"). If assuming the former, they must then decide whether to analyze all of their data or only putatively unlinked variable sites. Our goal in this chapter is to use simulated data to help guide these choices by comparing the robustness of different approaches to errors that are likely common in high-throughput genetic data sets.

Reduced-representation genomic data sets acquired from high-throughput instruments are becoming commonplace in phylogenetics [414] and usually comprise hundreds to thousands of loci from 50 to several thousand nucleotides long. Full likelihood approaches for inferring species trees from such data sets can be classified into two groups based on how they model the evolution of orthologous DNA sites along gene trees within the species tree—those that assume (1) each site evolved along its own gene tree (i.e., each site is "unlinked") [98, 164], or (2) contiguous, linked sites evolved along a shared gene tree [436, 298, 549, 831]. We will refer to these as unlinked- and linked-character models, respectively. For both models, the gene tree of each locus (whether each locus is a single site or a segment of linked sites) is assumed to be independent of the gene trees of all other loci, conditional on the species tree. Methods using linked-character models become computationally expensive as the number of loci grows large because of the estimation or numerical integration of all of the gene trees [831, 549]. Unlinked-character models on the other hand are more tractable for a large number of loci because estimating individual gene trees is avoided by analytically integrating over all possible gene trees [98, 164]. Whereas unlinked-character models can accommodate a larger number of loci than can linked-character models, most genetic data sets comprise linked sites, and unlinked-character models are unable to utilize the aggregate information about ancestry contained in such linked sites.

Investigators are thus faced with decisions about how best to use their data to infer a species tree. Should they use a linked-character method that assumes the sites within each locus evolved along a shared gene tree? Ideally, the answer would be "yes"; however, this is not always computationally feasible, and the model could be violated by intralocus recombination. Alternatively, should investigators remove all but one single-nucleotide polymorphism (SNP) from each locus and use an unlinked-character model? Or, perhaps they should apply the unlinked-character method to all of their sites, even if this violates the assumption that each site evolved along an independent gene tree. Important considerations in such decisions include the sources of error and bias that result from reduced-representation protocols, high-throughput sequencing technologies, and the processing of these data.

Most reduced-representation sequencing workflows employ amplification of DNA by means of polymerase chain reaction (PCR), which can introduce mutational error at a rate of up to 1.5×10^{-5} substitutions per base [585]. Furthermore, current high-throughput sequencing technologies have non-negligible rates of error. For example, Illumina sequencing platforms have been shown to have error rates as high as 0.25% per base [573]. In hope of removing such errors, it is common for biologists to filter out variants that are not found above some minimum frequency threshold [625, 431]. The effect of this filtering will be more pronounced in data sets with low or highly variable coverage. Also, to avoid aligning paralogous sequences, it is common to remove loci that exceed an upper threshold on the number of variable sites [293]. These processing steps can introduce errors and acquisition biases, which have been shown to affect estimates derived from the assembled alignments [293, 318, 431]. Given these issues are likely common in high-throughput genomic data, downstream decisions about what methods to use and what data to include in analyses should consider how sensitive the results might be to errors and biases introduced during data collection and processing.

Our goal is to determine whether linked- and unlinked-character models differ in their robustness to errors in reduced-representation genomic data and whether it is better to use all sites or only SNPs for unlinked-character methods. Linked-character models can leverage shared information among linked sites about each underlying gene tree. Thus, these models might be able to infer correctly the general shape and depth of a gene tree, even if the haplotypes at some of the tips have errors. Unlinked-character models have very little information about each gene tree and rely on the frequency of allele counts across many characters to inform the model about the relative probabilities of all possible gene trees. Given this reliance on accurate allele count frequencies, we predict that unlinked-character models will be more sensitive to errors and acquisition biases in genomic data. To test this prediction that linked-character models are more robust to the types of errors contained in reduced-representation data, we simulated data sets with varying degrees of errors related to miscalling rare alleles and heterozygous sites. Our results support this prediction but also show that with only two species, the region of parameter space where there are differences between linked- and unlinked-character models is quite limited. Further work is needed to determine whether this difference in robustness between linked- and unlinked-character models will increase for larger species trees.

10.2 Methods

10.2.1 SIMULATIONS OF ERROR-FREE DATA SETS

For our simulations, we assumed a simple two-tipped species tree with one ancestral population with a constant effective size of N_e^R that diverged at time τ into two

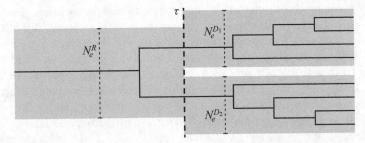

Figure 10.1. An illustration of the species tree model we used to simulate data. N_e^R, N_e^{D1}, and N_e^{D2} represent the constant effective population sizes of the root and each of the two terminal populations. τ represents the instantaneous separation of the ancestral population into two descendant populations. One hypothetical gene tree is shown to illustrate the gene trees simulated under a contained coalescent process for four haploid gene copies sampled from each of the terminal branches of the species tree.

descendent populations (terminal branches) with constant effective sizes of N_e^{D1} and N_e^{D2} (figure 10.1). For two diploid individuals sampled from each of the terminal populations (four sampled gene copies per population), we simulated 100,000 orthologous bi-allelic characters under a finite-sites, continuous-time Markov chain (CTMC) model of evolution. We simulated 100 data sets comprising loci of four different lengths—1000, 500, 250, and 1 characters. We assume each locus is effectively unlinked and has no intralocus recombination; that is, each locus evolved along a single gene tree that is independent of the other loci, conditional on the species tree. We chose this simple species tree model for our simulations to help ensure any differences in estimation accuracy or precision were due to differences in the underlying linked- and unlinked-character models and *not* due to differences in numerical algorithms for searching species and gene tree space. Furthermore, we simulated bi-allelic characters because unlinked-character multispecies coalescent models [98, 545] that are most comparable to linked-character models [298, 549] are limited to characters with (at most) two states.

We simulated the two-tipped species trees under a pure-birth process [846] with a birth rate of 10 by using the Python package DendroPy (version 4.40, Commit eb69003; [736]). This is equivalent to the time of divergence between the two species being exponentially distributed with a mean of 0.05 substitutions per site. We drew population sizes for each branch of the species tree from a Gamma distribution with a shape of 5.0 and mean of 0.002. We simulated 100, 200, 400, and 100,000 gene trees for data sets with loci of length 1000, 500, 250, and 1, respectively, using the contained coalescent implemented in DendroPy. We simulated linked bi-allelic character alignments using Seq-gen (version 1.3.4; [598]) with a generalized time reversible (GTR) model with base frequencies of A and C equal to 0 and base frequencies of G and T equal to 0.5. The transition rate for all base changes was 0 except for the rate between G and T, which was 1.0.

10.2.2 INTRODUCTION OF SITE PATTERN ERRORS

From each simulated data set containing linked characters described above, we created four data sets by introducing two types of errors at two levels of frequency. The first type of error we introduced was changing singleton character patterns (i.e., characters for which one gene copy was different from the other seven gene copies) to invariant patterns by changing the singleton character state to match the other gene copies. We introduced this change to all singleton site patterns with a probability of 0.2 and 0.4 to create two data sets from each simulated data set. The second type of error we introduced was missing heterozygous gene copies. To do this, we randomly paired gene copies from within each species to create two diploid genotypes for each locus; with with a probability of 0.2 or 0.4 we randomly replaced one allele of each genotype with the other. For the unlinked-character data set comprising a single site per locus, we simulated only singleton character pattern error at a probability of 0.4.

10.2.3 ASSESSMENT OF SENSITIVITY TO ERRORS

For each simulated data set with loci of 250, 500, and 1000 characters, we approximated the posterior distribution of the divergence time (τ) and effective population sizes (N_e^R, N_e^{D1}, and N_e^{D2}) under an unlinked-character model using ecoevolity (version 0.3.2, Commit a7e9bf2; [545]) and a linked-character model using the StarBEAST2 package (version 0.15.1; [549]) in BEAST2 (version 2.5.2; [85]). For both methods, we specified a CTMC model of character evolution and prior distributions that matched the model and distributions from which the data were generated. The prior on the effective size of the root population in the original implementation of ecoevolity was parameterized to be relative to the mean effective size of the descendant populations. We added an option to ecoevolity to compile a version for which the prior is specified as the absolute effective size of the root population, which matches the model in StarBEAST2 and the model we used to generate the data. The linkage of sites within loci of our simulated data violates the unlinked-character model of ecoevolity [98, 545]. Therefore, we also analyzed each data set with ecoevolity after selecting, at most, one variable character from each locus; loci without variable sites were excluded.

We analyzed the data sets simulated with one character per locus (i.e., unlinked data) with ecoevolity. Our goal with these analyses was to verify that the generative model of our simulation pipeline matched the underlying model of ecoevolity and to confirm that any behavior of the method with the other simulated data sets was not being caused by the linkage violation.

For ecoevolity, we ran four independent Markov chain Monte Carlo (MCMC) analyses with 75,000 steps and a sample frequency of 50 steps. For StarBEAST2, we ran two independent MCMC analyses with 20 million steps and a sample frequency of 5000 steps. To assess convergence and mixing of the ecoevolity and StarBEAST2 MCMC chains, we computed the effective sample size (ESS; [270]) and potential scale reduction factor (PSRF; the square root of equation 1.1 in [94]) from the samples of each parameter and considered an ESS value greater than 200 and PSRF less than 1.2 [94] to indicate adequate convergence and mixing of the chains. Based on preliminary analyses of simulated data sets without errors, we chose to discard the first 501 and 201 samples from the MCMC chains of ecoevolity and StarBEAST2, leaving 4000 and 7600 posterior samples for each data set, respectively.

10.2.4 PROJECT REPOSITORY

The full history of this project has been version controlled and is available at https://github.com/kerrycobb/align-error-sp-tree-sim and includes all of the data and scripts necessary to produce our results.

10.3 Results

10.3.1 BEHAVIOR OF LINKED (STARBEAST2) VERSUS UNLINKED (ECOEVOLITY) CHARACTER MODELS

The divergence times estimated by the linked-character method, StarBEAST2, were very accurate and precise for all alignment lengths and types and degrees errors, despite poor MCMC mixing (i.e., low ESS values) for shorter loci (figures 10.2–10.4). For data sets without error, the unlinked-character method, ecoevolity, estimated divergence times with similar accuracy and precision as StarBEAST2 when all characters are analyzed (figures 10.2–10.4). However when alignments contained errors, ecoevolity underestimated very recent divergence times with increasing severity as the frequency of errors increased (figures 10.2–10.4); estimates of older divergence times were unaffected.

The biased underestimation of divergence times by ecoevolity in the face of errors was coupled with overestimation of the ancestral effective population sizes (figures 10.5–10.7). When analyzing the alignments without errors, ecoevolity essentially returned the prior distribution on the effective size of the ancestral population (figures 10.5–10.7). Despite poor MCMC mixing, StarBEAST2 consistently estimated the effective size of the ancestral population better than did ecoevolity and was unaffected by errors in the data (figures 10.5–10.7). The precision of StarBEAST's estimates of N_e^R increased with locus length.

Estimates of the effective size of the descendant populations are largely similar between StarBEAST2 and ecoevolity; both methods underestimate the descendant population sizes when the data sets contain errors, and this downward bias is generally worse for ecoevolity (figures 10.8–10.10). The degree of underestimation increases with the rate of errors in the data sets for both StarBEAST2 and ecoevolity, and the results were largely consistent across different locus lengths. (figures 10.8–10.10).

When we apply ecoevolity to data sets simulated with unlinked characters (i.e., data sets simulated with one character per locus), we see the same patterns of biased parameter estimates in response to errors (figure 10.11) as we did with the linked loci (figures 10.2–10.4). These results rule out the possibility that the greater sensitivity of ecoevolity to the errors we simulated is due to violation of the method's assumption that all characters are unlinked.

10.3.2 ANALYSIS OF ALL SITES VERSUS SNPs WITH ECOEVOLITY

The unlinked-character model implemented in ecoevolity assumes that orthologous nucleotide sites evolve independently along separate gene trees. The data, however, were simulated under a model assuming that contiguous linked sites evolve along a shared gene tree. It would thus be a violation of the ecoevolity model to include all sites in the analysis. However, avoiding this violation by removing all but one variable site per locus drastically reduces the amount of data. When analyzing the simulated data sets without errors, the precision and accuracy of parameter estimates by ecoevolity was much greater when all sites of the alignment were used relative to when a single SNP per locus was used despite violating the model (figures 10.2–10.10). This was generally true across the different lengths of loci; however, the coverage of credible intervals is lower with longer loci. Analyzing only SNPs does make ecoevolity more robust to the errors we introduced. However, this robustness is due to the lack of information in the SNP data leading to wide credible intervals, and in the case of population

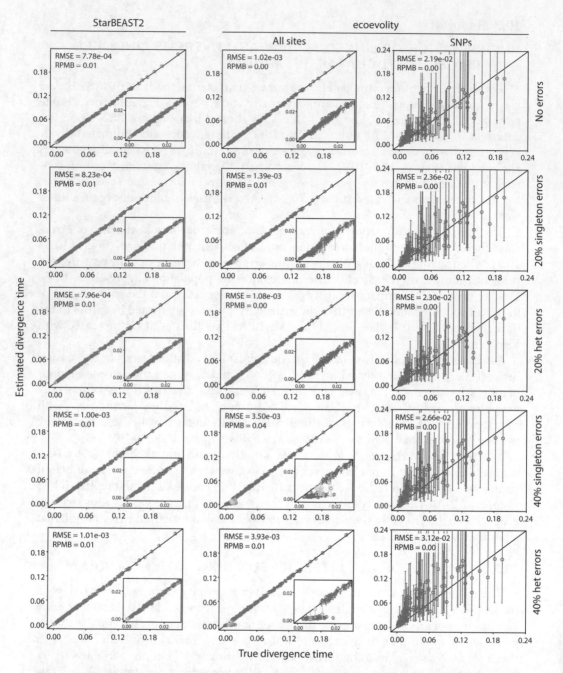

Figure 10.2. Accuracy and precision of divergence time estimates (in units of expected substitutions per site) with 1000 base pair loci. The left column shows estimates from StarBEAST2, and the center and right column shows estimates from ecoevolity using all sites and (at most) one SNP per locus. The top row shows estimates from 200 data sets simulated without character-pattern errors. Rows labeled 20% and 40% singleton errors show estimates from the same alignments after singleton site patterns were changed to invariant sites with probabilities 0.2 and 0.4, respectively. Rows labelled 20% and 40% het errors show estimates from the same (error-free) alignments after we randomly paired

size parameters, the marginal posteriors essentially match the prior distribution (figures 10.8–10.10).

10.3.3 COVERAGE OF CREDIBLE INTERVALS

The 95% credible intervals for divergence times and effective population sizes estimated from alignments without error in StarBEAST2 had the expected coverage frequency in that the true value was within approximately 95% of the estimated credible intervals. This was also true for ecoevolity when analyzing data sets simulated with unlinked characters (i.e., no linked sites). This coverage behavior is expected and helps to confirm that our simulation pipeline generated data under the same model used for inference by StarBEAST2 and ecoevolity. As seen previously [545], analyzing longer linked loci causes the coverage of ecoevolity to be lower because of the violation of the model's assumption that the sites are unlinked.

10.3.4 MCMC CONVERGENCE AND MIXING

Most sets of StarBEAST2 and ecoevolity MCMC chains yielded samples of parameters with a PSRF less than 1.2, indicative of convergence. However, we do see poor mixing (ESS < 200) of the StarBEAST2 chains as the length of loci decreases (figures 10.2–10.10). We see evidence of poor mixing and convergence for ecoevolity only when applied to data sets with errors. This is in contrast to StarBEAST2, for which the frequency and degree of poor MCMC behavior is largely unaffected by the type or frequency of errors. The proportion of simulation replicates in which StarBEAST2 had an ESS of the ancestral population size less than 200 was high across all analyses (figures 10.5–10.7). For the descendant population size, StarBEAST2 had better ESS values across all analyses, with the exception of rare estimates of essentially zero when analyzing 250 bp loci (figures 10.8–10.10).

10.4 Discussion

Phylogeneticists seeking to infer species trees from large, multilocus data sets are faced with difficult decisions regarding assumptions about linkage across sites and, if assuming all sites are unlinked, what data to include in their analysis. With the caveat that we only explored trees with two species, the results of our simulations provide

Figure 10.2. (*continued*) gene copies within each species into two diploid genotypes, and for each genotype we randomly replaced one allele with the other with probability 0.2 and 0.4, respectively. Each plotted circle and associated error bars represent the posterior mean and 95% credible interval. Circles and error bars are shaded dark grey if the effective sample sizes (ESS) of the estimate was greater than 200 and the potential scale reduction factor (PSRF) was greater than 1.2. If the ESS was less than 200, circles and error bars are shaded light grey. If the PSRF was greater than 1.2, circles and error bars are shaded intermediate grey. The root mean square error (RMSE) and rate of poor Markov chain Monte Carlo behavior (RPMB) is given for each plot, the latter of which is the proportion of estimates with ESS < 200 or PSRF > 1.2. We generated the plots with matplotlib version 3.1.1 [335].

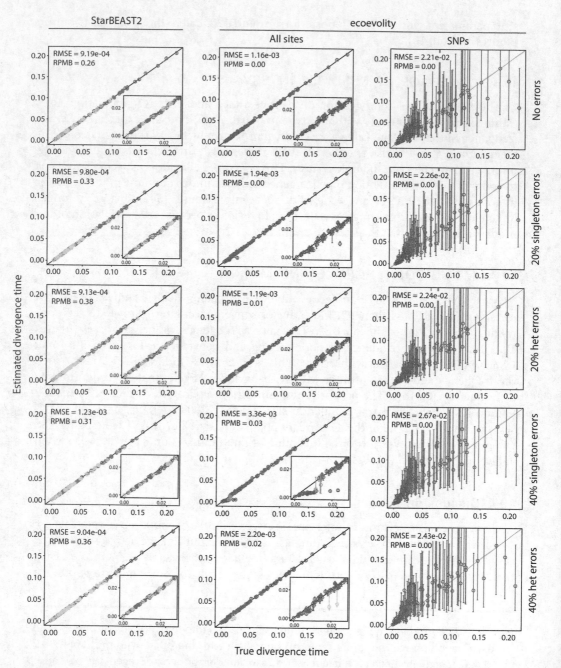

Figure 10.3. Accuracy and precision of divergence time estimates (in units of expected substitutions per site) with 500 base pair loci. See figure 10.2 for explanation.

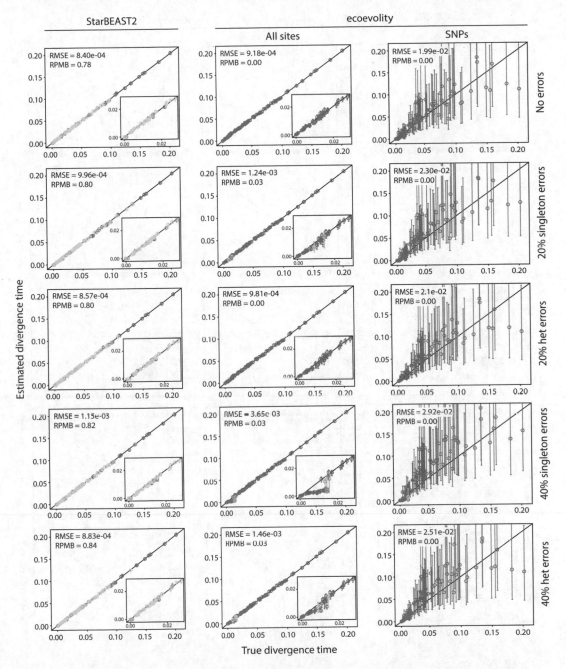

Figure 10.4. Accuracy and precision of divergence time estimates (in units of expected substitutions per site) with 250 base pair loci. See figure 10.2 for explanation.

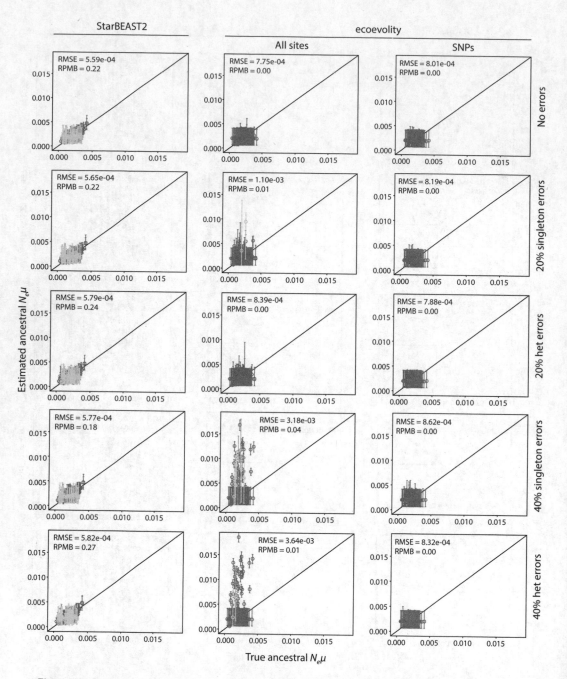

Figure 10.5. Accuracy and precision of estimates of root effective population size scaled by the mutation rate ($N_e^R \mu$) with 1000 base pair loci. See figure 10.2 for explanation.

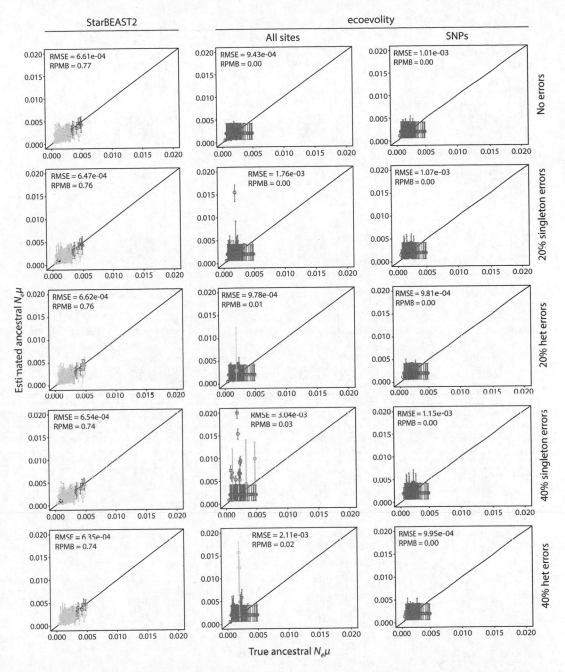

Figure 10.6. Accuracy and precision of estimates of root effective population size scaled by the mutation rate ($N_e^R \mu$) with 500 base pair loci. See figure 10.2 for explanation.

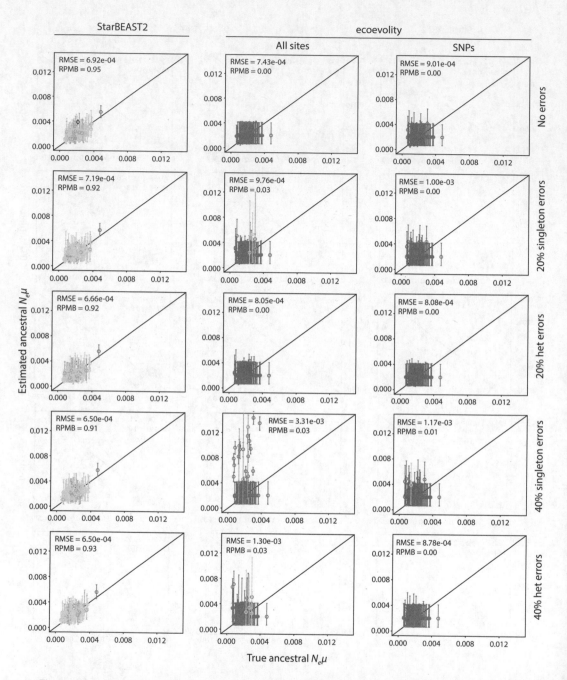

Figure 10.7. Accuracy and precision of estimates of root effective population size scaled by the mutation rate ($N_e^R \mu$) with 250 base pair loci. See figure 10.2 for explanation.

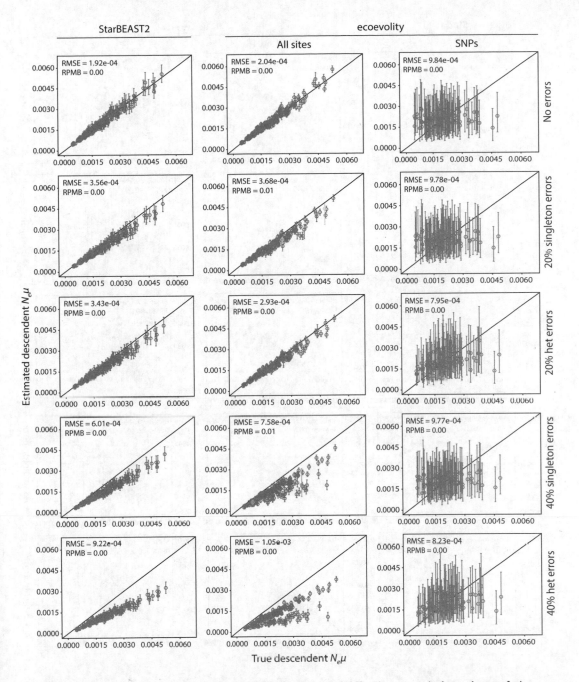

Figure 10.8. Accuracy and precision of estimates of effective population sizes of the descendant branches of the tree scaled by the mutation rate ($N_e^D \mu$) with 1000 base pair loci. See figure 10.2 for explanation.

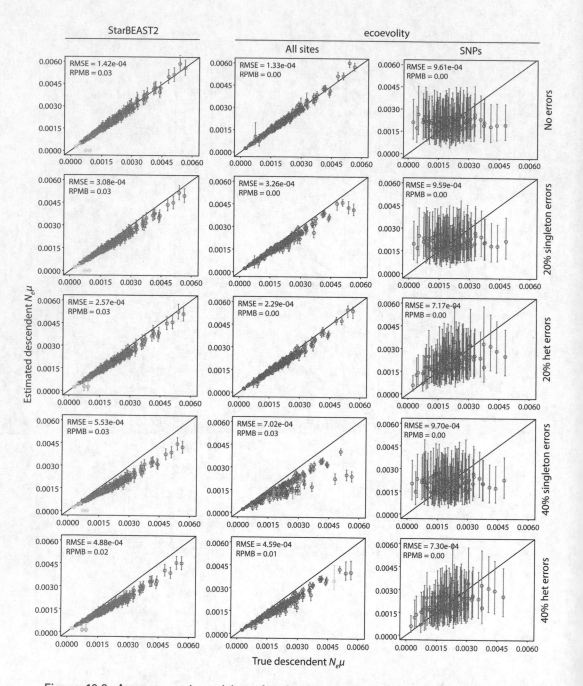

Figure 10.9. Accuracy and precision of estimates of effective population sizes of the descendant branches of the tree scaled by the mutation rate ($N_e^D \mu$) with 500 base pair loci. See figure 10.2 for explanation.

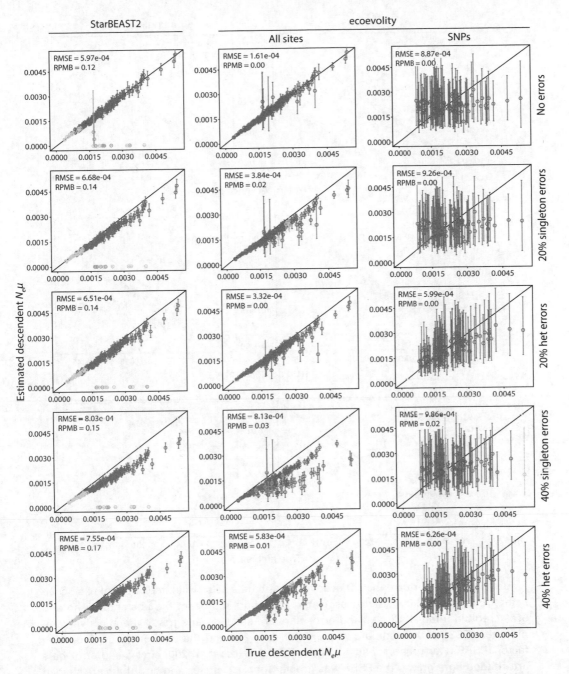

Figure 10.10. Accuracy and precision of estimates of effective population sizes of the descendant branches of the tree scaled by the mutation rate ($N_e^D \mu$) with 250 base pair loci. See figure 10.2 for explanation.

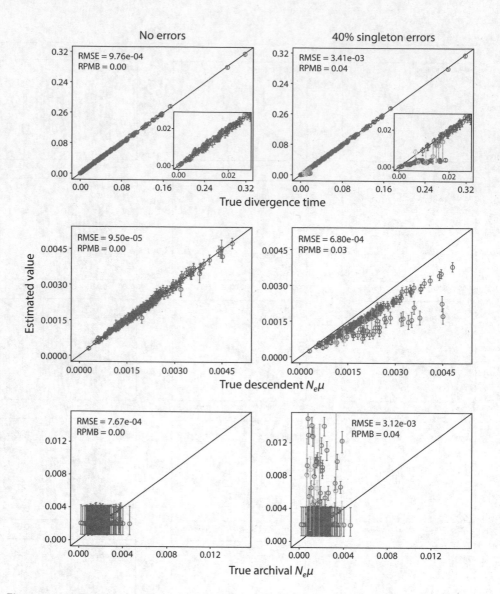

Figure 10.11. The performance of ecoevolity with data sets simulated with unlinked characters. Each plotted circle and associated error bars represent the posterior mean and 95% credible interval. Circles and error bars are shaded dark grey if the effective sample sizes (ESS) of the estimate was greater than 200 and the potential scale reduction factor (PSRF) was greater than 1.2. If the ESS was less than 200, circles and error bars are shaded light grey. If the PSRF was greater than 1.2, circles and error bars are shaded intermediate grey. The RMSE and RPMB is given for each plot, the latter of which is the proportion of estimates with ESS < 200 or PSRF > 1.2. See figure 10.2 for abbreviations. Inset plots magnify estimates of most recent divergence times. We generated the plots with matplotlib version 3.1.1 [335].

some guidance for these decisions. As we predicted, the linked-character method we tested, StarBEAST2, was more robust to the sequencing errors we simulated than was the unlinked-character method, ecoevolity. However, even with only two species in our simulations, the current computational limitations of linked-character models was apparent from the poor sampling efficiency of the MCMC chains, especially with shorter loci. For data sets with more species and many short loci, linked-character models are theoretically appealing, but current implementations may not be computationally feasible. The unlinked-character method, ecoevolity, was more sensitive to sequence errors but was still quite robust to realistic levels of errors and is more computationally feasible thanks to the analytical integration over gene trees.

Overall, for data sets with relatively long loci, as is common with sequence-capture approaches, it might be worth trying a linked-character method. If computationally practical, one stands to benefit from the aggregate information about each gene tree contained in the linked sites of each locus. However, if one's loci are shorter, as in restriction-site-associated DNA (RAD) markers, one is likely better off applying an unlinked-character model to all of one's data, even though this violates an assumption of the model. Below we discuss why performance differs between methods, locus lengths, and degree of error in the data and what this means for the analyses of empirical data.

10.4.1 ROBUSTNESS TO CHARACTER-PATTERN ERRORS

As predicted, the linked-character model of StarBEAST2 was more robust to erroneous character patterns in the alignments than was the unlinked-character model of ecoevolity. This is most evident in the estimates of divergence times, for which the two methods perform very similarly when there are no errors in the data (row 1 of figures 10.2–10.4). When errors are introduced, the divergence time estimates of StarBEAST2 are unaffected, but ecoevolity underestimates recent divergence times as both singleton and heterozygosity errors become more frequent (rows 2–5 of figures 10.2–10.4). However, ecoevolity divergence time estimates are only biased at very recent divergence times, and the effect disappears when the time of divergence is larger than about $8N_e\mu$.

These patterns make sense given that both types of errors we simulated reduce variation *within* each species. Thus, it is not too surprising that the unlinked-character model in ecoevolity struggles when there is shared variation between the two populations (i.e., most gene trees have more than two lineages that coalesce in the ancestral population). The erroneous character patterns mislead both models that the effective size of the descendant branches is smaller than they really are (figures 10.8–10.10). To explain the shared variation between the species (i.e., deep coalescences) when underestimating the descendant population sizes, the unlinked-character model of ecoevolity simultaneously reduces the divergence time and increases the effective size of the ancestral population. Despite also being misled about the size of the descendant populations (figures 10.8–10.10), the linked-character model of StarBEAST2 seems to benefit from more information about the general shape of each gene tree across the linked sites and can still maintain an accurate estimate of the divergence time (figures 10.2–10.4) and ancestral population size (figures 10.5–10.7).

This downward-biased variation within each species becomes less of a problem for the unlinked-character model as the divergence time gets larger, likely because the average gene tree has only a single lineage from each species that coalesces in the ancestral population. As the coalesced lineage within each species leading back to the ancestral population becomes a large proportion of the overall length of the average gene tree, the proportion of characters that either show fixed differences between the species or are invariant likely provides enough information to the unlinked-character model about the time of divergence to overcome the downward-biased estimates of the descendant population sizes.

From the ecoevolity results, we also see that when faced with heterozygosity errors, accuracy decreases as locus length increases. In contrast, accuracy of ecoevolity is not affected by locus length when analyzing data sets with singleton errors. This pattern makes sense in light of how we generated these errors. We introduced singleton errors per site and heterozygosity errors per locus. Thus, the same per-locus rate of heterzygosity errors affects many more sites of a data set with 1000 bp loci compared with a data set with 250 bp loci.

Unsurprisingly, the MCMC sampling performance of StarBEAST2 declines with decreasing locus length. There is less information in the shorter loci about ancestry and thus more posterior uncertainty about the gene trees. This forces StarBEAST2 to traverse a much broader distribution of gene trees during MCMC sampling, which is difficult due to the constraints imposed by the species tree. This decline in MCMC performance in StarBEAST2 does not appear to correlate with poor parameter estimates, and the distribution of estimates is generally as good as or better than those from eco-evolity. However, this might be because there is no uncertainty in the species tree in any of our analyses because there are only two species. As the number of species increases, it seems likely that the MCMC performance will further decline and start to affect parameter and topology estimates.

10.4.2 RELEVANCE TO EMPIRICAL DATA SETS

It is reassuring to see the effect of sequence errors on the unlinked-character model is limited to a small region of parameter space and is severe only when the frequency of errors in the data is large. Our simulated error rate of 40% is likely higher than the rate that these types of errors occur during most sample preparation, high-throughput sequencing, and bioinformatic processing. However, empirical alignments likely contain a mix of different sources of errors and biases from various steps in the data collection process. Also, real data are not generated under a known model with no prior misspecification. Violations of the model might make these methods of species tree inference more sensitive to lower rates of error.

The degree to which a data set will be affected by errors from missing heterozygote haplotypes and missing singletons will be highly dependent on the method used to reduce representation of the genome, depth of sequencing coverage (i.e., the number of overlapping sequence reads at a locus), and how the data are processed. To filter out sequencing errors, most pipelines for processing sequence reads set a minimum coverage threshold for variants or a minimum minor allele frequency. This filtering can result in the miscalling or removal of true variation, especially if coverage is low due to random chance or biases in PCR amplification and sequencing. Processing the data in this way can result in biased estimates of parameters that are sensitive to the frequencies of rare alleles [318, 431]. If the thresholds for such processing steps are stringent, these thresholds could introduce levels of error greater than our simulations.

10.4.3 RECOMMENDATIONS FOR USING
UNLINKED-CHARACTER MODELS

When erroneous character patterns cause ecoevolity to underestimate the divergence time it also inflates the effective population size of the ancestral population. We are seeing values of $N_e^R \mu$ consistent with an average sequence divergence between individuals *within* the ancestral population of 3%, which is almost an order of magnitude larger than our prior mean expectation (0.4%). Thus, looking for unrealistically large population sizes estimated for internal branches of the phylogeny might provide an indication that the unlinked-character model is not explaining the data well. However, there is little information in the data about the effective population sizes along ancestral branches, so the parameter that might indicate a problem is going to have very large credible intervals. Nonetheless, many of the posterior estimates of the ancestral population size from our data sets simulated with character-pattern errors are well beyond the prior distribution.

Whether using linked- or unlinked-character models with empirical high-throughput data sets, it is good practice to perform analyses on different versions of the aligned data that are assembled under different coverage thresholds for variants or alleles. Variation of estimates derived from different assemblies of the data might indicate that the model is sensitive to the errors or acquisition biases in the alignments. This sensitivity is especially true for data for which sequence coverage is low for samples and/or loci. Given our findings, it might be helpful to compare the estimates of the effective population sizes along internal branches of the tree. Seeing unrealistically large estimates for some assemblies of the data might indicate that the model is being biased by errors or acquisition biases present in the character patterns.

Consistent with what has been shown in previous work [545, 546], ecoevolity performed better when all sites were utilized despite violating the assumption that all sites are unlinked. This suggests that investigators might obtain better estimates by analyzing all their data under unlinked-character models rather than discarding much of it to avoid violating an assumption of the model. Given that the model of unlinked characters implemented in ecoevolity does not use information about linkage among sites [98, 545], it is not surprising that this model violation does not introduce a bias. Linkage among sites does not change the gene trees and site patterns that are expected under the model, but it does reduce the variance of the those patterns owing to them evolving along fewer gene trees. As a result, the accuracy of the parameter estimates is not affected by the linkage among sites within loci, but the credible intervals become too narrow as the length of loci increase [545, 546]. However, it remains to be seen whether the robustness of the model's accuracy to linked sites holds true for larger species trees.

10.4.4 OTHER COMPLEXITIES OF EMPIRICAL DATA
IN NEED OF EXPLORATION

Our goal was to compare the theoretical performance of linked and unlinked character models, not their current software implementations. Accordingly, to minimize differences in performance that are due to differences in algorithms for exploring the space of gene and species trees, we restricted our simulations to a two-species model and a small number of individuals. Nonetheless, exploring how character-pattern errors and biases affect the inference of larger species trees would be informative. The species tree topology is usually a parameter of great interest to biologists, so it would be interesting to know whether the linked model continues to be more robust to errors than the unlinked model as the number of species increases. We saw the MCMC performance of Star-BEAST2 decline concomitantly with locus length in our simulations because of greater uncertainty in gene trees. Given that data sets frequently contain loci shorter than 250 bp, it is important to know whether good sampling of the posterior of linked-character models becomes prohibitive for larger trees. Also, ecoevolity greatly overestimated the effective size of the ancestral population in the face of high rates of errors in the data. Exploring larger trees will also determine whether this behavior is limited to the root population or is a potential problem for all internal branches of the species tree.

Exploring other types of errors and biases would also be informative. To generate alignments of orthologous loci from high-throughput data, sequences are matched to a similar portion of a reference sequence or clustered together based on similarity. To avoid aligning paralogous sequences it is necessary to establish a minimum level of similarity for establishing orthology between sequences. This can lead to an acquisition bias due to the exclusion of more variable loci or alleles from the alignment [318]. Furthermore, when a reference sequence is used, this data filtering will not be random with respect to the species, but rather there will be a bias toward filtering loci and alleles with greater sequence divergence from the reference. Simulations exploring the affect of these types of data acquisition biases would complement the errors we explored here.

In our analyses, there was no model misspecification other than the introduced errors (except for the linked sites violating the unlinked-character model). With empirical data, there are likely many model violations, and our prior distributions will never match the distributions that generated the data. Introducing other model violations and misspecified prior distributions would thus help to better understand how species tree models behave on real data sets. Of particular concern is whether misspecified priors will amplify the effect of character-pattern errors or biases.

We found that character-pattern errors that remove variation from within species can cause unlinked-character models to underestimate divergence times and overestimate ancestral population sizes in order to explain shared variation among species. This raises the question of whether we can explicitly model and correct for these types of data collection errors in order to avoid biased parameter estimates. An approach that could integrate over uncertainty in the frequency of these types of missing-allele errors would be particularly appealing.

PART III

Beyond the Species Tree

With many estimation techniques available for inferring species phylogenies, such analyses are now routinely carried out for species that span all of life. However, species trees are often a starting point, rather than the focus, in various endeavors to understand major principles and test hypotheses about the evolutionary forces at work within species. The diverse chapters included in this part of the book are intended to highlight how species trees can be used as a basis for studying such processes and for untangling other aspects affecting the evolutionary trajectory of species.

Chapter 11, written by Digiacomo, Cloutier, Grayson, Sackton, and Edwards, demonstrates the utility of comparative genomic study within a species tree framework and the interplay of the challenges associated with these two efforts. The authors begin by describing the process of inferring a species tree for a group as large and diverse as birds, focusing in particular on the paleognathous birds and the challenges this group poses. By contrasting summary methods for species tree estimation with a phylogeny for the birds inferred using concatenated genomic data, the authors find a common observation of strong branch support for the latter estimate based on the standard bootstrap compared with the weak support of more appropriate methods of bootstrapping based on phylogenomic subsampling. Furthermore, despite the loss of information associated with summary methods, and despite gene tree estimation error, the authors demonstrate that the estimated phylogenetic relationships from the species tree are confirmed by analysis of rare genomic changes, such as patterns of CR1 retroelement insertions. In the second half of the chapter the authors use the inferred species tree as a foundation for downstream comparative genomics analyses. However, nontrivial challenges had to be addressed to avoid misleading inferences. Specifically, the authors address the assessment of uncertainty in whole-genome alignments and its effect on species tree estimation. With the goal of identifying genes with potentially functional roles, they use a comparative framework that accommodates the differences in the gene trees across loci—that is, a species tree framework—rather than pooling the genomic data across the genome (i.e., concatenating the data) to infer a single tree. The authors then describe a procedure for detecting variation in the evolutionary rate of divergence across genes to identify regions of the genome with putative functional roles, as well as an analysis of transcriptomics and epigenetics as quantitative traits, all of which rely on a well-supported phylogeny as the basis for analysis. Overall, the chapter provides an excellent example of the importance of the estimated species tree in downstream analyses with clear evolutionary significance for understanding the loss of flight by drawing on lessons learned from the detailed analysis of a clade of paleognathous birds.

The theme of studying phylogenetic heterogeneity continues in chapter 12, written by Pease and Weinheimer, which considers the importance of heterogeneity and discordance in making conclusions about the underlying evolutionary processes by drawing on both philosophical ideas and empirical observations. Beginning from a historical perspective on discordance, the authors consider three interrelated questions. The first addresses what a species tree represents conceptually in the context of heterogeneity and discordance. The second focuses on methods that quantify, analyze, and visualize discordance in empirical settings. Finally, the authors consider methodology that can be used to learn about phylogenetic relationships for a collection of species in the presence of discordance. This last section separately examines several common sources of conflict, such as introgression and hybridization, selection, trait evolution, and coevolution. Empirical examples are used to illustrate these processes as well as to highlight the important caveats that must be considered in analyses of genomic data with various approaches to contend with such processes. The chapter as a whole provides a careful examination of how to reconcile phylogenetic discord, and uncertainty, with the goal of describing evolutionary relatedness as a framework for studying variation among species.

In chapter 13, McKenzie and Eaton consider an aspect of genome-scale phylogenetic inference that is often overlooked, namely the correlations among gene genealogies that arise because of spatial proximity of genes along a chromosome. The authors note that current approaches tend to ignore such correlations, instead focusing on one of two extremes: either genealogies are viewed as independent, unlinked phylogenies sampled from a species-level phylogeny or no such species phylogeny is assumed and genealogical variation is measured in sliding windows along a chromosome with no species-level constraints imposed. To examine the more realistic situation of local correlation among genealogies subject to species-level constraints, the authors use simulation to explore the effects of distance between genealogies, effective population sizes, and the generation time represented by branch lengths in the underlying species trees, leading to some interesting observations. The first is that with a larger number of generations along a species tree branch, the size of nonrecombined blocks is smaller. This observation has obvious implications for many of the methods described in part 1 of this volume for inferring species trees from multilocus data under the multispecies coalescent. They also show that similarity of trees for a sample of linked versus unlinked genes depends upon tree size. The authors go on to propose a method that applies a weighting scheme to improve local gene tree inference, showing that models that incorporate spatial characteristics of sample genes hold a lot of promise for improving phylogenetic inference overall.

The final chapter of this section (and the book) is a contribution by Brown, Mount, Gallivan, and Wilgenbusch focused on the important issue of visualizing and describing sets of trees. In addition to describing a set of techniques for exploratory analyses based on techniques for nonlinear dimension reduction, as implemented in the TreeScraper software package, the chapter introduces the display of relationships both among trees and among bipartitions within trees using networks. The chapter also describes applications for detecting communities of trees within these networks. The authors motivate this discussion by emphasizing the importance of developing methodology to enable exploratory analyses of sets of trees, particularly as both the magnitude and complexity of data sets available for analyses at this scale continues to grow at a rapid pace. They highlight this point by reference to studying the relationships among sets of gene trees and the species tree in phylogenomic data, especially with respect to effects related to the substitution model applied in a phylogenomic study, to the assumption of a common underlying species tree when estimating gene trees (as opposed to the lack of constraint imposed by a common species tree), as well as to differences in genetic variation across genomes.

Taken together, these four chapters exemplify the breadth of study topics that are beyond a focus on the species tree itself but that represent the varying possibilities that are enabled by species-level phylogenomic frameworks. That is, it is the inferred species tree that forms the foundation of downstream functional analysis, as in chapter 11, that allows for development of techniques for quantifying and studying discord in the evolutionary history of different genes or sets of genes, as in chapters 12 and 14, and that enables incorporation of spatial relationships at the chromosomal level into phylogenomic inference, as in chapter 13. Species tree inference, beyond the goal of estimating the species tree, has even more diverse applications than covered here, and in the future, the purview to which it is applied will no doubt expand in new directions. For example, whether species trees or gene trees, or phylogenetic networks [50], provide the proper framework for comparative analyses and related tree-based analyses has already become a subject of debate, as has the subject of the multispecies coalescent [601] for inferences of species boundaries [737, 305]. Advances in coalescent-based models for applications to serially collected samples, such as those involved in studying viral evolution, will no doubt continue to expand (see, e.g., [39]). An additional area for future development is the incorporation of other processes leading to phylogenetic discord, such as gene duplication and loss, into models for species-level phylogenetic inference. Current models (e.g., [604]) primarily assume a fixed species tree and seek to estimate gene-level relationships as a means of elucidating the history of duplication and loss within a genome (but see, e.g., [88, 514]), though clearly these processes are informative about the species-level phylogeny as well. Last, as phylogenetic methods are being increasingly applied in settings in which evolution is occurring at different scales, such as the study of the progression of cancer within a single individuals (e.g., [847]), such applications will push methodological developments that include models and algorithms for capturing the underlying processes involved in such histories, as applications of multilocus data once spurred the development of species tree inference itself to capture the biological process underlying the divergence history of species across the tree of life.

CHAPTER 11

The Unfinished Synthesis of Comparative Genomics and Phylogenetics: Examples from Flightless Birds

Alexandria A. DiGiacomo, Alison Cloutier, Phil Grayson, Timothy B. Sackton, and Scott V. Edwards

11.1 Introduction

A major aim in evolutionary biology is to understand the evolution of complex traits, and currently there is increasing interest in understanding the contribution of regulatory evolution to phenotypic change. A major hypothesis of modern genomics is that the noncoding portion of genomes—constituting the vast majority of DNA of eukaryotic organisms—is a vast regulatory network [479, 478]. Deciphering this network is one of the signature goals of diverse research programs and major international consortia such as the ENCODE project [527, 751, 313]. Although genomic regulatory landscapes are invariably complex, ongoing advances in sequencing technologies, statistical modeling, and epigenetic profiling are facilitating our ability to identify regulatory elements such as enhancers and understand the effect of regulatory change on phenotypic evolution [119, 127, 643].

These technologies, applied in the context of a robust phylogeny, open the door to studying the genotype-to-phenotype map even in nonmodel species for which population analyses or controlled breeding are not possible—the emerging PhyloG2P paradigm [703]. A robust phylogeny is crucial to detecting and analyzing when traits and genomic features change in tandem and is particularly powerful in instances of convergent evolution [302, 466, 669]. Phylogenetic relationships form the foundation of how we think about differences between organisms and how we analyze those differences in a statistically sound manner [242]. Yet, as the integrated use of genomic data, phylogenetics, and functional genomics has become more common, full integration of the phylogenetic perspective is lacking in many bioinformatics tools central to the PhyloG2P enterprise. As we aim to show in this chapter, the integration of phylogeny and coalescent theory in many downstream analyses in comparative genomics is still a work in progress.

There are now a number of examples illustrating the promise of the integration of comparative and functional genomics (reviewed in [703]). The successes of these fields have emerged despite the fact that many of the models and protocols employed to understand links between genotype and phenotype still ignore fundamental frameworks such as phylogenetic relationships or fundamental processes such as incomplete lineage sorting. These fields have made great strides in recent years to incorporate a phylogenetic perspective into genome analysis, yet they lag far behind, for example, phylogenetic and comparative methods themselves, which have for years acknowledged the importance of phylogeny and coalescent processes [217, 370, 823, 602] in phylogenetic analysis of genetic and phenotypic evolution. Sources of uncertainty are generally not propagated appropriately through the ensemble of bioinformatic protocols that compose a typical analysis pipeline [204, 282]. However, researchers continue to seek ways to incorporate biological realism into comparative genomics and thereby inform downstream functional genomics experiments. In this chapter we review our recent work on paleognathous birds and present it as a test case for how phylogenetics and coalescent methods can set the stage for downstream functional analyses designed to decipher phenotypic novelty. We also illustrate how many methods in comparative genomics, including many of the fundamental tools for comparing and analyzing whole genomes, have yet to embrace the complexity of phylogenetic and coalescent variation.

11.1.1 PHYLOGENETICS OF MODERN BIRDS

With greater than 10,000 species, modern birds are rivaled only by fishes and nonavian reptiles as the most diverse extant vertebrate clade [231, 48, 758]. Their impressive range of morphological, behavioral, ecological, and locomotory diversity has captured the interests of evolutionary biologists since Darwin [400]. Furthermore, convergent phenotypes across distantly related avian lineages are common, as seen in examples such as the specialized flipper-like wings of great auks and penguins; the increased body size of many flightless species; and the similar body plans and beak morphologies of the diurnal birds of prey, such as hawks and falcons, which now comprise unrelated phylogenetic lineages [296, 226, 690, 643]. The identification and study of convergent traits inherently relies upon robust hypotheses of phylogenetic relationships [635, 450, 4, 401]. Although resolving the phylogeny of birds has garnered much recent attention, and massive data sets have now been applied to its resolution, significant debate still clouds many regions of the avian tree, particularly among the deeper, interordinal branches [287, 483, 343, 589, 728]. The three major lineages of birds registered by the divergence of the Palaeognathae and Neognathae, and the subsequent divergence within the Neognathae of the Galloanserae and Neoaves, have been consistently well supported by numerous pre-genomic and genomic studies [150]. However, relationships at many taxonomic levels, particularly within the Neognathae, have proven difficult to resolve.

Early branching patterns within the major avian lineages, such as the paleognaths, have been especially contentious, in part due to the burst of short branches early at the base of the tree and subsequent series of long branches leading to modern lineages [796, 288]. One suggestion for why these deep branches are so challenging to resolve is that these branches are short and likely are accompanied by substantial gene tree variation. Such short branches would be difficult to resolve even if incomplete lineage

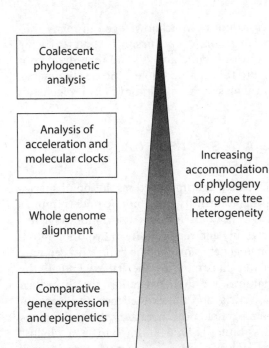

Coalescent phylogenetic analysis

Analysis of acceleration and molecular clocks

Whole genome alignment

Comparative gene expression and epigenetics

Increasing accommodation of phylogeny and gene tree heterogeneity

Figure 11.1. Schematic showing how different aspects of the phylogenomics pipeline have accommodated phylogenetic analysis and gene tree heterogeneity. Four different components of the pipeline (gene expression, whole-genome alignment, analysis of acceleration and rates, and phylogenetic analysis) have incorporated a phylogenetic perspective and gene tree heterogeneity to an increasing extent from bottom to top.

sorting (ILS) were not an issue because such branches present few opportunities for phylogenetically informative molecular changes to accumulate between divergences. When such short branches in the species tree are combined with the compounding effects of short and varying branches in gene trees, the challenges for resolution become exacerbated. Finally, additional challenges arise when such deep, rapid divergences are followed by long branches, which can result in undetected saturation of some sequence sites [239, 796, 288]. In the avian tree, many of these terminal long branches will no doubt be subdivided through additional taxon sampling, but so far, the structure of the avian tree mirrors those trees often used in simulation for the most challenging test cases of phylogenetic inference [581, 562, 437].

Our experience of analyzing the paleognath data set to probe for genetic signatures of loss of flight laid bare the many discrepancies in logic between phylogenetic analysis on the one hand and downstream analyses relying on this tree on the other. Whereas phylogenetic analysis has clearly embraced issues forced upon it by the multispecies coalescent model, most downstream applications either ignore this fundamental new perspective or accommodate it only partially (figure 11.1). In this chapter we outline these discrepancies and ask what sorts of consequences this shortfall may have for our analyses of genome evolution and comparative genomics. We find that some types of comparative genomics methods, such as simple mapping of read data to a reference genome and comparative analyses of transcriptome and ATAC-seq data, have few tools tailored to accommodate phylogenetic relationships of species as a framework (but see [205, 628, 206]). Our experience of analyzing the paleognath data sets thus convinced us that the comparative genomics revolution is still an unfinished synthesis. Although issues of gene tree heterogeneity across the genome have been appreciated for some time (e.g., [582]), this awareness has in some cases not been acknowledged in many routine aspects of phylogenomic and comparative genomic pipelines. After first reviewing the paleognathous birds as a test case for phylogenomic analysis, we then divide our downstream functional genomics analyses into three parts: whole-genome alignments, detection of rate variation on gene trees, and analysis of transcriptomics and epigenetics as quantitative traits.

11.1.2 PALEOGNATHOUS BIRDS AS A TEST CASE FOR POST-GENOMIC PHYLOGENETICS

The increased availability of genome-scale data and improved resolution of species trees have proved invaluable to studies seeking to identify the genomic underpinnings of convergent traits. Such studies yield insight not only into the association of genetics and phenotypes but also into the predictability and repeatability of evolution [401]. The combined application of multilocus phylogenetics on the one hand, and genomic and developmental tools on the other, is well illustrated by the Palaeognathae, the clade of birds consisting of the volant tinamous and flightless ratites, which in turn include ostriches, emus, cassowaries, rheas, kiwis, and the extinct moa and elephant birds [288, 145, 643]. Early studies based on morphological characters placed tinamous as sister to a monophyletic clade of flightless ratites [149, 643], implying a single loss of flight in the common ancestor of ratites. However, more recent phylogenetic work based on molecular data has recovered the ratites to be paraphyletic, implying up to six independent losses of flight within the clade [292, 40, 509, 533, 276, 839, 643]. New phylogenetic analyses based on large numbers of noncoding elements have yielded new insight into paleognath relationships [145] and have laid the foundation for functional genomic work investigating the genomic basis for convergent phenotypes in this group [643]. Our studies in paleognaths illustrate the promise of genome-scale analyses of phylogenetically challenging clades and also provide useful insight into how comparative and developmental genomics can be improved by acknowledging heterogeneous signals in the data, especially those stemming from gene tree variation.

11.2 Building of a Whole-Genome Species Tree for an Ancient Radiation of Birds

Recent molecular studies have clarified some aspects of paleognath phylogeny, perhaps most notably in recovering the extinct, wingless moa as sister to the volant, or flighted, tinamous embedded within a paraphyletic ratite clade [576, 288, 40]. However, other relationships were poorly resolved, in particular the placement of rheas, which in previous work are variously recovered as the sister to the remaining nonostrich taxa, to tinamous, or to a clade containing emu, cassowaries, and kiwi [292, 576, 288, 689, 509, 589]. These topological differences are accompanied by short internal branches deep in the species tree, which suggest a rapid ancient divergence among paleognath lineages. The small number of extant paleognath species prohibits additional sampling to subdivide long terminal branches but facilitates de novo whole-genome sequencing across the focal clade, which can increase phylogenetic confidence not only by employing many nuclear loci but also by allowing underlying patterns of sequence variation to be considered at a genome-wide scale. To that end, data sets were compiled from 10 new high-quality paleognath genomes [643] and a new reference-based nuclear genome assembly for the extinct little bush moa (*Anomalopteryx didiformis*,

[146]) alongside existing publicly released assemblies [854, 412]. The resulting data sets incorporate sequence from all extant ratite genera, as well as four representatives of the more species-rich tinamous.

Because a robust phylogeny is necessary to further downstream analyses of genome evolution, comparative genomics, and functional genomics, these data were used to estimate the paleognath species tree and assess its accuracy. The species tree was estimated using both coalescent and concatenation methods. The results were further assessed using a combination of rare genomic changes to provide additional topological support, phylogenetic subsampling to analyze topological conflicts, and topology tests to determine if gene tree error or real signal was the root cause of observed gene tree heterogeneity.

The contrasting results of summary species tree versus concatenation methods can be seen in the example of the placement of rheas in the paleognath tree. We used three classes of noncoding nuclear markers to investigate paleognath relationships: conserved nonexonic elements (CNEEs, $N = 12{,}676$ loci), introns ($N = 5016$), and ultraconserved elements (UCEs, $N = 3158$; see Cloutier et al. [145] for details). Each marker panel was analyzed with two summary species tree methods, MP-EST [440] and ASTRAL-II [507], which take as input estimated gene trees for each locus (in [145] and [643]; inferred under maximum likelihood with RAxML, [720]). Maximum-likelihood topologies of partitioned concatenated sequence alignments were also inferred for each marker type using ExaML [379]. Many relationships are congruently recovered by the three methodologies, for instance relationships among kiwi species and among tinamous and the recovery of a paraphyletic ratite clade in which the earliest split occurs between ostrich and all other paleognaths and in which moa and tinamous form sister lineages within the non-ostrich clade (figure 11.2). MP-EST and ASTRAL species tree methods recover fully congruent topologies across all data sets and place the rheas as sister to an emu/cassowary + kiwi clade (figure 11.2a). In contrast, analysis of concatenated alignments with ExaML instead places the rheas sister to the other non-ostrich paleognaths for all marker types and additionally recovers emu + cassowary as sister to moa + tinamous for CNEEs and UCEs (figure 11.2b, c).

Observed topological conflicts between the two species tree methods and analysis of concatenated supermatrices are accompanied by moderate to high bootstrap support. However, there is a growing awareness that traditional bootstrap resampling procedures might provide an inadequate measure of confidence in topologies recovered from large phylogenomic data sets (e.g. [220, 657, 262]). Phylogenomic subsampling offers an alternative approach to characterize the underlying support for conflicting clades recovered by different data sets or methodologies. This approach employs a variation of double bootstrapping [668] in which input data is resampled with replacement both at the level of entire genes and also across sites within genes. In phylogenomic subsampling, this double bootstrapping procedure is applied to replicate data sets of increasing numbers of loci (for example, 10 replicates each of 100 loci, 200 loci, etc.). Recovered clades that are robustly supported by the underlying data are expected to produce consistent results for random subsamples of loci, and bootstrap support is expected to rise with an increasing number of analyzed loci [220]).

The effects of subsampling are clearly illustrated with respect to estimating the sister group to rheas (figure 11.3; see [145] for full subsampling results). For example, replicates analyzed with MP-EST or ASTRAL consistently recover rheas as sister to emu/cassowary + kiwi, with support for alternative topologies declining sharply for moderately sized replicates (figure 11.3a, b). In contrast, the same data analyzed under

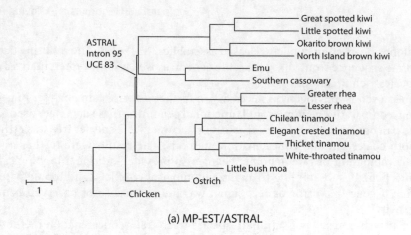

(a) MP-EST/ASTRAL

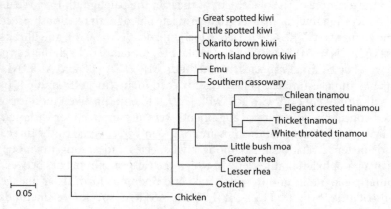

(b) ExaML-introns

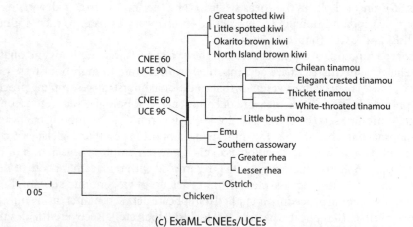

(c) ExaML-CNEEs/UCEs

Figure 11.2. Species trees inferred using (a) MP-EST or ASTRAL for all marker types; (b) ExaML on concatenated data for introns; and (c) ExaML for conserved nonexonic elements (CNNEs) or ultraconserved elements (UCEs). Bootstrap supports are indicated for clades with <100% support. Branch lengths in (a) are drawn in coalescent units, with values shown for the MP-EST total evidence tree combining all loci. Terminal branches are uninformative and drawn as a constant value. Branch lengths in (b) and (c) are in units of substitutions per site and are shown for the ExaML UCE tree in (c).

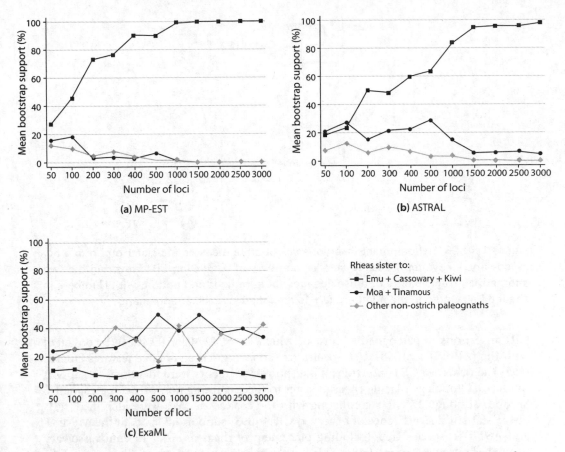

Figure 11.3. Phylogenomic subsampling showing support for alternative hypotheses for the sister group to rheas from (a) MP-EST (b) ASTRAL, and (c) ExaML. Plots show the average bootstrap support from 10 replicates of randomly sampled loci within each size category.

concatenation with ExaML exhibit unstable behavior that alternates between recovering rheas as sister to moa + tinamous or to all other non-ostrich paleognaths, although with low bootstrap support (figure 11.3c). This inconsistent behavior indicates substantial conflict in the underlying data that is not accommodated within a concatenation framework.

An alternative approach to resolving conflicting topologies estimated from sequence-based data involves analysis of rare genomic changes, such as patterns of CR1 retroelement insertions. Retroelement insertions are usually scored as binary presence/absence characters. As a result, they may have a fairly simple underlying model, although many of the complexities of retroelement evolution, such as sudden bursts of element transposition, have not been incorporated into models describing their evolution. Retroelements are also largely homoplasy-free characters [606], although, assuming their neutrality, they are equally subject to ILS as any other genomic regions. Analysis of retroelements can therefore serve the dual purpose of corroborating inferred relationships while also highlighting regions of the species tree where extensive ILS is likely, as demonstrated in Cloutier et al.'s [145] identification of 4301 informative

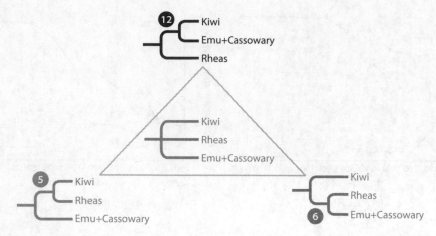

Figure 11.4. CR1 retroelement insertions corroborate rheas as the sister group to emu/cassowary + kiwi (top), despite smaller numbers of conflicting CR1s favoring the two alternative arrangements of these lineages (bottom right and bottom left). Numbers in shaded balls indicate the number of CR1s identified for each arrangement.

CR1 insertions in paleognaths, most of which (N = 4274, or 99.4%) are consistent with the MP-EST and ASTRAL species tree topology. Three notable features emerge from the observed CR1 insertions. First, no retroelements were identified to support the ExaML topology placing rheas as sister to the remaining non-ostrich paleognaths. Second, although 27 CR1s conflicting with the coalescent trees were identified, tests of the statistical significance of observed CR1s [395] support all splits in the MP-EST and ASTRAL species tree, including placement of rheas as sister to emu/cassowary + kiwi (figure 11.4, p = 0.048). Third, although the placement of rheas is corroborated, conflicting CR1 insertions among the rheas, emu + cassowary, and kiwi lineages indicate differing evolutionary histories in this region of the species tree (figure 11.4). The roughly equal numbers of conflicting CR1s for the two alternative topologies further implicate ILS as the underlying process generating heterogeneity [193, 395, 186].

Phylogenomic subsampling and CR1 retroelement insertions thus both support the MP-EST/ASTRAL species tree topology. Conflicting CR1s additionally suggest substantial ILS, which can produce an erroneous species tree topology in sequence-based analyses if loci are modeled within a concatenation framework that does not accommodate this variation in gene tree histories. Estimated gene trees indeed showed not only a large amount of topological heterogeneity primarily involving rearrangements of the short internodes along the backbone of the species tree but moreover that the most common gene tree topology does not match the species tree—an empirical anomaly zone [145]. Coalescent branch lengths for both MP-EST and ASTRAL species trees for each marker type were also consistent with values expected to generate an anomaly zone across the successive short internodes forming the common ancestor of emu/cassowary and kiwi, and with this clade to rheas [176, 145].

A much-debated issue in the current literature is the extent to which gene tree estimation error (GTEE) affects the performance of summary species tree methods and, by extension, the inference of an empirical anomaly zone [321, 624, 818, 823]. Simulations show high precision for both MP-EST and ASTRAL in estimating the true species tree even in the presence of uninformative or poorly estimated gene trees as long as enough loci are used and estimation error is unbiased [818, 504, 823]. The use of topology tests on gene trees to ask whether gene tree error could be responsible for the gene tree heterogeneity is demonstrated in the paleognath data set. We found that, depending on the test, up to 80% of gene trees could reject the topology of the coalescent species tree, suggesting that much of the heterogeneity in gene tree topologies and departure from the species tree we observed was, in fact, real signal [145]. An issue distinct from but related to the issue of GTEE is branch length error in the species tree. It has been suggested that estimated branch lengths of species trees may be more affected by GTEE than the species tree topology itself [657]. On this basis, Gatesy et al. (2019) [262] recently argued that the proportion of observed gene tree heterogeneity attributed to ILS would be inflated in coalescent simulations owing to underestimated coalescent branch lengths in the species tree from which gene trees are simulated. We therefore reassessed the paleognath data by generating 10 replicates of 1000 randomly chosen loci for each of the three marker panels, without site-wise bootstrapping [683], and submitting each replicate to the same MP-EST inference originally used for all loci. Because several authors (e.g., [657, 683]) express reservations about using bootstrap replicate gene trees as input to species tree methods, we here present results from "optimal" gene trees (gene-wise resampling of Simmons et al. [683]): gene trees inferred under maximum likelihood from the full input alignment without bootstrapping (complete methods, scripts, and input gene trees can be found in the Dryad data release accompanying [145], and raw results for the replicates discussed here are available from the authors upon request). It should be noted the practice of gene-wise resampling recommended by [683] was already in practice by several groups working on large data sets, driven primarily by its reduced computational time and the knowledge that gene-wise resampling is sufficient for data sets with many loci (e.g., [343]).

The application of gene-wise resampling to measure branch lengths in the species tree and to assess the proportion of GTEE due to ILS can be seen in the paleognaths for which all replicates recovered the same species tree topology obtained using MP-EST or ASTRAL on the full set of loci. Decomposing gene trees into sets of rooted triplets for the emu + cassowary, kiwi, and rhea lineages also demonstrates remarkably consistent proportions of gene trees recovering the major, or species tree, topology and the two minor, or alternative, topologies across random samplings of input loci and across marker types (figure 11.5a). We would likely not expect such consistency, or for results to so closely mirror those from retroelement insertions, if observed gene tree heterogeneity predominantly arose through estimation error. Instead, with a sufficiently large number of loci, empirical gene trees appear to provide an adequate and unbiased representation of the true gene tree distribution in this case. Despite gene tree estimation error that is almost certain to exist in large phylogenomic data sets, the symmetry and relatively high frequency of the two alternative minor topologies further support a clear and consistent signal of ILS across the two short internodes separating these lineages. Accordingly, coalescent branch lengths estimated from the full set of loci fall within the range recovered from smaller random samplings of input gene trees (figure 11.5b, c). MP-EST branch lengths for each replicate also fall within the range encompassing an empirical anomaly zone across this region of the species tree [176]. Taken together, these results indicate that extensive ILS has contributed to some historic difficulties in resolving paleognath relationships but that robustly supported results are obtained for this group when genome-level data is combined with species tree methods that account for coalescent variation in gene tree histories.

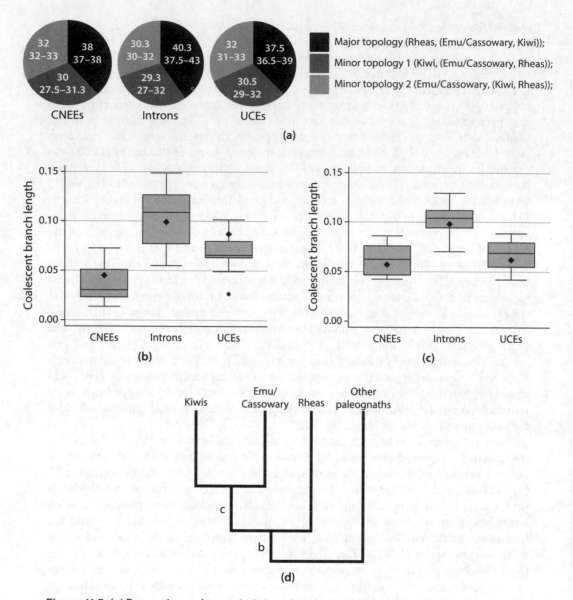

Figure 11.5. (a) Proportions of rooted triplets showing the major and two minor topologies in empirical gene trees for the rhea, emu + cassowary, and kiwi lineages. Large pie charts at top show values for the full set of loci in each marker panel, with the range in values found in 10 replicates of 1000 randomly chosen loci shown below. (b) and (c) Coalescent branch lengths for the full set of loci in each marker panel (diamond symbols) and from 10 replicates of 1000 randomly chosen loci for each marker type (boxplots). Values for (b) the common ancestor of rheas, emu, cassowary, and kiwi and for (c) the common ancestor of emu, cassowary, and kiwi are shown according to the (d) tree at right.

11.3 The Unfinished Synthesis of Comparative Genomics and Genomic Heterogeneity

11.3.1 A SPECIES TREE FOR PALEOGNATHOUS BIRDS AS A FOUNDATION FOR COMPARATIVE GENOMICS

Our species tree for paleognathous birds was a starting point for many downstream analyses of genome evolution and comparative genomics in this group. Building the tree for paleognaths caused us to grapple with many issues surrounding the application of coalescent methods to large data sets: was the gene tree heterogeneity we observed, and the evidence for an anomaly zone in the paleognath tree, real or due to gene tree error? Can summary ("two-step") coalescent methods, in which gene trees are estimated separately from the species tree, recover a reliable species tree? Would the estimated tree be different if we used full Bayesian methods, which cannot easily accommodate data sets as large as the paleognath data set? During this process we learned that, although phylogenetic analysis has been infused with such issues largely because of the introduction of coalescent methods to the field, many of the downstream applications of comparative genomics that use the estimated species tree have so far ignored these issues.

11.3.2 ACCOMMODATION OF UNCERTAINTY INTO WHOLE-GENOME ALIGNMENTS

Alignments are of course the bedrock of comparative genomics and represent the first crucial step in many downstream analyses of divergence and variation. For decades the phylogenetics community has proceeded largely on the assumption that the alignment for a given data set is fixed, despite that many other parameters in downstream analyses, such as models of substitution and trees and branch lengths, are estimated and uncertainties measured with varying degrees of success. Guang et al. [282] argued that uncertainty should be incorporated into all aspects of phylogenomic workflows, including alignments, and Howison et al. [316] and others have suggested ways to incorporate uncertainty into alignments by means of posterior distributions of a Bayesian analysis. Such methods, however, have been applied only to organisms such as viruses with very small genomes, and it is unclear how practical they will be when applied to eukaryotic genomes. Some software, such as BaliPhy [610] and PASTA [505], simultaneously estimate phylogeny and sequence alignment. The concept of pangenome graphs, which are built on reference-free alignment methods that attempt to capture both sequence and structural variation in genomes, are also an exciting new development.

New reference-free whole-genome alignment methods promise a new era of alignment that is much better suited to comparative genomics [563, 32]. Previous methods of aligning whole genomes, such as MULTIZ, first specify a reference species to which all other species in the analysis are aligned [75, 501]. Such methods have the undesirable property that any genomic regions or sequence not in the reference are discarded from further analysis. Additionally, it is widely appreciated that the alignment quality will degrade as the focal taxon is more removed phylogenetically from the reference, a trend also widely known in BLAST and other analyses requiring alignments at varying phylogenetic depths [188, 523, 187, 524, 695, 121, 525]. By using a phylogenetically based, reference-free whole-genome aligner (ProgressiveCactus, [32]), we were able to explore evolutionary patterns across the entirety of all the bird genomes in our analysis, including regions that were unique to particular lineages and not found in chicken (e.g., figure 11.6). (In practice, many of our analyses were still confined to regions present

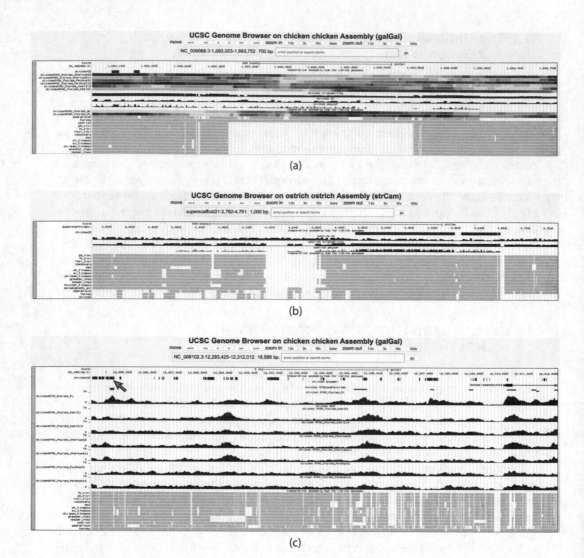

Figure 11.6. Screenshots of the ratite genome browser illustrating the power of phylogeny-aware whole-genome alignment [32]. (a) A genomic region uniquely absent in paleognaths at position 1,693,348–1,693,458 bp based on the chicken genome as reference sequence. From top to bottom, tracks show conserved elements (CE) in chicken; tracks with "ATAC" and "Pooled" indicate intensity of ATAC-seq peaks in various developing organs (PecMus = pectoral muscle; S signifies the Illumina sequencing index); next are alignability, GC content, and phyloP score, a measure of local sequence conservation (Pollard et al. [583]). Tracks at the bottom in light grey represent genomic alignments. The paleognath deletion is the large white space at the bottom. Chicken E stages for each ATAC track are as follows: SternumS2: E9, SternumS11: E10, PecMusS3: E9, PecMusS12: E10, KeelS10: E10,KeelS1: E9, chickenATAC_Pooled_HL: E4.5, chickenATAC_Pooled_FL: E4.5. (b) A genomic region uniquely present in in paleognaths at 8,825–3,868 bp based on ostrich as reference sequence. Tracks shown are a subset of those in panel (a), with GC, alignability, and phyloP based here on ostrich. (c) Visualization of quantitative ATAC-seq peaks in multiple organs, aligned with conserved nonexonic element mCE1623056 near the TBX5 gene and accelerated convergently in ratites (arrow) and in a region of open

in the chicken genome because we projected the ProgressiveCactus alignment on the chicken genome as a reference when calling CNEEs with phyloP [583]). Progressive-Cactus requires a guide tree on which it computes pairwise alignments among sister lineages, computing ancestral genomes in much the same way as the first generation of ancestral state methods are used to compute ancestral nucleotides and amino acids.

Although methods like ProgressiveCactus have a clear phylogenetic approach to whole-genome alignment, their ability to accommodate uncertainty in the phyloge-netic relationships of species is somewhat rudimentary, and there is little possibility of incorporating gene tree heterogeneity into the alignment process. For example, early in our analysis of paleognath genomes, before our phylogenetic analyses were begun, the position of rheas was uncertain among previously published analyses [292, 689, 40]. Therefore, in our alignment of paleognath genomes, we incorporated uncertainty in the phylogenetic position of rheas by inputting a polytomy in their placement on the guide tree. Though the impact of this uncertainty on downstream analyses is unclear, this approach is much better than previous reference-based methods in at least acknowl-edging some uncertainty in the guide tree. However, depending on how far one wants to push the gene tree paradigm, the tree for various segments of the genome is likely to vary on different scales and in many cases may vary from nucleotide to nucleotide if the underlying species history facilitates extensive ILS. Recombination is another factor that will cause different regions of the genome, and even adjacent nucleotides, to differ in their underlying history and branch lengths.

As useful as it is to play devil's advocate in such situations, it is unclear how important issues of ILS and recombination will be when attempting to align whole genomes, although very little work has examined this issue. The phylogenetic relation-ships of genes and species is central to identifying single copy orthologs, and many pipelines have been proposed for making this selection more rigorous [204, 18]. A single phylogeny is also required for inferring the locations of CNEEs in genomes, usu-ally done with a hidden-Markov model such as implemented in programs like phyloP [680, 583, 327]. For CNEE detection one requires a tree with a fixed topology and "neu-tral" branch lengths—usually based on fourfold degenerate sites of genes as a proxy that is also easily alignable across species. In this case the important metric is total branch length: for inference of conserved regions, phyloP assumes uniform scaling of the neu-tral rate across lineages, which again is likely violated extensively in genome-wide data. However, we suspect that the effect of ILS in such cases is minimal, although further research is required. Incomplete lineage sorting may prevent some CNEEs from being detected, if the wrong gene tree is used for a region and SPILS (substitutions produced by ILS, the phenomenon in which failure to incorporate gene tree heterogeneity can induce apparent increases in substitution rate in gene trees discordant with the species tree [495]); however, it should not create false positive CNEEs (inference of CNEE where one does not exist). Indeed, it is still unclear whether the biases incurred by ignoring gene tree heterogeneity are substantial enough to warrant extensive development of

Figure 11.6. (*continued*) chromatin in the developing forelimb. Chicken E stages are as in panel (a). The TBX5 gene is in the upper right labeled with its chicken Ensemble gene number. ATAC tracks are as in panel (a). The identification of lineage-specific genomic regions was made possible by using the phylogenetically aware whole-genome aligner ProgressiveCactus ([32]; see text).

ILS-aware alignments methods, especially when such methods are likely to be extremely slow and computationally intensive.

11.3.3 GENE TREE HETEROGENEITY AND DETECTING RATE VARIATION IN GENES AND NONCODING REGIONS

Tests of the molecular clock have a long history in molecular evolutionary studies, but even today, statistical tests for changes in evolutionary rate assume a single gene tree for all loci in the genome. We now know that gene tree heterogeneity around the genome could undercut traditional tests for rate changes if the tree used to test for rate changes (the species tree) is an incorrect description of the locus under study (the gene tree), SPILS as described above [495]. Many tests involving molecular clocks and estimation of divergence times have still not come to grips with gene tree heterogeneity. The only software we are aware of that can accommodate gene tree heterogeneity and estimate lineage- and gene-specific rates on different branches is *BEAST [549]. This is extremely powerful software that, as a Bayesian method, accommodates many sources of uncertainty appropriately, unlike so-called two-step species tree methods that can sometimes misallocate variance in the hierarchy of gene trees and species trees [823]. As of now, most methods for estimating divergence time based on relaxed and other clock models do so by working on concatenated data, assuming that a single gene tree describes the data set and that the gene coalescence time is synonymous with the species divergence time, which we have known for decades is usually not true [28, 194] (see [443] for a recent empirical example). Below, we describe how our paleognath data set uses rate variation among lineages to link potential candidate genomic regions to convergent genomic traits. However, as our example shows, estimating rate variation and molecular clocks in the context of gene tree heterogeneity is still an imperfect art and represents a major gap in the unfinished synthesis of comparative genomics and coalescent theory.

Previous genomic studies of convergent evolution typically focused on protein-coding regions in the genome [163, 68, 28, 194, 235] (reviewed in [401]), leaving the role of regulatory regions in convergent evolution relatively unexplored. In our analysis of paleognath genomes we used a program called PHAST [679] to identify CNEEs with possible regulatory functions from a whole-genome alignment. Overall, we delimited ∼1.3 million CNEEs but restricted attention and analysis to 284,001 putative regulatory elements that were greater than 50 bp. We developed a novel Bayesian method called PhyloAcc (pronounced phylo-ay-cee-cee) [317] to identify 2355 (∼0.83%) CNEEs exhibiting accelerated rates of evolution within one or more lineages of flightless ratites but not in any other analyzed birds, with 256–630 ratite accelerated regions (RARs) independently accelerated in at least two ratite lineages, depending on filtering parameters. These RARs present candidates for regulatory regions whose function has changed in ratite lineages (figure 11.6c). When combined with epigenetic marks, such as open chromatin signatures detected by ATAC-seq, and near to genes of developmental interest, such as TBX5 [68], convergent acceleration of noncoding regions is a powerful tool for prioritizing regions to study functionally [643]. Although the original PhyloAcc could not accommodate gene tree heterogeneity, preliminary tests of PhyloAcc-GT, which does accommodate gene tree heterogeneity, shows that it has higher accuracy and a lower false positive rate than PhyloAcc (unpublished). Preliminary investigations suggest that accommodating gene tree heterogeneity into the analysis of the paleognath data set with PhyloAcc-GT does not substantially change the number of CNEEs detected as accelerated, either convergently or in single lineages. Still, in other phylogenetic contexts, accommodating such heterogeneity might make a larger difference in which regions are estimated to be accelerated and by how much. Accommodating gene tree heterogeneity has also not been applied to efforts to estimate divergence times of ratite lineages [839], and it is unclear how much divergence times will change when such heterogeneity is accommodated.

Issues of gene tree heterogeneity will also affect analysis of protein-coding regions. In our analysis of the paleognath data [643], as well in taxonomically broader analyses of adaptive evolution across birds [678], such gene tree heterogeneity could present false positives for the same reasons SPILS could affect estimated rates for noncoding regions. In this regard, like PhyloAcc, methods such as PAML, a popular software for measuring adaptive evolution in proteins [830], are also susceptible to unforeseen effects of gene tree variation. Shultz and Sackton [678] recently examined rates of adaptive evolution across ~11,000 gene alignments across 48 species of birds. Their analysis suggested that, as in primates, genes involved with the immune system had widespread signals of adaptive evolution on multiple avian lineages. They tested whether assuming a single gene tree such as the estimated species tree, had an effect on their analysis, and saw very little difference in conclusions when using the species tree as a constant across genes or the inferred gene tree for each alignment, which effectively accommodates gene tree heterogeneity. We suspect that regional rate heterogeneity across the genome, which is a well-known phenomenon, at least in bird and other vertebrate genomes [225, 118], may be a more serious concern for methods examining rate heterogeneity among lineages in a tree. Methods like PhyloAcc can effectively incorporate variation among lineages in genome-wide substitution rates; for example, in our and previous paleognath data sets, it is clear that tinamou genomes evolve much faster than their nonvolant relatives [145, 643]. Here, however, we refer to heterogeneity in substitution rate across individual genomes, which, like gene tree heterogeneity, is not accommodated in most methods for estimating rates of evolution. This issue is another way of incorporating increased biological realism into statistical models used on genome-wide data.

An issue arising out of our paleognath data set, and which is an important component of incorporating gene tree heterogeneity into clock analyses, is that of the low phylogenetic information content of many short loci of interest in functional genomics. For example, in addition to being candidates for regulatory activity, CNEEs have also been proposed to serve as useful markers for phylogenetic analysis. Like UCEs, CNEEs are relatively easy to identify in eukaryotic genomes, but, unlike UCEs, it is the core regions of CNEEs rather than the flanking regions that have been proposed as useful phylogenetic markers. Additionally, CNEEs are likely much more numerous than at least traditional UCEs because they are not ultraconserved. Edwards et al. [219] showed that CNEEs performed almost as efficiently as introns and UCEs when resolving avian phylogeny. However, as in the paleognath data set, gene trees derived from CNEEs show somewhat lower bootstrap support and likely are more error prone than for other markers. In this case it may be that forcing the species tree on such genes may be a useful stopgap approach (or adopting the most common gene tree, owing to the anomaly zone in this data set), although a fully Bayesian approach should be able to accommodate such error more elegantly. Conserved nonexonic elements present a particularly acute gray zone where an individual locus may have enough substitutions to be classed as accelerated, perhaps even enough for SPILS to be an issue, but not enough to infer a robust gene tree. However, in the case of the paleognath data set, we know from previously published analyses that RARs are not more likely to occur in regions of gene tree discordance, suggesting that ILS does not grossly distort these results (table S13 of [643]).

Another approach to analyzing loci with low phylogenetic signal might be to run rate analyses on a small set of likely gene trees for loci like CNEEs. For the paleognath data set, we have a fairly good idea of what the set of plausible gene trees is because large amounts of gene tree heterogeneity are confined primarily to two internodes, those linking rheas to emus and kiwis. Cloutier et al. [145] used the approximately unbiased test [677], as implemented in IQ-Tree [536] to check each alignment against a small set of plausible gene trees. Their analysis showed that, depending on the criterion used, ~25% to ~80% of loci could reject the species tree topology, indicating that another tree is a better description for that locus. Such approaches lend some skepticism to claims that gene tree error alone can explain gene tree heterogeneity [264] and provide a temporary way forward before more comprehensive statistical approaches are developed to handle gene tree heterogeneity.

11.3.4 PHYLOGENETIC ANALYSIS OF QUANTITATIVE 'OMICS DATA: GENE EXPRESSION AND EPIGENETICS

A final area in which comparative genomics is still coming to grips with heterogeneity of biological processes is in gene expression and epigenetics, both of which are emerging as central tools in the discovery of loci influencing phenotypes. In our work on paleognaths, we used both of these approaches in combination with analyses of rate variation to prioritize loci to characterize functionally [643]. The convergent acceleration of CNEEs in flightless paleognaths implicated multiple putative regulatory regions associated with flight loss within the clade. In an effort to narrow down our options for functional analysis even further, and to determine which CNEEs might also serve as functional enhancers, we performed the epigenetic approach called ATAC-seq [102], a technology that identifies areas in the genome found in open chromatin, which, unlike DNA tightly bound around nucleosomes ("closed" chromatin), are available for transcription and the binding of transcription factors. ATAC-seq peaks have previously been associated with an enrichment of regulatory regions and transcription start sites, providing a helpful result toward identifying and characterizing regulatory region convergence [102, 266]. We performed ATAC-seq on eight tissues, including hindlimb and forelimb buds, sampled at early developmental stages of chickens. The ATAC-seq peaks combined with the identified convergent RARs and previously published ChIP-seq peaks for embryonic chicken [666] yielded 54 candidate enhancers believed to be active in the developing chicken forelimb. An electroporated beta-actin/GFP enhancer construct assay was used to screen candidates for activity during early chicken forelimb development. A convergent RAR, mCE967994, was found to exhibit consistent and strong activity during chick forelimb development, making it a prime candidate for further study. Enhancer activity of mCE967994 was tested for in the homologous region of the volant elegant crested tinamou, which was identified as conserved in our analysis, and the flightless greater rhea, which was identified as accelerated. We found the tinamou version of the enhancer to consistently drive GFP expression in the developing chicken limb bud, whereas the rhea version did not. These results not only provide evidence for a loss of constraint of a regulatory element associated with loss of flight but also demonstrate the necessity of species trees in answering biological questions.

There are numerous methods for analyzing epigenetic and transcriptomic data, but from the standpoint of phylogenetics, such methods are still in their infancy. Over the past decade, a number of approaches to analyzing gene expression data have been suggested to help overcome the phylogenetic interdependence of species [59, 205, 628, 82, 206]. Such approaches often have as their basis the classical quantitative genetic models that provided a framework for early analyses of quantitative trait variation within and among species [403, 241]. Dunn et al. [206] showed that ignoring phylogeny when analyzing comparative gene expression data could result in false positive and unsubstantiated claims. There are still relatively few published studies robustly incorporating a phylogenetic perspective when analyzing gene expression or epigenetic data [765, 135], likely because of cost and complexity of obtaining appropriate tissues for analysis. As with other forms of genomic data analysis, there is a tendency for geneticists to first focus on models incorporating increasing genomic complexity rather than increasing phylogenetic complexity [139]. This bias is likely because 'omics data still comes primarily in the form of single-species studies, with relatively few studies comparing gene expression or epigenetic marks across multiple species [821, 104, 430]. The first wave of such studies, at least in vertebrates, tended to compare species broadly across big clades, such as mammals, with little phylogenetic context (e.g., mouse, dog, human, etc.) or well-characterized ecological variation [91]. Some new tools have emerged that explicitly incorporate the phylogenetic relationships of species into the analysis of gene expression and identification of regulatory modules [639, 367]. In principle, gene expression patterns could be influenced by the fact that genes with varying expression also have varying gene trees. Indeed, coalescent theory and gene tree heterogeneity is just now making inroads into our thinking about quantitative traits generally [494]. It is still unclear how important such heterogeneity is in influencing the variation in gene expression and epigenetic marks we observe across extant species today.

11.4 Conclusions

Our phylogenomic analysis of paleognathous birds, followed by a variety of downstream analyses involving statistical analysis and functional assays, has provided a glimpse of the promise and challenges of post-genomic phylogenetics. The radiation of flightless birds, an ancient, rapidly diverging radiation, is a challenging phylogenetic problem and one that we believe coalescent methods are essential to resolve. Even so, the phylogenetic analysis was challenged even further by the difficulties of distinguishing true biological signal from gene tree error, especially given the extremely short internodes of the species tree and the low resolving power of some of the noncoding markers we used. We also identified a number of gaps in current implementations of downstream analyses of these data: whole-genome alignments, analyses of rate variation among lineages, and analyses of quantitative 'omics data, such as transcriptomics and epigenomics. Throughout these analyses we were confronted with myriad shortcomings of existing models, particularly around the issues of gene tree heterogeneity and phylogenetic uncertainty. Yet, the accuracy of the species tree was integral to identifying convergently accelerated regions in ratites, which subsequently were assessed functionally. Inaccuracy in this critical early step could have resulted in false positives of convergent rate acceleration of elements that do not affect the trait of interest, or conversely a false negative preventing detection of CNEEs. Furthermore, the extent of the effect that these inaccuracies have on downstream analyses results frequently cannot be understood without repeating the analyses using different trees and/or branch lengths. We believe the next phase of comparative genomics methods will be toward increased integration of complex biological signals, gene tree heterogeneity foremost among them, into comparative analyses. The rich history of evolutionary models available from classical comparative methods and evolutionary biology generally will provide a robust foundation for this next phase of comparative genomics models.

CHAPTER 12

Phylogenetic Analysis under Heterogeneity and Discordance

James B. Pease and Ellen I. Weinheimer

A *system* is an arrangement or a general ordering based on the consideration of a single part, such as the calyx or fruit in plants, or the teeth, claws, number of digits, etc., in the quadrupeds. A *method*, to the contrary, is not restricted to the consideration of a single part, but summons others to its aid.
— Jean-Baptiste Lamarck, Méthode, in *Encyclopédie méthodique: Botanique* (Paris, 1796), IV, 129.

12.1 Introduction

Phylogenetic methods attempt to construct orderly models of biodiversity free from lingering ambiguity and uncertainty. Simultaneously, we appreciate the aesthetic disorder of nature, a beauty and richness of random evolutionary walks. No observable set of biological characters precisely indicates an unambiguous, universal scheme of relationships for any group of organisms. An idealized world of biodiversity in which all characters fit a precisely defined system with clocklike regularity would be uncanny and deeply unfamiliar. The absence of this natural irregularity would also nullify most of the contrast and variation that motivate inquiries across the biological sciences.

Here, we first examine a history of ideas surrounding evolutionary heterogeneity. What does the widespread existence of genotypic and phenotypic heterogeneity mean conceptually for the phylogenetic study of organisms? Second, we describe methods for assessing, quantifying, and visualizing phylogenetic heterogeneity. How do we appropriately assess phylogenetic heterogeneity and use it to inform questions and hypotheses? Finally, we outline current approaches for grappling with various analyses when phylogenies are heterogeneous and make recommendations for improvement of these analysis techniques. The full consideration of these issues shows that overcoming apparent impasses caused by phylogenetic discordance and incongruence requires expanding our conceptual perspectives even more than refining our quantitative methods.

12.2 The Origin of Discordance
12.2.1 A HISTORY OF SYSTEMS AND METHODS

In 1690, John Ray reflected, "The correct and philosophical division of any genus into species is by essential differences. But the essences of things are unknown to us.

Therefore, in place of these essential characters, some characteristic accidents should be used" [607]. The date 1690 is notable because it means Ray's words were as old to Darwin at the 1859 publication of *On the Origin of Species* as Darwin's work is to current scholars. This meditation on uncertainty in biological data (and Lamarck's above) would be equally at home in contemporary discussions of molecular phylogenomics as they were in the writings of seventeenth- and eighteenth-century natural philosophers like John Locke, Lamarck, and Linnaeus [806]. These pre-Darwininan philosophical considerations of discordance challenge a conventional narrative that Charles Darwin, Alfred Russel Wallace, and their mid-nineteenth-century contemporaries marked a transition between Linnean typology to "polytypic" methods that drew from many lines of evidence centered around inheritance.

Recently, science historians have offered revisions to this narrative ([687, 805, 807]; see also counterpoints in [722]). Careful examination of pre-Darwinian writings finds the progress narrative is oversimplified, and appreciation of underlying incongruity in biological characters dates back at least to the early Enlightenment. A revised history indicates the omnipresence of both "monotypic" systems that center on key characters and "polytypic" methods that draw consensus from heterogeneous information. The relative dominance of these perspectives among practicing biologists has waxed and waned over time, and their dialog forms an important tension in debates over concepts of organismal relationships and the evolution of populations and species.

The contemporaneous developments of the modern evolutionary synthesis and modern statistics in the early twentieth century vastly improved understanding of the relationships between genotype, phenotype, and inheritance. However, the fundamental argument continues over the relative utility and importance of genotypic or phenotypic characters in modeling evolutionary relationships. Quantitative debates and new methods for coping with "incongruence" among different characters can be found in a chain of dialog stretching across the mid-twentieth century ([684, 709, 708, 237, 500, 510] provide some useful signposts). Farris [237] makes a particularly clear distinction between "internal" molecular characters (directly inherited) and "external" phenotypic characters (indirectly inherited but directly under selection).

Meanwhile, biotechnology advanced rapidly in the form of DNA structure, nucleic acid sequencing, and the central dogma of molecular biology, which facilitated a return to more monotypic thinking. Molecular phylogenies based on key genes (e.g., organelle genes and ribosomal RNA) were held up as overarching lenses for "resolving" the tree of life. This revolution was effective in reshaping our views of biodiversity, particularly in uniting the microbial and multicellular domains, expanding our understanding of the diversity of microbial lineages, and clarifying many phylogenetic relationships where phenotypic data had been ambiguous. However, even some of the earliest molecular data sets found incongruence [705, 704] and presaged the next stage of the debate.

Comparative multigene studies in the 1990s and 2000s began the return of the cycle toward polytypic thinking with the widespread finding that phylogenies from various molecular gene sequences appeared to disagree. This ante-genomic period of molecular evolution conversations about genuine phylogenetic heterogeneity were often lost among debates over methodological bias, error, and model choice. Discussions (sometimes acrimonious) persist on a variety of method conflict points, including concatenation versus coalescent-like methods [385, 263, 817], missing and biased data [801, 650, 66, 320, 609, 760], phylogenetic inference error "zones" [239, 236, 176], and other model misspecifications ([726, 359]; and reviewed by [553]). The underlying issue is that if different phylogenetic methods infer various tree topologies inconsistently from the same data, then we may question the importance, quantifiability, or even the very existence of phylogenetic heterogeneity generated by biological processes. In short, we doubt whether heterogeneity is a fault of models or a feature of data.

The genomic revolution was supposed by many to resolve much of the prior evolutionary molecular conflict. Unsurprisingly, in hindsight, genomes were found far more functionally complex and phylogenetically diverse than was previously imagined by most. More data and computational power have not "resolved" conflicting characters so much as settled most of the earlier doubts that heterogeneous phylogenies could be attributed completely to model error [630, 582, 307]. Debates from the 1960 and 1970s over trait "incongruence" have been renewed as "phylogenomic discordance," with an accompanying efflorescence of polytypic methods and the ascendance of coalescent theory. Multigenomic data has rendered testable a range of new patterns and processes, including incomplete lineage sorting [328, 743, 559, 456, 770], introgression [1, 174], gene duplication–loss [5], composition and rate biases [253, 695], hemiplasy [37, 283], and others (e.g., [496, 299, 617]).

Looking at the near future, rapid advances in molecular biology continue to improve our connections between genotype, phenotype, and environment. As we connect the biochemical steps that mediate genotype and phenotype, a range of new characters are available for phylogenetic modeling (e.g., methylation states, protein binding sites, structural variants, expression levels, and gene interaction networks). Digitization has also transformed trait models with a range of new "phenomic" approaches, including high-dimensional morphometrics [147], metabolomic profiles [128, 286, 665], and textual metasearches [357]. New criteria and compound metrics based on them (e.g., principal components) are being accompanied by a range of new unsupervised machine-learning approaches to cope with this increasing scope and complexity of data [664, 824]. While the immediate present debates continue to focus largely on intragenomic molecular conflict among loci or genotype–phenotype conflict, we stand poised to engage in far broader debates about how new strands of evidence can be woven together to infer more comprehensive evolutionary histories and "the tree of life."

As we examine this revised history of thinking on systems and methods of biodiversity, we can thematically connect seventeenth-century arguments over "essential" characters, twentieth-century character "incongruence," and twenty-first-century phylogenomic "discordance" to a future of biological data that is increasing in both heterogeneity and dimensionality. Technological developments in statistical methods, DNA sequencing, and now machine learning have led to hopes of finding "the key" to unlock the tree of life and clarify its apparent heterogeneity. But history shows the benefit of an ongoing and productive coexistence of our dual desires for orderly, simple systems for studying biodiversity and messy, comprehensive methods for encompassing the stochastic reality of nature.

12.2.2 CONCEPTS OF HARMONY AND DISCORDANCE

The Ship of Theseus is an ancient paradox that asks whether a ship whose parts are all slowly replaced during a voyage remains the same ship upon return to harbor. This philosophical paradox of identity has been raised in the context of the relationship between species and constituent individuals [267, 52] and between microbial genomes under horizontal gene transfer and component genes [191]. Inference of species phylogenies and population histories in relation to gene and trait phylogenies is a similar yet distinct problem. Is the evolutionary history of a species or population separate from the histories of the genes, traits, and other characters carried by their constituent members? By analogy, is the provenance of each wood plank, sail panel, and iron nail different from the journey of the ship as a whole? In applying this paradox's principle to phylogenomics, we must consider not only physical belonging and identity but also temporal congruence and shared historical paths. Individual organisms can have phylogenetic relationships independent of overall species relationships, and genes within the genomes of individuals can have phylogenetic relationships independent of even these histories of individual organisms. We also continue to discover even more ways to measure biological relatedness through our increasingly complex understanding of both genomes and ecosystems. This means the "parts" of our species "ships" are becoming ever more numerous and layered. How then should we view the relationship between the histories of these parts to the history of the whole? As with most paradoxes, reflection on the imperfections of possible solutions is more important than arrival at a resolution.

Phylogenetic studies, in practice, generally start by inferring a species tree model. The species tree is usually approximated by a consensus species tree (CST) model, inferred from the total molecular or phenotypic character data (or some designated subset). The CST is used as an estimator of the "true species tree," a topology describing the actual order of species divergences through time (the statistical "unknown"). The CST and "true species tree" nonterminal nodes represent speciation events, and nonterminal branches represent ancestral lineages. We can then sample what we will generically label "mereotic trees" (MTs; from Greek μέρος "part," cognate with mereology as the philosophical study of parts to a whole). Mereotic trees are trees inferred from some subunit of the whole data. We purposefully avoid "gene trees" as a term in order to include phylogenies inferred from any data subcomponent (traits, nongenic loci, etc.). However, molecular gene trees will be a common type of MT discussed. When these individual MT models have topologies that differ either from the CST or each other, they are said to be "in conflict," "incongruent," or "discordant." In practice, discordance generally means incongruence of unrooted topologies but can include differences in rooting point, branch lengths, or node rank, when specified.

We can gain perspective on some curious aspects of the conceptual framework we have just described if we attempt to apply it to a univariate distribution of data. We do not generally say that any given univariate data point is "in conflict" with the other data points simply because they do not share the same value. Nor would we find it remarkable that data points are not the same value as the mean, median, or any other metric of the distribution. Studies have shown that the sheer size of the n-dimension tree space means that a consensus tree may be unlikely to be congruent with any of the constituent trees for even a moderate number of taxa with modest heterogeneity [83, 175]. Why then does the nature of phylogenetic discourse lead us so often to talk rigidly about "conflict" instead of viewing MTs as a distribution? Why then do we call phylogenies "discordant" but other types of natural data "distributed"?

Genome-wide sets of MTs often exhibit irregular distributions, which means any "averaged" consensus tree may not be identical to any of the MTs [492, 336]. In part, this perspective is informed by the knowledge that branches in a species phylogeny are supposed to represent actual biological lineages. The idea that a species might have a variable set of closest relatives for different loci sits uncomfortably in the tradition of evolutionary discourse, particularly systematics and taxonomy. Even more disconcerting might be confronting the idea that traits can have different closest relatives ([552], see discussion in [799]). Consider a polygenic trait controlled by hundreds of genes.

Molecular evolutionary processes for individual genes are neither fully independent from nor dependent on other genes or the process of speciation. Positive epistasis and physical linkages are correlative forces (though see [686]), while recombination or negative epistasis are dispersive. Recombination and assortative mating mean that the birth–death process of gene alleles can proceed independently from species birth–death (i.e., incomplete lineage sorting). We also know that a polygenic trait influenced by 100 genes might have a trait phylogeny congruent with loci of major effect or perhaps an average of the trait-associated loci that resembles none of the gene trees [427, 494, 497, 702].

From this logic, we can conclude that the following phylogenies could all be accurately inferred but have distinct true topologies: (1) the molecular phylogenies of trait-associated genes of major/minor effect, (2) a trait phylogeny inferred from phenotypic characters, (3) a total molecular consensus tree, (4) a trait-associated consensus tree from just genes of major effect for the given trait, and (5) the species branching order ("true species tree"). Acknowledging that MTs can biologically disagree among themselves and with the CST and that CSTs are an imperfect and conditional estimator of the "true species tree" are both important. However, recognizing MTs as a distribution raises new questions about species, genes, and traits.

12.2.3 THE SPECIES TREE

Related to a discussion of heterogeneous trees is the question of what we mean by "the species tree." George Gaylord Simpson in 1943 eloquently encouraged the field to give up trying to develop a universal definition of "species" [685]. The relative importance, prevalence, and probabilistic conditions of various speciation mechanisms are unsettled matters and are not likely to be universally resolved in the near future. Demarcation of taxonomic species units is still largely done by specialist systematists who now may use newer genetic species determination techniques [834, 551, 712].

If concepts of "species" are tenuous and conditional, then it logically follows that uncertainty translates through to a decision of whether two tips on a phylogeny are or are not part of the same species. Establishing species identity is generally a matter of taxonomic codes or clade-specific standards of genetic, ecological, or phenotypic differentiation. Additionally, the internal nodes on a phylogeny are ostensibly "speciations," but speciation is a process not an instantaneous event. Molecular trees place these nodes by fitting character substitutions on branches, which means they are a rough average of molecular divergence. These observations are not meant to imply that we cannot model species trees without fully understanding speciation but rather to affirm that species trees are modeling a particular species birth–death process to which the birth–death processes of mereotic trees may be only indirectly related.

Following from many authors [455, 633, 179, 456, 90], we will outline a framework of three general perceptual stances on species trees and mereotic trees. First, a strict speciocentric stance (figure 12.1a) views the species tree as the focal tree used for analyses and assumes that MTs have the same topology as the species tree. Under this framework, all data are modeled on a single topology regardless of whether separate inference of an MT's topology might be concordant or discordant. This envisions the species tree in the narrowest possible way by constraining all traits, loci, and molecular processes to exactly the same evolutionary history (aka the "Procrustean bed" [290]). The practical benefit of this approach is that all data subsets are directly comparable branch for branch because the tree shape does not change. However, this simplicity comes at the

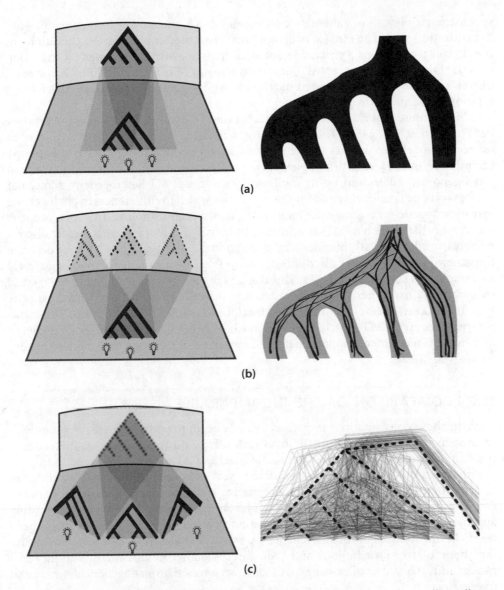

Figure 12.1. Conceptual diagrams visually representing the fundamental paradigm direct-ing analyses for each stance (foreground), and secondary frameworks for consideration (shadows).(a) A strict speciocentric stance for which no discordance is considered. (b) A relaxed speciocentric stance for which discordance is primarily considered in relation to a consensus. (c) A mereocentric stance for which underlying tree heterogeneity from data subsets is the focus and consensus trees are emergent.

potentially high cost of ignoring interesting variation or other biases, which we explore more fully in the later sections on individual analyses.

Second, a relaxed speciocentric stance (figure 12.1b) still views the species tree as the primary paradigm, but allows for MT variation. This stance is exemplified by the species tree–gene tree coalescent and reticulation diagrams found in many sources. A relaxed speciocentric view allows alternative topologies due to incomplete lineage sorting, but hybridization should be rare (MTs do not often cross species boundaries). Additionally, the species tree remains a generally good descriptor of a low-heterogeneity distribution of MTs. Discordance is accepted, but the diversity of MTs is more akin to "error bars" in tree space and is largely viewed in the context of a central, overarching species tree topology that maintains primacy.

Third, a mereocentric stance (figure 12.1c) views the MT distribution as the primary data (e.g., [630, 157, 248]). This perspective is common, sometimes necessary, in work on recent species complexes or metapopulations for which the species tree is poorly identifiable, in lineages with complex or frequent introgression or hybridization, and in phylogenomic studies addressing multimodal sources of MT heterogeneity. Molecular studies on gene families often adopt this stance, for which the differences in phylogenetic histories of gene paralogs are the focus instead of species orthologs. In a mereocentric stance, the distribution of MTs is nebulous and birth–death of species is largely disconnected from birth–death of individual gene and trait lineages. The CST may describe the general shape of the MT distribution, and MTs can adopt a range of topologies freely and transgress species boundaries. This concept of species trees is similar to the relaxed speciocentric, except here the heterogeneous MT distribution is primary and the species topology is seen as emergent from this distribution, rather than vice versa. These two perspectives on the MT–species tree relationship are not necessarily mutually exclusive and may even accurately approximate different parts of the same evolutionary history for a given lineage.

12.2.4 COMPARISON OF THE INCOMPARABLE

With these considerations unpacked, we can begin to address the main issues with heterogeneity for the phylogenetic investigator. How do we integrate an understanding of phylogenetic heterogeneity into approaches that require phylogenetic trees as parameters? If MTs evolved along different sets of ancestral lineages (i.e., differ in internal branch configuration), then comparing processes across individual MTs becomes difficult, if not completely intractable. Particularly challenging are the existing approaches that necessitate prior specification of a tree topology (often a single fixed tree). In the remainder of this chapter, we will first describe how to characterize the shape of tree distributions and then discuss both new discordance-aware techniques and adaptations of existing methods to accommodate phylogenetic diversity and uncertainty.

12.3 Characterization and Quantification of Phylogenetic Heterogeneity

12.3.1 QUANTIFICATION AND VISUALIZATION OF DISCORDANCE

Quantification and visualization of the underlying heterogeneity of phylogenetic signal establishes the breadth of genetic distinctiveness among lineages. Phylogenomic data sets can have distributions of gene trees with nearly negligible variation or moderate heterogeneity or can be so diverse that individual trees are all unique and distinct ([630, 568, 206]; see figure 12.2). However, even data sets with high levels of overall MT discordance can sometimes show broad agreement on particular lineages (figure 12.2d, e). Discordance metrics thus track the shape of and variation in phylogenomic sets.

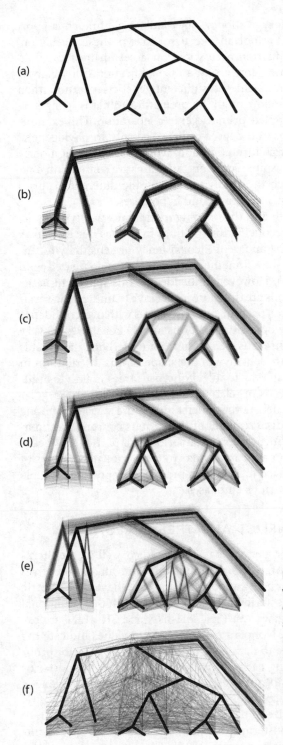

Figure 12.2. (a) A consensus tree shown with varying levels of underlying heterogeneity as cloudograms (b–f). As the series progresses, the consensus tree remains the same but more heterogeneity is added: (b) minor variation in branch lengths only changes to (c) a secondary phylogeny, (d, e) increasing intragroup heterogeneity, and finally (f) total heterogeneity, whereby all underlying trees are unique.

Accurately charting the degree of lineage support (beyond simple branch support scores related to a consensus) can be an important precursor to hypothesis formation. For example, lack of clear genetic distinction across loci for a set of lineages could be either an indication of introgression or, conversely, a disqualification from analysis of introgression itself. If lineages have not achieved sufficient genetic separation, then molecular analysis of introgressive gene flow might be premature. What is being perceived as secondary flow is merely incomplete primary genetic separation. This example illustrates how understanding the scope and degree of heterogeneity in phylogenetic relationships can be crucial to hypothesis formation and avoidance of setting poorly informed expectations of species relationships. Such contradictions are common in phylogenetic discordance analyses, in which the relativism among the different histories makes it difficult to establish expectations, contrasts, and hypotheses.

Related to this topic is a current conflict in the ongoing monotypic–polytypic cycle, namely concatenation *versus* gene tree reconciliation methods in species tree inference. Supporters of concatenation point to the apparent inconsistency or sensitivity to different weighting schemes in consensus-based methods as a weakness compared with full concatenation methods (e.g., [262]). However, embedded in this statement is the converse truth that concatenation methods are therefore insensitive to underlying variation and can present marginally supported parts of consensus trees with overconfidence [675]. Synthesis conclusions from these opposing positions are (1) gene trees do vary and so both methods approximate a consensus tree often from underlying data that is discordant, and (2) different methods are sensitive to the shape of this distribution in different ways. This debate over the "best" way to infer consensus trees is akin to arguing over geometric *versus* harmonic means in relation to a true mean for an irregular univariate distribution. Refocusing the debate on understanding the variation among gene trees and the biological causes of disagreement of the various consensus estimators seems preferable to continued argument over the superiority of particular "mean" consensus tree estimators. This situation appears similar to the various estimators of population genetic diversity, which have led to a modern practice of reporting multiple diversity metrics and contrasting their various emphases.

12.3.2 QUANTIFICATION OF CONFLICT AND TREE EVALUATION

Inference of phylogenies under heterogeneity has been studied well in the literature, but the effect of heterogeneity on branch/bipartition support evaluation receives less attention. The two most common branch support methods are the nonparametric bootstrap (NBS; [241]) and Bayesian posterior probabilities (BPP; [406]). Additional methods such as the Shimodaira-Hasegawa test (SH), SH-like, and SH-aLRT are expansions of these methodologies [284]. Nonparametric bootstrap values indicate the sensitivity of each bipartition on the tree to data resampling; Bayesian posterior probabilities are the probabilities of observing each particular bipartition in the posterior distribution of trees. Neither NBS or BPP was designed to characterize phylogenetic diversity among loci but rather to assess confidence in the single CST inference.

The sensitivity of NBS drops as the size of data increases, leading to a saturation of the NBS values at 100 [423]. Bayesian posterior probabilities can similarly gain confidence in its original conclusions as the data increases [836]. These are not deficiencies of these methods but do, in practice, lead to two common misinterpretations: (1) high NBS or BPP scores imply lack of MT heterogeneity, and (2) NBS and BPP quantify the relative level of MT heterogeneity. Certainly, strong discordance will eventually cause a drop in the consistency of inference, leading to lower confidence in the consensus. But these methods primarily are designed to report whether the given CST represents a majority of the data (even if a slim majority) and not to characterize the relative degree of underlying heterogeneity for a branch among MTs.

Concordance scores have been developed to characterize MT heterogeneity, though they have not yet been widely adopted outside genome-wide phylogenomic surveys. These include quartet puzzling [733, 734], Bayesian concordance factors [27], internode and tree certainty (IC/TC) scores [647, 373, 858], ASTRAL [657, 850], SplitScores [17], and quartet sampling [567]. These methods sample different subsets of the data to highlight the relative heterogeneity of data subsets at nodes. Collectively these have revealed some striking features of CSTs. In particular, they have shown more clearly that deeper relationships with less phylogenetic signal might be less accurately represented as a result of trade-offs in the joint likelihood optimization of the total tree.

Phylogenetic branch concordance scores help highlight the variation in concordance among branches in a CST. Nonparametric bootstrap and BPP scores are based more on the question: "How often does the majority of the resampled data support a given CST branch?" Concordance scores instead ask: "What proportion of my data support various clade relationships?" The first question is conditioned on an implicitly majority-rule view, while the latter simply measures the distribution of support. Concordance factors can therefore more specifically differentiate some lineages with broad support from other lineages with weaker support across many MTs. This means lineages can be classified as more or less identifiable across MTs instead of more or less weakly supported by an overall majority of MTs.

12.3.3 VISUALIZATION OF CONFLICT

Visualizing discordance can also be useful as a diagnostic (figure 12.3). Reticulated networks are a common representation of phylogenetic signal distribution ([45, 311, 337, 338, 710]; see also [519, 752, 73]). However, few current practical methods can address an acyclic network structure [749, 711, 790]. "Cloudograms" (visual overlays of MTs [455] created by the program Densitree [86]) provide another useful way to diagnose and showcase relative levels of discordance. Cloudograms can be useful in showing the degree of heterogeneity in various clades or, in some cases, visualizing strong secondary phylogenies due to introgression (figure 12.2).

With networks or cloudograms, the tendency is to assume automatically that the CST runs like a skeleton through the densest part or visual middle of these diagrams. However, this can be false for a variety of reasons, including weighting parameters and particular biased processes that may produce asymmetric distributions of trees around the CST. Such asymmetry in the distribution of inferred MTs is, in fact, predicted in many cases by coalescent models [179, 294]. Care must be taken also to use appropriate shading or other visual cues to indicate the relative strengths of various edges, lest highly supported branches are lost in a morass of reticulations.

STRUCTURE plots and principal components analyses derived from (usually biallelic) genetic data are another way to approximate heterogeneity among closely related populations and species [499, 588, 410]. Careful interpretation is needed here since these analyses are cluster based, instead of formal hypothesis tests. The tendency can be to overinterpret any apparent intermediacy in the frequencies of individual alleles or overall patterns of allele similarities as evidence of hybridization or introgression (see commentary in [542, 409]). With careful interpretation and consultation with

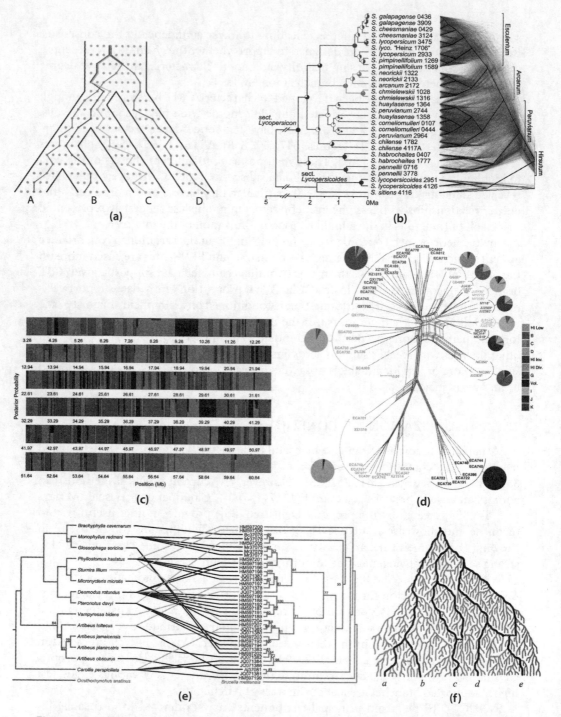

Figure 12.3. Example visualizations of gene tree variation. (a) Gene-tree species tree coalescent diagram [176]. © 2006 J. H. Degnan & N. A. Rosenberg. (b) Consensus phylogeny with node concordance pie charts and gene tree cloudogram [568]. © 2016 J. B. Pease et al. (c) Gene tree distributions by shade on the chromosomes of the mouse genome [795]. © 2009 M. A. White et al. (d) Phylogenetic network diagram showing potential relationships as reticulations [154]. © 2022 eLife Sciences Publications Ltd.

other methods, however, these nonphylogenetic approaches can complement tree-based analyses in which allele patterns are inconsistent with a single tree topology.

When scaffold- or chromosome-mapped data are available, spatial mapping of a smaller number of trees on chromosomes has been used to visualize various phylogenies. Studies in a variety of organisms show chromosomes "painted" with different evolutionary histories [307, 795, 258, 434, 248, 568]. Several tools can also do calculations of discordant site patterns, introgression patterns, quartet painting, or other systems of visualization that show the spatial distribution of different gene trees or multispecies allele patterns on the chromosomes [410, 211, 473, 64, 228, 570]. These chromosome-anchored approaches can be powerful for identifying chromosomal blocks with enrichment of alternative evolutionary histories that can be identified as regions of introgression, inversion, or strong selection [248, 394, 568, 402]. Spatial approaches can also make functional connections to loci contained in distinct genomic blocks with different histories. Additional tools also exist that test molecular recombination within loci by means of phylogenetic methods [792, 434, 471]. Collectively, these tools create diagrams that are useful for visually assessing phylogenetic heterogeneity and suggesting possible further hypothesis testing.

12.4 Analysis under Phylogenetic Heterogeneity

12.4.1 TESTING OF INTROGRESSION AND HYBRIDIZATION UNDER PHYLOGENETIC HETEROGENEITY

A natural place to start a discussion on phylogenetic heterogeneity is hybridization since studying homoploid hybrid lineages or introgression fundamentally requires at least two phylogenetic histories (though for different reasons). For homoploid hybrid formation, we expect to sample two phylogenies in different MTs: one in which the hybrid lineage is most closely related to one parental lineage and another in which the hybrid lineage is related to the other parental lineage. The challenge is to estimate the proportion that each parental lineage contributes to the homoploid hybrid lineage since the contributions from parental lineages generally will not be equal after even a few generations. This can make lineages of hybrid origin challenging to distinguish from those that introgressed later or with high levels of background incomplete lineage sorting, though several methods exist that address this question, including the f-statistics [612], PhyloNet [749, 790], and HyDe [77].

Postspeciation introgression means sampling a mix of two primary groups of MTs: those representing the speciation phylogeny and those representing the introgression phylogeny. The introgression phylogeny shows closest relationship between donor and recipient of gene flow joined at a node that represents an introgression (not a speciation). In this case, a determination must be made as to which trees are speciation and which introgression. In general, various metrics have relied on the logic that introgression trees

---

Figure 12.3. (*continued*) (e) Tanglegram showing topological differences between two phylogenies [419]. © 2014 B. R. Lei & K. J. Olival. (f) Representation of a phylogenetic network with reticulations [560]. © 2015 F. Pardi & C. Scornavacca. All images are from open access articles distributed under the terms of the Creative Commons Attribution License, which permits unrestricted use, distribution, and reproduction in any medium, provided the original author and source are credited.

should have more recent divergences than speciation trees since, by definition, introgression must occur after speciation. These metrics include Patterson's D-statistic (the "ABBA-BABA" test; [277]), d_{XY} [471], D_{FOIL} [569], D_{min} [636], PhyloNet [749, 790], and D_1/D_2 [301].

Another common assumption is that the most frequent tree in the set of MTs is the true branching order, and a less common tree is the introgression tree. However, introgression has been found to affect a substantial portion of the genome in some cases [248]. Additionally, the presence of introgression itself may interfere with the inference of a consensus tree, and multiple introgressions in a clade can cause substantial issues as well (see discussions in [569, 174]). While two trees are the necessary minimum, most species that are biologically capable of interspecific gene flow or hybrid offspring generation are usually relatively recently diverged. Therefore, incomplete lineage sorting is usually also present and leads to a more diverse array of topologies within which the introgression and speciation topologies are nested. This topological diversity often means the task is far more complicated than deciding which of only two phylogenies are the speciation and introgression trees.

As we have discussed previously, a clear species phylogeny is not strictly necessary to test for introgression [566]. The role of a species phylogeny in introgression testing is to establish baseline expectations of relatedness. Introgressing loci stand in contrast to this expectation as an alternative phylogenetic model. Lineages that are genetically heterogeneous in their phylogenetic relationships may be indicating that divergence is still ongoing and incomplete in nascent species–population complexes. If the lineages in question lack specific genetic distinctiveness necessary to establish their separateness, then confidently establishing their unexpected similarity as evidence of introgression can be difficult, if not impossible.

Therefore, the most general approach to introgression testing is to take care to establish that the lineages being investigated are genetically separate enough that the incursion of alleles is relevant and measurable. Otherwise, the late stages of an ongoing speciation process could be interpreted as introgression or hybridization, when in fact genetically intermediate signals are simply an incomplete divergence rather than a postspeciation convergence. In other words, lineages will be labeled as admixing when in reality they are in a late stage of divergence. The fact that both processes can look similar from allele pattern and frequency data is at the heart of the challenge in studying postspeciation gene flow.

12.4.1.1 Example: Introgression under Uncertainty in Birds

An example of a procedure to test for introgression under heterogeneity comes from a study of Central and South American siskins (*Spinus*) by [58]. In this study, 31 distinct phylogenetic topologies were reconstructed using RAxML [720] and SNAPP and SVDquartets [140]. The D-statistic test requires selection of two sister taxa that are assumed not to introgress, a third taxon that is a candidate for introgression with one of the first two, and an outgroup to polarize ancestral/derived alleles. However, four possible consensus trees were inferred for this clade, and certain quartets of putatively introgressing taxa were not in precisely the same configuration in all four trees. Instead of simply selecting one tree (based on a standard convention or cutoff), the authors instead conducted introgression tests using all four possible trees. Once D-statistics for the clades of interest had been calculated using various tree topologies, the results were compared. Collectively, the analysis found some patterns of introgression were consistent regardless of the species tree used, while others were inconsistent.

This example is instructive as an heterogeneity-aware approach in two ways. First, [58] inferred a variety of phylogenetic trees by means of a range of methods. They then analyzed the diversity of these trees and clades to shape their hypothesis formation for tests of introgression. This is an example of using the analysis of phylogenetic heterogeneity to guide the hypothesis testing. Second, the authors conducted their analyses under a range of possible trees and then concluded that some introgressing pair inferences were insensitive to the phylogenetic topology used (and others were not). The approach uses the heterogeneity of results to further analyze the relative strength or likelihood of introgression rather than merely assuming a fixed tree for analysis. In this case, the approach is helped by the relative similarities of the 31 possible trees, which shared some clades in common while other taxa were variable in their relationships. While effective for siskins, such an approach would likely lose efficacy if the pool of plausible trees was so variable as to lack any consensus lineages.

12.4.2 TESTING OF SELECTION UNDER PHYLOGENETIC HETEROGENEITY

Phylogenetic tests of selection generally involve analyzing particular branches, sites, or branch–site combinations for evidence of an excess nonsynonymous substitutions (d_N/d_S test) by using popular software like PAML [539, 830] and HyPhy [377]. Phylogenetic tests of selection are more challenging than tests of introgression under conflict for two reasons. First, tests of selection are often carried out gene by gene whereas introgression tests are often done on large regions or for the whole-genome level [832, 540]. Second, unlike introgression where we expect tree heterogeneity, tests of selection typically focus on particular ancestral branches. If protein-coding genes are evolving under different histories, then the ancestral branches are not the same for all gene trees.

When testing selection under MT heterogeneity, should we proceed from a strict speciocentric view and use the consensus tree for selection tests across all loci? Or should we adopt a mereocentric stance and analyze each gene under its own inferred gene tree topology? If a fixed tree is used for all loci, this can cause substitution overcounting, since forcing data on a suboptimal phylogeny may cause falsely convergent substitutions to be inferred [496]. However, fitting each gene to its own phylogeny risks (1) testing nonoverlapping sets of branches that cannot be directly compared, and (2) overfitting the tree to its data, including the selective increase in molecular rates [833, 835]. Therefore, if an MT and species tree are genuinely discordant, then an inherent trade-off exists between overcounting substitutions by underfitting the data to a consensus tree or undercounting substitutions by overfitting each gene to its own MT.

One solution is to perform only pairwise measurement of d_N/d_S between pairs of species, which avoids the tree model entirely (e.g., [227, 393, 774]). This approach avoids the problem of phylogenetic discordance by drawing contrasts nonphylogenetically (or strategically using unambiguous tree relationships). However, this approach cannot address questions that require joint inference of ancestral lineages and can become cumbersome when the species and genes are numerous. Additionally, two species is the minimum amount of data one can use to calculate d_N/d_S, which substantially limits the power of the test. Despite potentially severe limitations, pairwise approaches can at least give some validation of phylogenetic d_N/d_S results if the phylogenetic heterogeneity brings the results from individual gene trees or certain key taxa into question.

Branch tests examine molecular evolution across specific branches on a phylogeny in a likelihood ratio test (LRT), with null models usually specifying more uniform rates and alternate models incorporating single or multiple elevated-rate branches [830]. Early applications of PAML to comparative genome data sets tended to use a fixed tree topology for all loci (strict speciocentric; [540, 29, 202, 378, 767, 755]). Noting this use of a fixed tree is merely factual and not critical. The sparsity and phylogenetic scope of the genomes, relatively early state of discussion about the multispecies coalescent and discordance, and computational limitations meant that assuming a fixed species tree was a rational and practical choice for these studies.

As genomic data sets have grown, considerable confusion has arisen regarding how to conduct selection branch tests effectively when the branches vary. One practical workaround limits analyses to branches that appear in a large proportion of MTs (e.g., present in >70% of gene trees). As with introgression, this means we will end up focusing on the parts of the overall CST that are relatively unambiguous in their congruence with the distribution of MTs. Practical examples of this kind of "conservative branch" approach appear in several studies [534, 568, 148]. The limitation here is that inference may be restricted to a subset of genes for a particular branch or that key branches of prior interest may be consistent for only a small number of genes.

We might also consider whether genes whose reconstructed phylogeny does not implicate the branch in question would return reliable or significant signals of selection under d_N/d_S. Certainly if there is not enough signal even to infer a branch from the gene's alignment data, one would not expect an LRT of the branch test to be significant or meaningful. For the purposes of selection tests, the absence of a branch in a tree derived from the locus in question may be relevant evidence, if indirectly. This phenomenon may arise more frequently when branches deeper in time are tested since there is more potential for homoplasy to mask or distort the ancient selective signal. More work is needed to unpack these questions or develop alternative ways of integrating selective signals from incongruent trees.

Site tests look for repeated substitutions in codons as a sign of codon-specific positive selection. Similar to branch tests, setting a fixed consensus topology can cause inference of substitutions on branches that are not necessarily present in a gene tree inferred from species trees [496]. These tests are also sensitive to the presence of multi-nucleotide mutations that violate the assumed independence of substitutions in a codon [663, 764].

Overall, testing for selection under phylogenetic heterogeneity is problematic in that direct comparison of the same lineage branch in a tree may not be possible across the entire genome. As with hybridization, context is important and focusing on high-concordance clades, raising the minimum cutoff for LRT expectations to accommodate discordance, and selecting candidate genes can alleviate this problem in many cases. Some of the suggestions above provide practical workarounds but do not resolve this core issue.

12.4.2.1 Example: Selection under Heterogeneity in Wild Tomatoes

An example of heterogeneity-aware selection testing comes from a study of wild tomato species (*Solanum* sect. *Lycopersicon*) by [568]. Phylogenomic study of 29 transcriptomes found 2744 distinct topologies in a set of 2745 100 kb window phylogenies. The concatenated and coalescent-like consensus phylogenies were also topologically distinct from all of these 100 kb phylogenies. However, several branches of the major subclades (corroborated also in this case by trait and habitat information) were consistent across a large proportion of both fixed-sized windows and gene partitions. The red-fruited clade that includes domesticated tomatoes was monophyletic in 98% of trees from individual genes, with other major clades being monophyletic in 78%, 34%, and 6% of genes. Despite the strong degree of overall phylogenetic heterogeneity (nearly all 100 kb trees unique), several key clades were represented across a substantial proportion of genes.

Branch tests of selection (d_N/d_S) using PAML [830] were conducted for branches leading to key clades only for genes for which the gene tree contained the key branch. For the red-fruited clade, nearly all genes were suitable for testing under a common branch test because they all contained the same ancestral branch. The major subclades with 78% and 34% support were also testable across thousands of genes. The most heterogeneous clade with only 6% of support had to be excluded, though better supported groups within this clade were analyzed. This conservative strategy bypasses the problem of gene tree–consensus tree disagreement by conducting analyses only on targeted branches present in a substantial number of genes. The obvious limitation is that not all branches are testable, and this approach will not be possible when gene tree heterogeneity is extreme. However, it would be unusual for a gene alignment to have sufficient synonymous and nonsynonymous substitutions to test d_N/d_S for a branch but the same branch not to have enough substitutions to be inferred in the gene tree. This study shows a conservative approach to gene-by-gene tests of selection focusing on only branches where there is strong concordance, bypassing the problem of heterogeneity to test sets of congruent branches in spite of overall tree heterogeneity.

12.4.3 TESTING OF TRAITS UNDER PHYLOGENETIC HETEROGENEITY

Studies that model trait phylogenies often have discussed that choice of the "wrong tree" might lead to inaccurate inference of trait models (see discussion in [800]). However, a different problem to consider is whether the tree is constant across traits or whether traits trees themselves are genuinely heterogeneous. The answer may, in part, depend on the genetic basis of the trait itself. For a Mendelian trait controlled by a single locus of major effect, a simple answer might be that the gene's tree should correspond to the trait tree. However, a growing sense of broad interconnection among genes (e.g., "omni-gene hypothesis" among others; [89]) poses a challenge to this simple framing. Can traits with apparently simple genetic basis truly evolve independently of other genes in the genome?

For polygenic traits, the complex epistatic functional effects of many alleles mean the phylogenetic patterns of the trait itself can be independent of the genes influencing that trait. Analyses of gene modules or pathway-based phylogenetic analyses have attempted to bridge the genotype–phenotype gap [251, 250, 144, 233, 494]. As with the relationship between the overall CST and MTs, we should consider the potential for MT diversity in genes underlying a polygenic trait. A consensus of the gene subset that affects the trait of interest might have a topology that differs from all the underlying MTs and possibly the CST as well. A tree model constructed from the trait data is thus itself another phylogenetic model, distinct from the molecular total CST, molecular MTs, or other consensus trees from sets of trait-associated loci.

What does gene/trait discordance mean conceptually and how might it affect analyses? Trait data are often mapped to a consensus molecular tree to test for the fit. When trait data poorly fit the consensus molecular tree (using metrics like Pagel's λ; [558]), this is used as evidence for either increased rate of change or multiple gains and/or losses of the trait [343, 827, 397, 815]. However, when the underlying diversity of molecular gene trees is strong, the assumption might be tenuous that trait-altering events would evolve according to a phylogeny consistent with either the consensus species tree or even a consensus from genes of major effect.

As recommended for other techniques, quantifying the level of phylogenetic heterogeneity is an important precursor. If a particular lineage is poorly defined across gene segments (i.e., is a "fuzzy" branch with poor concordance; figure 12.2), then inference of trait changes involving that branch become more suspect. In other words, if a short branch in a molecular consensus tree deep in time is particularly difficult to infer from the molecular data, then we should take care in taking all branches in the topology of a CST as stable and fit for inference of a major trait transition.

Gene expression and gene pathway evolution are emerging frontiers in understanding trait evolution as we continue to try to clarify the connections between the genome and organismal traits. Models have been proposed to analyze individual gene expression values and transcriptome-wide profiles in a phylogenetic context, yet the challenges in this area are high given they compound the uncertainty and complexity in phylogenetic methods and transcriptome expression quantification [544, 91, 169, 206]. Additionally, the network of gene interactions itself can become a multivariate trait modeled using a combination of phylogenetics and systems biology [342, 631, 641]. The challenges here can be severe since we still poorly understand the functional connections between genes in even model lab systems (and less in wild species in situ). Regardless, these new opportunities to study the joint evolution of molecular characters, processes, and systems have potential to inform genotype–phenotype functional connections as we model the evolutionary processes that change them.

12.4.3.1 Example: Coevolution of Multiple Genes and Traits in *Jaltomata*

An example of trait evolution under consideration of heterogeneity was conducted in the flowering plant genus *Jaltomata* (Solanaceae) by [815]. This analysis explicitly acknowledges the evolution of key fruit color and corolla shape traits probably occurred under two different phylogenetic histories. The potential relationship of the apparent polyphyletic trait distribution is attributed to the underling incomplete lineage sorting, shared ancestral variation, and introgression. In this way, genetic hemiplasy is explicitly connected to trait hemiplasy in interpreting the results. The analysis thus recognizes that variation in gene trees might also have an effect on the variation in trait trees, rather than relying only on a fixed molecular tree and assuming convergent or parallel evolution of the same traits repeatedly. Also notable is that Wu et al. [815] proceed with analyses after quantifying the phylogenomic heterogeneity in a manner that informs both subsequent hypothesis formation and interpretation of the results.

12.4.3.2 Example: Coevolution of Oyster Genes with Bacterial Virulence

Another example comes from [785] in a study of *Vibrio* and oyster larvae. The topology of each bacterial gene tree was compared with a pathogen virulence trait tree to identify individual genes that were concordant with the trait-derived tree's topology. The degree of congruence of each host gene tree with the virulence trait phylogeny was also used as a means to identify the degree to which specific genes coevolved with the pathogen.

12.4.4 TESTING OF COEVOLUTION UNDER PHYLOGENETIC HETEROGENEITY

Coevolutionary studies often draw conclusions from the discordance of the two organismal phylogenies in relationships such as plant–pollinator, parasite–host, and others [632, 557, 200, 168, 474]. "Tanglegrams" [557, 477, 167] that visually compare two or more trees to highlight points of difference can be constructed by several software packages, such as Dendroscope and Mesquite [340, 458]. Points of topological discordance (often measured by some variant of Robinson-Foulds distance) between the two CSTs are interpreted as evidence for or against coevolution. Generally, phylogenetic coevolution studies compare the two CSTs and draw biological conclusions from their disagreement or agreement in the context of a particular pair of coevolving traits or biomolecules mapped onto that CST [599, 419, 216]. The coevolution of different haplotypes within polyploid lineages also make use of these techniques [353, 77]. Other methods exist that improve phylogenetic inference based on coinference of two trees [41, 201]. Some, in part, base coinference on gene tree information, but largely the complexity of reconciling the two consensus trees is what dominates the considerations [555] rather than the diversity of MTs within each consensus.

What happens when we consider the underlying diversity of MTs? Charleston [129] outlined multispecies coevolutionary networks of trees ("*jungles*") and explicitly modeled lineage sorting and hybridization events (see also [106]). Page [557] also discussed the implications of added complexity of MT heterogeneity for cophylogeny inference. Note that there are three separate applications of cophylogeny reconciliation that are common: (1) intragenomic comparisons of gene trees, particularly paralogs, (2) nuclear-organellar genome coevolution, and (3) full coevolution/cospeciation between separate, free-living clades. In the case of molecular coevolution, usually differences in topology and rates are analyzed for two coevolving genes. When we consider coevolution of polygenic traits or whole species, can we confidently assert patterns of coevolution based on differences in CSTs when strong underlying heterogeneity is present in one or both MT distributions? Should we be analyzing coevolution of individual genetic functional points of interaction? Should we compare distributions of gene phylogenies? Should we look for correlations by dating the phylogenies in absolute time?

For now, these questions of how to handle discordance in coevolutionary studies remain largely open, but we raise them here for two reasons. On a technical level, all of the previously mentioned concerns about gene tree/species tree reconciliation still apply, but with the uncertainty compounded to the power of the number of coevolving groups. On a conceptual level, the presence of MT heterogeneity (both molecular and trait) underlying those consensus trees creates a complication even for how we think about the evolutionary pressures that drive coevolution. Considering the levels of population, gene, trait, and genome, care must be taken to circumscribe exactly what is coevolving. We highlight them here to recommend careful future consideration of how two species trees, particularly ones supported by genome-wide data, should be compared.

12.4.4.1 Example: Coevolution of Figs and Wasp Pollinators

An example of an approach to coevolution with discordance considered comes from [653] in a study of figs and their associated wasp pollinators. In this study, the authors used methods to infer a posterior distribution of trees based on Bayesian Markov chain Monte Carlo approaches that take underlying molecular gene tree diversity into account. They then performed their analyses of speciation, cospeciation, host switching, and extinction using trees drawn from the posterior distribution. This approach therefore integrated a cophylogenetic reconciliation approach over the uncertainty from a genome-wide data set. As with the other examples discussed, a distribution of gene trees is drawn and assessed before hypothesis formation, and then (as with the introgression study from [58]) integration is performed over various possible trees to check the consistency of results.

12.5 Conclusion

As the many cited studies and many more unanswered questions in the previous sections highlight, much remains to be considered both conceptually and practically as we adapt tree models and systematics perspectives to "clouds," "forests," and "jungles." Appreciating the history, characterizing the shape, and anticipating the presence of phylogenetic heterogeneity are steps to enhancing phylogenomic inquiries as character types and their interactions and contrasts all multiply .

While specifics of the methods mentioned here will change faster than books can be written, three general thematic recommendations can be drawn from this analysis of the state of phylogenomic discordance. First, we can safely reject a historical progressionary narrative whereby modern phylogenetic methods entirely replaced systematic thinking. Rather we can see that "systems" based around key factors and integrative "methods" have coexisted conceptually through many technological revolutions and have formed a useful, contentious dialog.

Second, appraisal of the distribution of trees in a phylogenomic or multitrait data set is important for understanding the degree to which key clades are supported genome-wide. The consistency of identifiability for lineages among data partitions is both an initial result and a guide for hypothesis formation in molecular evolution tests. We should take caution in treating any consensus tree (concatenated or coalescent-like) as "fixed" with all branches equally reliable but also avoid declaring a phylogeny too chaotic and unknowable in the absence of broad agreement. Most empirical phylogenomic data sets sampled have shown a mix of broadly defined and poorly supported clades, and knowing which clades are which can be crucial to avoiding setting poor expectations in molecular evolutionary tests.

Finally, while each phylogenetic analysis highlighted here has its own caveats, a common thread is to be aware and skeptical of any approach involving the mapping of one data type onto trees inferred from another. As we continue to expand data sets and construct more phylogenies in new ways, the contrasts and distributions among them can offer information if we listen. Keeping the paradox of the Ship of Theseus clearly in mind will prevent holding too tightly to a singular tree of life or surrendering too easily to false patterns in the chaos of intertwined trees.

CHAPTER 13

The Multispecies Coalescent in Space and Time

Patrick F. McKenzie and Deren A. R. Eaton

13.1 Introduction

The multispecies coalescent (MSC) is a model for inferring a phylogenetic tree from a distribution of sampled genealogies—or, in practice, a distribution of empirical gene trees inferred from multilocus genetic data [455, 456, 177]. By integrating over genealogical variation, the MSC improves estimation of both tree topologies and divergence times in addition to providing estimates of other demographic parameters of interest, such as population sizes [218, 234]. Its influence on phylogenetics has been broad and pervasive, as is evident in the many extensions that have been developed for incorporating the MSC into studies of gene duplication and loss [604], introgression [845], and even character evolution [283]. As we approach one decade since the publication of *Estimating Species Trees* [372], it is valuable to re-examine the MSC, and its assumptions, to ask how we can best approach new challenges and opportunities in the coming era of ubiquitous whole-genome data sets. One area in which we believe the MSC has great potential is in improving the inference of local genealogical variation across whole genomes.

A key distinction between species tree inference under the MSC and the inference of genealogies sequentially distributed across genomes is the effect of genetic linkage. The MSC explicitly assumes that genealogies are unlinked, that is, statistically independent, whereas genealogies distributed across a contiguous genomic region are not independent and are expected to be spatially autocorrelated. This correlation (linkage disequilibrium) decays over time as recombination causes samples within different genomic regions to trace back to different sampled ancestors [330]. While this decay function has been well studied in the context of single populations [490], its effect on the similarity of genealogies constrained by a species tree model is poorly understood, including the influence of species tree parameters. Recent algorithmic advances have now made it possible to simulate efficiently entire chromosomes with recombination to produce correlated tree sequences [360], which presents a powerful new opportunity to investigate the relationship between species tree parameters and sequential genealogical patterns across genomes.

Genome-wide phylogenetic inference is currently approached from two methodological extremes: either (1) a single species tree is inferred as a hierarchical model to describe the expected distribution of unlinked genealogies across the genome; or (2) no hierarchical model is assumed, and gene trees are inferred independently in sliding windows of concatenated sequences along the genome [473]. The latter approach is often applied to identify introgressed regions based on their deviation from a genome-wide average [776]. However, the dearth of information contained within small genomic windows can cause high gene tree estimation error in this approach, and similarly, increasing window size to be too large will cause errors from concatenation of multiple distinct histories. The MSC provides a potential path forward. A parameterized species tree inferred from unlinked locus data may be able to provide priors on the expected distribution of genealogies both globally across the genome as well as spatially among linked trees.

In this chapter we explore this concept by using simulations to estimate the effect of species tree parameters on the rate of decay of phylogenetic similarity across the spatial extent of a chromosome with a uniform recombination rate. We show that a decay function can be estimated to describe the spatial autocorrelation of genealogies and that by incorporating this function into gene tree inference, accuracy can be significantly improved compared with existing sliding window methods.

13.2 Coalescent Simulations

To investigate genealogical variation along chromosomes we simulated genealogies under a range of species tree models in Python using the *ipcoal* package [489]. This package takes as input a tree topology and demographic parameters (divergence times, effective population sizes, mutation rate, and recombination rate) to generate a parameterized simulator for the program *msprime* [360]. Using this model we then simulated coalescent genealogies constrained by a species tree topology. To generate linked trees we simulated a 1 Mb chromosome and recorded the true genealogy spanning each position of its length since different genealogies span different intervals along the chromosome between recombination locations. To generate unlinked trees we simulated 1000 independent, nonrecombined loci and stored the single observed genealogy from each locus. Species trees and genealogies were plotted and manipulated using the Python package *toytree* [212]. Annotated code to reproduce all analyses in this chapter is organized into jupyter notebooks and available via https://press.princeton.edu/ISBN/9780691207605.

The distributions of linked and unlinked genealogies simulated on the same species tree are easy to distinguish when visualized: linked genealogies exhibit significant auto-correlation whereas unlinked genealogies exhibit greater variation (figure 13.1c, d). We explored a range of parameters to describe realistically linked and unlinked genealogical variation in genome-wide phylogenetic data sets. To focus our analyses on fewer total parameters we performed all simulations on completely imbalanced tree shapes (but different tree sizes) in which internode lengths and effective population sizes of internal edges are all set to be equal. All simulations were performed using a per-site per-generation recombination rate of 1e−9, and in the case when sequence data was generated, a per-site per-generation mutation rate of 1e−8 applied under the JC69 substitution model. The parameters we investigated for their effect on the distribution of genealogies include tree size (number of tips), the probability of incongruence (internode edge lengths in coalescent units), and tree height (the number of generations between internodes).

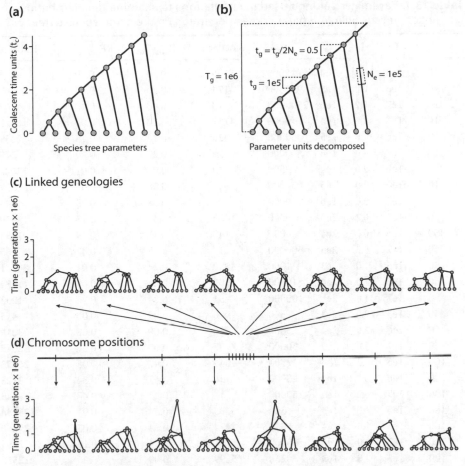

Figure 13.1. The effect of genetic linkage on the spatial distribution of genealogies across a chromosome. (a) A species tree topology with edge lengths in coalescent units can fully describe the probability of incongruence among unlinked genealogies. (b) Coalescent units (t_c) are a composite of time in generations (t_g) and the effective populations size (N_e). The extent of linkage among genealogies is influenced by both t_g and N_e and thus not fully explained by t_c alone. (c–d) Genealogies are plotted with tips in the same order as the species tree topology to highlight incongruence. Arrows indicate the positions of genealogies on a chromosome; linked genealogies are close together and unlinked genealogies are far apart. (c) Linked genealogies are spatially correlated because many samples share the same ancestors until a recombination event occurs. (d) Unlinked genealogies are independent and exhibit greater variation among a sampled set than do linked genealogies.

13.2.1 UNITS, SPACE, AND TIME

The effect of time, measured in units of generations, is not typically of interest for studies of the MSC since the probability of incongruence (among unlinked genealogies) can be explained entirely by internode lengths measured in coalescent units (t_c), which are calculated as $t_c = t_g/(2N_e)$, where t_g is time in generations and N_e is the

Table 13.1. Parameter settings used in simulations to examine the distribution of linked versus unlinked genealogies generated on the same species tree.

	Ntips	T_g	t_c	t_g	N_e	Block size	$RF_{unlinked}$	$RF_{linked-5K}$	Half-life
0	10	1e5	0.2	1e4	25000	1706	0.40	0.19	6576
1	10	1e6	0.2	1e5	250000	174	0.40	0.40	607
2	10	1e7	0.2	1e6	2500000	18	0.40	0.40	67
3	10	1e5	1.0	1e4	5000	4761	0.22	0.05	16715
4	10	1e6	1.0	1e5	50000	525	0.22	0.15	2359
5	10	1e7	1.0	1e6	500000	56	0.22	0.20	270
6	10	1e5	2.0	1e4	2500	8849	0.08	0.01	22627
7	10	1e6	2.0	1e5	25000	906	0.08	0.05	4659
8	10	1e7	2.0	1e6	250000	93	0.08	0.08	443
9	50	1e5	0.2	2e3	5000	1485	0.47	0.13	15890
10	50	1e6	0.2	2e4	50000	149	0.47	0.39	1838
11	50	1e7	0.2	2e5	500000	15	0.47	0.47	185
12	50	1e5	1.0	2e3	1000	4830	0.24	0.01	84731
13	50	1e6	1.0	2e4	10000	509	0.24	0.09	11884
14	50	1e7	1.0	2e5	100000	51	0.24	0.22	1233
15	50	1e5	2.0	2e3	500	9900	0.09	0.00	124415
16	50	1e6	2.0	2e4	5000	866	0.09	0.02	23019
17	50	1e7	2.0	2e5	50000	87	0.09	0.08	2048
18	100	1e5	0.2	1e3	2500	1345	0.47	0.08	26873
19	100	1e6	0.2	1e4	25000	138	0.47	0.34	3128
20	100	1e7	0.2	1e5	250000	14	0.47	0.47	345
21	100	1e5	1.0	1e3	500	5405	0.24	0.01	178444
22	100	1e6	1.0	1e4	5000	474	0.24	0.05	22688
23	100	1e7	1.0	1e5	50000	51	0.24	0.19	2499
24	100	1e5	2.0	1e3	250	7299	0.09	0.00	218558
25	100	1e6	2.0	1e4	2500	877	0.09	0.01	39616
26	100	1e7	2.0	1e5	25000	85	0.09	0.06	4169

Note: All simulations were performed on an imbalanced species tree with uniform internode edge lengths. Three free parameters were explored: the number of tips (Ntips) on the tree, total tree height in generations (T_g), and internode edge lengths in coalescent units (t_c). Two additional parameters are shown for which values were determined entirely by values of the free parameters: the internode length in units of generations (t_g) is determined by T_g and Ntips, and effective population size (N_e) is determined by t_c and t_g. Results are reported as the mean values calculated from 1000 simulated genealogies. The size of nonrecombined genomic blocks (block size) decreases with time in generations. This affects the RF distance between linked genealogies but not unlinked genealogies. $RF_{linked-5K}$ is the RF distances among linked trees separated by 5 Kb on a chromosome. The phylogenetic half-life was calculated from fitting an exponential curve to the rate of decay of phylogenetic linkage.

effective diploid population size. Because t_c is a ratio of time and population size, the absolute value of t_g has not been of interest, only its relation to N_e (figure 13.1a, b). However, in the context of a sequential coalescent process it turns out that t_g does matter, since recombination is modeled as a per-generation process, and so both t_g and N_e affect the number of recombination events and thus the similarity of neighboring genealogies.

The effect of time in units of generations is demonstrated in table 13.1. Here we simulated linked and unlinked genealogies on the same species trees and over a range of parameters. In each data set we measured the average Robinson-Foulds (RF) distance between all pairwise unlinked genealogies and, in the case of linked trees, between 1000 pairs of genealogies randomly sampled from positions that are spaced 5 Kb apart. Robinson-Foulds distances are reported here with normalized (scaled) values to account for differences in tree size, but non-normalized RF distances show the same qualitative results (not shown). In the unlinked data sets t_g has no effect on the similarity of genealogies—only t_c is relevant—as has been traditionally recognized in the MSC. However, for linked genealogies t_g has a large effect. When edge lengths are longer in units of generations, and N_e is similarly scaled to retain the same probability of incongruence (t_c), the size of nonrecombined blocks becomes smaller, and the average RF distance between neighboring genealogies is greater.

A notable result of these simulations is the observation that the size of non-recombined genomic blocks becomes very small in certain regions of parameter space, particularly when the internode lengths in units of generations are very long. This is troubling for the MSC, which requires that loci represent a single genealogical history as opposed to multiple concatenated genealogies. The effect of recombination within loci has been investigated previously, both in the first edition of this book [123] as well as in a series of critical examinations of the effect of "concatalescence" [716]. At issue is whether gene tree estimation error is elevated when concatenating data from multiple genealogies into a single locus. The results from our simulations suggest there are some regions of parameter space where the size of nonrecombined blocks becomes quite small, and so this issue may warrant further examination. In this chapter, however, rather than examine recombination and its effects as a critique on existing MSC approaches, we aim instead to explore how linkage among recombined genealogical blocks of the genome can possibly be a useful source of information when analyzing whole genomes.

13.2.2 TREE SIZE, TREE SPACE, AND PHYLOGENETIC DECAY

The enormous size of phylogenetic tree space is a constant source of computational burden in phylogenetics, but, intriguingly, it may actually provide a source of information in the context of the sequential coalescent process. This is because as the size of tree space grows in larger data sets so too does the expected RF distance between any two random unlinked genealogies. This increase in RF distance is particularly true when t_c is very small, such that all coalescent events occur deeper than the root of the species tree. In this case the topology of unlinked genealogies is hardly constrained by the species tree at all, and almost any genealogy can be observed. However, adjacent genealogies on the same chromosome are still expected to share significant similarity since few recombination events are likely to have occurred between them. Consequently, the degree to which linked genealogies are more similar to each other, *relative* to the similarity among unlinked genealogies, is a function of parameters of the species tree, including the tree size.

This type of relative measurement provides a means to develop a statistic to describe the rate of decay of spatial autocorrelation in genealogies across a genome. We propose the term "phylogenetic linkage" (*PL*) to describe the ratio of RF distances among linked genealogies separated by some genetic distance in the genome relative to the average RF distance among unlinked genealogies.

$$PL(d) = 1 - (RF(d)_{linked} / RF_{unlinked}).$$

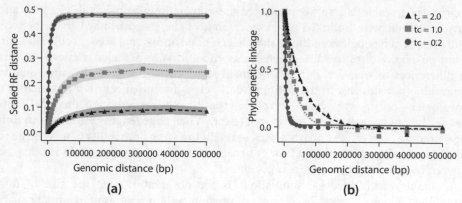

Figure 13.2. (a) The Robinson-Foulds (RF) distance between genealogies separated spatially on a chromosome plateaus as linkage decays by recombination and approaches the average RF distance between unlinked genealogies. (b) The ratio of RF distances between linked and unlinked genealogies (phylogenetic linkage) measured at different genetic distances approximates an exponential decay function. Results are shown for data simulated on a 100 tip species tree with total tree height of $T_g = 1e6$ generations, and N_e of 2.5e4, 5e3, or 2.5e3, corresponding to edge lengths in coalescent units (t_c) of 0.2, 1.0, and 2.0, respectively.

In other words, if two genealogies spaced d distance apart on a chromosome are as different from each other as two randomly sampled unlinked genealogies are on average, then they are effectively unlinked. By measuring phylogenetic linkage at increasing genetic distances between genealogies we can infer a rate of decay of phylogenetic linkage across the genome. For each simulated data set we then fit an exponential decay function using the *scipy* package in Python. From the estimated decay rate parameter (λ_d) we estimated a phylogenetic linkage half-life, representing the distance in base pairs at which two genealogies are expected to lose half of their phylogenetic linkage (table 13.1; figure 13.2).

When t_g is larger, phylogenetic decay occurs faster since more recombination events are possible over each internal edge of the tree (e.g., compare rows 0 and 7 in table 13.1). Similarly, when N_e is greater, recombination events are more likely to cause a change in the topology, and thus phylogenetic linkage decays faster (e.g., compare rows 0, 3, and 6 in table 13.1). Finally, when the total tree size (Ntips) is greater, the decay of linkage occurs more slowly since the average difference between unlinked genealogies is greater, and thus it takes longer for sufficient spatial information to decay to approach the unlinked mean RF distance (e.g., compare rows 0, 9, and 18 in table 13.1).

13.3 Linked Genealogies and Gene Tree Inference

Unlike in simulations, the true genealogical history for any region of the genome is an unknown and unobservable variable. It is something we must infer based on the signal left by the mutational process. So how can our understanding of the decay of phylogenetic linkage be useful in the context of gene tree inference? One way to approach this problem is to ask what is the expected length over which a site supporting a bipartition in one position of the genome continues to be true in neighboring regions of the genome?

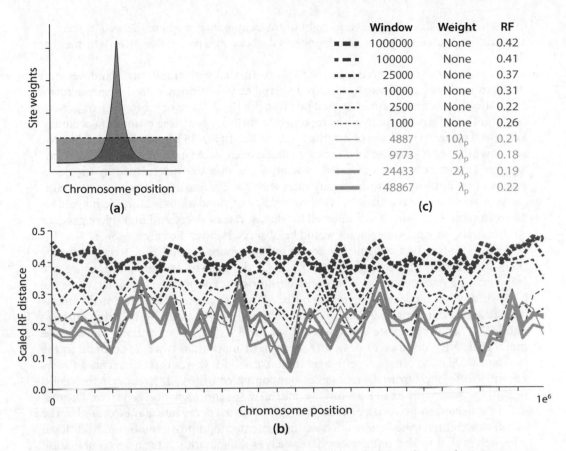

Window	Weight	RF
1000000	None	0.42
100000	None	0.41
25000	None	0.37
10000	None	0.31
2500	None	0.22
1000	None	0.26
4887	$10\lambda_p$	0.21
9773	$5\lambda_p$	0.18
24433	$2\lambda_p$	0.19
48867	λ_p	0.22

Figure 13.3. The accuracy of gene tree inference in sliding windows along a chromosome. (a) We compared windows that were of a fixed length and with uniform site weights to windows with lengths and weights determined by a function of exponentially decaying phylogenetic linkage inferred from simulations under the species tree parameters. (b) The scaled RF distance between inferred gene trees and the true simulated genealogy at 50 positions across a genome measured with different window sizes and weightings. (c) Windows with uniform (no) weights inferred less accurate gene trees than those with weights distributed by a phylogenetic decay function, as measured by the mean scaled RF distance to the true simulated genealogies.

The standard sliding window approach gives equal weight to all sites within an alignment window. An alternative approach could be to extend the size of the window to ensure that there is sufficient information to infer a resolved gene tree but to apply variable weights to sites in the alignment such that those near the center of the window have greatest weight and can override alternative signals (figure 13.3a). This weighting has the effect that if no local information exists to support the true local genealogy then data from more distant regions can inform that part of the tree but with decreasing weight as their probability of representing the same genealogy decays with distance.

We implemented a weighted approach to local gene tree inference by using the "-a" weights file argument during maximum likelihood tree estimation in RAxML v.8.2.12 [720]. A distribution of site weights was generated to give exponentially decreasing weight to sites on either side of a central position. For computational efficiency we cut off the window size to the left and right of the center at sites where the weight reached 1/10,000 of the center given the exponential decay rate parameter estimated for that data set.

Decay-weighted gene tree inference was compared with traditional windows with uniform (no) weights. Gene trees were inferred at 50 positions spaced evenly across a 1 Mb simulated chromosome. At each position the RF distance between the true genealogy and the inferred gene tree was recorded to measure gene tree estimation accuracy. We tested uniform windows of lengths 1 Kb, 2.5 Kb, 10 Kb, 25 Kb, 100 Kb, and 1 Mb, the last of which represents the total concatenated chromosome gene tree. Decay-function-weighted gene trees were estimated in windows with a size determined by the decay rate, and we additionally tested decay rates with $2\times, 5\times$ and $10\times$ faster rates to examine sensitivity to rate estimation. We show results for simulations performed on data set 10 from table 13.1, which was selected for its fast rate of decay and high incongruence so that many distinct genealogies would be observed across the chromosome.

The decay-function-weighted windows inferred more accurate gene trees on average than did uniform windows (figure 13.3b, c). Of the uniform windows, the largest size (representing concatenation of the entire chromosome) performed the worst, while the best window size appears to be near 2.5 Kb. All four decay-function-weighted window sizes tested had lower mean RF scores (greater accuracy) than the best scoring uniform window. The best estimate was observed for the $5\times$ decay rate window, which had a mean scaled RF distance of only 0.18, making it more than twice as accurate as the genome-wide concatenation gene tree. In nonscaled RF scores this represents an average of 35 differences from the true genealogy compared with 43 differences in the 2.5 Kb uniform windows, 51 in 10 Kb windows, and 80 in the concatenation gene tree. The reason a $5\times$ decay rate performed better than the estimated decay rate may be caused by the cutoff to weighted window sizes that we implemented to improve run times. Additional parameters that we did not explore here, such as the mutation rate and recombination rate, are likely to be important factors as well since they affect the information content within each nonrecombined genomic block.

13.4 Conclusions

For over a decade the goal of phylogenomic analyses has primarily focused on inferring a single species tree to represent the distribution of genealogical variation across the entire genome. However, as whole-genome data becomes available there is increasing interest in the spatial distribution of genealogies at specific locations across the genome. This type of local ancestry information can be useful for testing evolutionary questions about patterns of hemiplasy versus convergence [283], identifying introgressed regions [234], and testing hypotheses about adaptation [472]. Despite the development of advanced hierarchical models for inferring species trees, such methods have yet to be developed for spatially linked gene tree estimation.

Here we have demonstrated that the probability of incongruence described by the edge lengths of a species tree in coalescent units does not capture the expected spatial similarity of genealogies across chromosomes. Instead, in addition to the ratio of t_g to N_e, which describes the probability of incongruence, it becomes necessary to consider the magnitudes of these parameters as well. This observation presents an interesting scenario: imagine a balanced tree with two clades where every edge has the same t_c edge lengths. In one clade these edges are composed of high N_e and t_g values, while in the other clade edges have low N_e and t_g values. Despite having the same probability of incongruence, the two clades would exhibit very different rates of change in their topology per unit length spatially across the genome. Unlike in our simulations, the rate of decay would likely not be uniform and would covary more among some edges than others.

In theory, this expectation could be built into sliding window analyses based on a parameterized species tree inferred from unlinked loci. The simple approach that we implemented here, applying weights to alignment windows, is only a first step. A more appropriate direction to focus in the future would be to use species tree information to establish tree topology priors in a Bayesian context that could be used to improve local gene tree estimation by combining both the expected genome-wide distribution of genealogies as well as the expected similarity among neighboring genealogies. In contrast to treating recombination as a source of error for phylogenetic analyses, this direction of research aims to accommodate recombination as a source of historical information. There is no doubt that the MSC will continue to be extended to meet the needs introduced by new types of data and the many questions that they inspire.

CHAPTER 14

Tree Set Visualization, Exploration, and Applications

Jeremy M. Brown, Genevieve G. Mount,
Kyle A. Gallivan, and James C. Wilgenbusch

14.1 Introduction to Visualizing and Exploring Tree Sets

All phylogenetic studies are built around sets of trees. Tree sets carry different kinds of information depending on the data and approaches used to generate them, but ultimately the variation they contain and their structure is what drives new phylogenetic insights. However, tree sets also present particular challenges. Trees are complicated objects that are meant to be interpreted visually, and each tree, on its own, can carry a huge amount of information. Trees also naturally exist in a high-dimensional space that is challenging to conceptualize.

In some cases, tree sets can be understood and explored in formal statistical frameworks. For instance, readers of this book will undoubtedly be familiar with the multispecies coalescent model (MSC) that describes how stochastic coalescent processes explain variation in the histories of individual genes and how these histories differ both from each other and from the overarching species tree. Models like the MSC are elegant and incredibly valuable for testing different hypotheses, but their application is specific to those cases in which we can reasonably expect the variation in trees to be explained by the process(es) they include. Some tree sets may not have a natural process-based model to describe their variation. For instance, sets of trees meant to represent the uncertainty in a phylogenetic estimate (i.e., bootstrap sets or those drawn from a posterior distribution) have variation that is not described by a biological process. Other variation in trees may result from problems with data quality or poor fit between our observed data and the models we have available (e.g., misalignment, unintended paralogy, or heterogeneous evolutionary processes).

In order to better understand the variation in and structure of tree sets generally, we need tools that are generic, flexible, and exploratory. These tools can serve as natural complements to more formal, statistical investigations and allow us to flag surprising or unexpected observations, to better understand the results of model-based studies (e.g., by comparing the trees generated from replicate runs or assuming different kinds of models), as well as to build intuition.

In this chapter, we describe a set of tools for understanding variation and structure in tree sets and provide some examples of how they can be applied to questions related to phylogenetics and species tree inference. These tools include both visualization techniques and quantitative summaries, and they are currently implemented in the CloudForest computational platform [768], which is built around the TreeScaper software package [322]. We start with a brief overview of the approaches, then move to examples of how they can be applied.

14.1.1 TREE SET VISUALIZATION

Nonlinear dimensionality reduction (NLDR) is a general class of approaches for projecting the position of data that exist in a high-dimensional space onto a lower-dimensional space that can be visualized and explored more easily (typically, in two or three dimensions; [416]). This projection is done in a way that attempts to preserve the pairwise distances among data points as much as possible, to minimize distortions, and to provide insight into the relationships between data points in their original, high-dimensional representation. However, some distortion may be inevitable since visualization is restricted to no more than three dimensions, but the intrinsic dimensionality of the data (i.e., the number of dimensions required to maintain the original distances between data points) may be larger. CloudForest includes tools to estimate the intrinsic dimensionality of the pairwise distances among a set of phylogenetic trees [802]. When this intrinsic dimensionality is much greater than the number of dimensions used for visualization, the visualization should be interpreted with caution (though the projection may still be useful for other downstream analyses).

Nonlinear dimensionality reduction techniques have a long history of development in mathematics and other applied fields but were first used to visualize phylogenetic trees by Amenta and Klingner [20], who developed the TreeSetViz module within Mesquite [457]. Since then, several studies have explored applications of NLDR in phylogenetics (e.g., [304, 350, 782]). More recently, [802] addressed some outstanding questions about how different NLDR approaches perform in the phylogenetic context. In particular, they compared different stress functions as well as different algorithms for optimizing projections. While a variety of tree-to-tree distances, stress functions, and algorithms are available in CloudForest, we will focus here on examples of how NLDR and complementary techniques can be applied to questions in phylogenetics and species tree inference, and we refer readers to other publications for more technical detail.

Before we apply NLDR, a few important points should be mentioned. First, NLDR projections can be incredibly valuable for the intuitive and visually appealing summaries they provide, but in most phylogenetic cases the low-dimensional projections will distort the original tree-to-tree distances to varying degrees and should be interpreted with caution. Second, there are several stress functions available in CloudForest for performing NLDR, which, in our experience, can sometimes meaningfully affect the visualization of tree space. No single function will necessarily always perform best, although [802] suggest that curvilinear components analysis (CCA; [181]) best preserved the original tree-to-tree distances in their tests with mitochondrial data. Given these potential differences in results, one approach is to focus on those relationships that remain constant across different functions. Third, NLDR visualizations can be a powerful exploratory technique that suggests further avenues of investigation in more formalized statistical frameworks, and this exploratory role should be included in descriptions of a phylogenomic or species tree workflow.

14.1.2 DETECTION OF STRUCTURE IN TREE SETS

Interpretations of tree space as a network have a deep history in phylogenetics, largely in the context of exploring these networks to find the optimal tree (or set of trees) given some criterion—like a parsimony score, likelihood score, or posterior probability (see Felsenstein [242] for an excellent overview of approaches for exploring tree space and [794] for a recent study on the properties of these searches in Bayesian analyses). In this context, networks are conceptualized with trees as nodes and edges connecting "neighboring" trees. Whether or not trees are considered neighbors will depend on the type of tree alteration that is used to move through tree space. Note that this type of network is distinct from the networks that are used to describe relationships among species, where nodes represent splitting or fusion events between lineages and branches represent the evolution of different lineages.

Here, we also use networks, but we are not primarily interested in finding the "best" tree. Instead, we are interested in how the tree set itself is structured. For instance, are there regions of tree space where trees in a set are more abundant or dense? Do trees form distinct groups? To investigate these properties, we construct networks with trees as nodes and we add edges between all pairs of trees in the set (figure 14.1). These edges are then weighted as a function of a pairwise tree distance (several distance options are available in CloudForest). More specifically, edges are weighted with the affinity between two trees, which, roughly speaking, is the inverse of distance. Trees that are more similar (have a low pairwise distance) will have a higher pairwise affinity and will be connected by an edge with a higher weight. If trees in a set are clumped in tree space, we expect to find regions of the network where some trees form groups, such that there are large edge weights inside the groups and small edge weights between groups (figure 14.1). Note that while we use these topological networks to detect structure in tree sets, they can be very challenging to visualize for large sets of trees. Therefore, we often display results from these networks by coloring points (trees) in an NLDR plot.

In addition to tree-based networks, we also construct bipartition-based networks. In this second type, bipartitions are nodes, and the edges connecting pairs of bipartitions indicate how often they are found together in the same trees. More formally, the edges that connect the bipartitions are weighted by their covariances in presence/absence across trees in a tree set (figure 14.2). Bipartitions that are found together in the same tree more often than expected by chance will have large, positive covariances, and those that are found together less often than expected by chance will have large, negative covariances. If there is very little structure in the tree set, the bipartition network should contain only edges with low weights (both positive and negative; figure 14.2a). However, if the tree set is highly structured (for instance, when combining trees inferred from two genes with strongly conflicting phylogenetic signal), then bipartitions should form strong associations (figure 14.2b). Those bipartitions found in trees inferred with gene 1 will all have strong, positive covariances with each other, as will those found in trees inferred with gene 2. However, bipartitions from gene 1 trees should have strong, negative covariances with bipartitions from gene 2 trees.

A major advantage of conceptualizing both trees and bipartitions as parts of a network is the ability to take advantage of the substantial progress mathematicians have made in formalizing the study of networks [535]. Here, we specifically focus on a class of

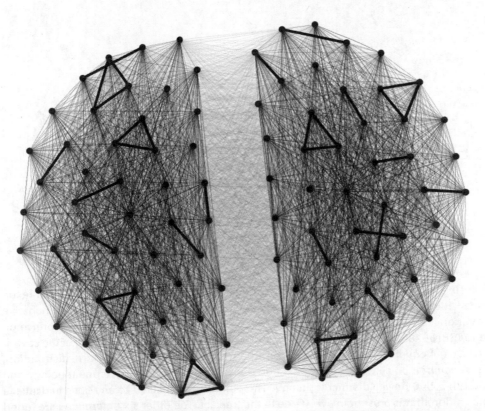

Figure 14.1. An example topological affinity network with 100 25-taxon trees grouped into two distinct communities. Each node is a different tree topology, and the edges between nodes are weighted by the affinity of each pair of trees. Larger affinities (smaller distances) are represented by thicker lines. The arrangement of points was chosen simply to make edges visible, and the placement of nodes is not optimized for two-dimensional representation as it is in an NLDR plot.

techniques called community detection methods [269, 252]. Roughly speaking, the goal of community detection is to find sets of nodes that form distinct communities in which nodes in a community are tightly connected by edges with large, positive weights and nodes in different communities are weakly connected or are connected by edges with large, negative weights. (Note that topological networks will have only positive weights on edges, while bipartition networks will have both positive and negative edge weights.) Community detection methods are able to look for this type of structure in networks without the need to specify the number or size of these groups a priori. Many different community detection models can be applied to networks, and CloudForest implements several different options [322]. Here, we focus on use of the constant Potts model [756] because it is better able to detect communities of widely varying sizes than many other models (a property known as being resolution limit free).

Many community detection methods, including constant Potts, include tuning parameters that can adjust the focus of the model on communities of varying size and number. At one extreme of the tuning parameter values, the model will prefer to put all nodes in one large community. At the other extreme, the model will prefer to place every node in its own community. Since these results are not biologically interesting, we focus

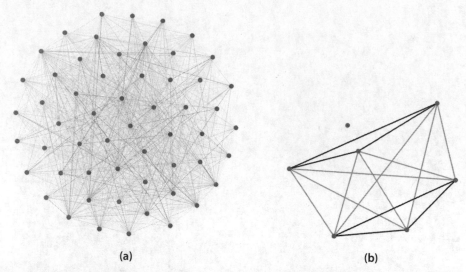

(a)   (b)

Figure 14.2. Examples of bipartition covariance networks based on sets of seven-taxon trees. Blue branches indicate positive covariances and red branches indicate negative covariances. (A color version of this figure is available through https://press.princeton .edu/ISBN/9780691207605.) The line weight indicates the relative magnitude of the covariance. (A) A covariance network based on 1,000 trees sampled from a uniform distribution. (B) A covariance network based on 1,000 trees, where 500 trees have one topology and the other 500 have a distinct topology. The point not connected to any lines indicates a bipartition that is present in every tree in the tree set. The other six bipartitions are found in only one topology or the other.

on the communities identified by the model with intermediate parameter values. While different potential tuning parameter values give us a great deal of flexibility, we generally do not know ahead of time which parameter values we should prefer. Instead, we allow the network to tell us. By adjusting the tuning parameter in small increments, we can look for regions of parameter values that all return the same community structure (called plateaus in parameter space). These regions of stability indicate that the detected community structure is a natural property of the network. A given network may display zero, one, or multiple of these intrinsic community structures.

14.2 Applications to Gene Trees, Species Trees, and Phylogenomics

Here, we provide examples of ways in which combinations of tree set visualization and community detection can be applied to questions about gene tree variation, species tree inference, and phylogenomics. These tools are intended to allow exploration, provide new intuition, and facilitate more detailed investigation. As a result, many of the examples we outline below do not overturn previous understanding but rather provide a new perspective on it.

14.2.1 SENSITIVITY TO MODELS OF SEQUENCE EVOLUTION

Species tree inference can be compromised by systematic errors in gene tree inference, which can occur when models of sequence evolution do not fit the data well. Many studies have explored how inferred gene trees, or the distribution of uncertainty

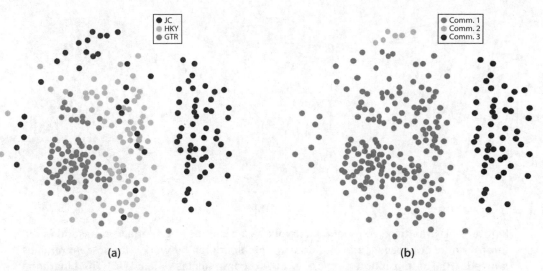

Figure 14.3. Two-dimensional visualizations from non-linear dimensionality reduction (NLDR) using weighted RF (wRF) distances of primate *cytb* trees. (A) Trees are colored by the model assumed in different analyses. (B) Trees are colored based on communities detected from a topological affinity network.

in these trees, may vary depending on which model of sequence evolution is used in an analysis (e.g., [421, 331]). However, the comparison of gene trees resulting from analyses that assume different models can be challenging, especially as trees become large. Also, in some cases, we would like to gain intuition for how the tree changes with model assumptions, especially when the best model is not clear.

Tree set visualization and community detection methods can provide intuitive summaries for how results change across analyses and focus attention on those parts of the tree that strongly conflict across analyses. These methods can also highlight regions of trees that are consistently uncertain in all analyses. To illustrate how these tools can be used in this context, we conducted Bayesian phylogenetic inference using RevBayes [341] for an alignment of *cytb* sequences from primates. We conducted three separate analyses that assumed a Jukes–Cantor (JC), Hasegawa–Kishino–Yano (HKY), or generalized time reversible (GTR) model (see [829] for an overview of these and other related models).

We sampled 80 trees from the posterior distributions of each analysis and then visualized these trees in two dimensions using NLDR based on weighted Robinson-Foulds (RF) (wRF; [619]) distances (figure 14.3). Note that we use weighted (incorporating branch lengths) and unweighted (not incorporating branch lengths; [87, 620]) RF distances frequently in this chapter for explicative purposes because of their familiarity to most readers and the parallelism that exists between the two forms. However, other distances (e.g., geodesic; [69]) may offer a richer perspective on the relationships between trees and are worth exploring in empirical studies. When trees are shaded based on the assumed model, a cursory visual inspection reveals that the results of our analysis are sensitive to the assumed model (figure 14.3a): HKY and GTR produce credible sets of trees that are closer together in tree space compared with trees from the JC credible set. A few trees from the JC analysis are similar to trees sampled with HKY but generally not those sampled with GTR. The space occupied by HKY trees tends to sit in between those from JC and GTR, which suggests that an HKY analysis may support some relationships in common with JC and some in common with GTR.

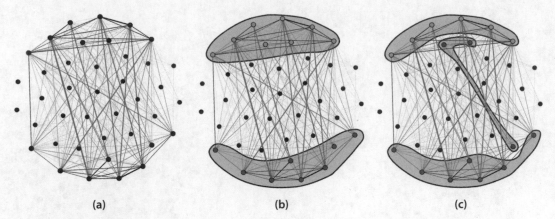

(a)   (b)   (c)

Figure 14.4. Bipartition covariance network based on trees from analyses assuming different models of sequence evolution. (A) The bipartition network with no communities labeled. (B) The bipartition network with two communities labeled. (C) The bipartition network with three communities labeled. Note that nodes in different communities are colored differently.

Community detection based on topological affinities shows intrinsic structure with three communities. These three communities reinforce the distinction between trees sampled from JC analyses and those from GTR and HKY (figure 14.3b). One community includes all trees from HKY and GTR (and a few from JC), while the other two communities are solely trees from the JC analysis. These communities highlight the wide dispersion and structuring of the JC credible set as well as the greater similarity between HKY and GTR results.

The results of community detection on bipartition covariance networks identifies the major topological areas of conflict between results from JC and GTR and reinforces the intermediate position of HKY trees in tree space (figure 14.4). These networks exhibit intrinsic structuring with both two and three communities. When the two communities are mapped onto the network (figure 14.4b), several large, positive covariances are obvious within each community, as are many large, negative covariances that separate the two. Interestingly, the three-community results show how the third community is composed of bipartitions originally included in both of the two, larger communities (figure 14.4C). However, these bipartitions generally had weak covariances (both positive and negative) connecting them to any other bipartition and are themselves connected only by weakly positive edges (thin, black lines). Therefore, this third community tells us less about major topological conflict in our tree set, so we focus on the two, larger communities.

The nature of the conflict between tree sets from different analyses is clarified by mapping bipartitions from different communities to the maximum a posteriori (MAP) trees from each analysis (figure 14.5). The first community (light shading) from the two-community analysis contains bipartitions that map primarily to the GTR MAP tree. These bipartitions either strongly conflict between the GTR and JC/HKY analyses or have intermediate posterior probabilities across all three analyses. In both cases (strong conflict between analyses or consistent uncertainty across them), there are identifiable sets of trees with alternative resolutions for these clades. The second community (dark shading) largely contains those bipartitions that conflict with the GTR tree and

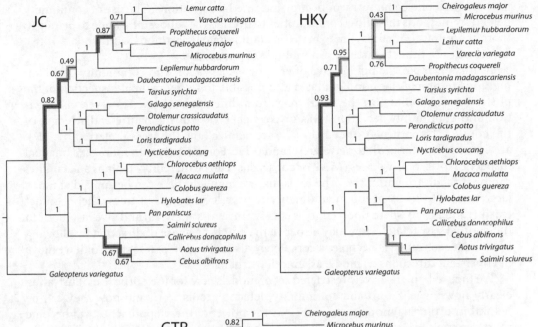

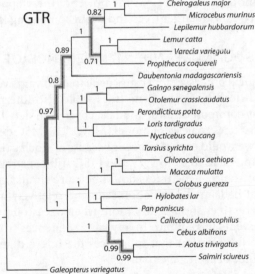

Figure 14.5. Maximum *a posteriori* trees from analyses of primate *cytb* sequences, assuming different models of sequence evolution. Branches are labeled with posterior probabilities and colored branches correspond to those bipartitions found in the two communities in Fig. 14.4B.

are primarily found in the JC tree. One exception is the single darkly shaded bipartition that appears in the GTR MAP tree (figure 14.5). This bipartition is found on the far left of the darkly shaded community in figure 14.4b and is connected by positive covariances to nodes in both the lightly and darkly shaded communities, indicating that it does not strongly conflict with other bipartitions. The community detection methods that we currently use for analysis of these networks do not allow nodes to be assigned to more than one community or occupy intermediate positions. The use of community detection methods that allow communities to overlap (i.e., allow for a node to belong to more than one community) may better accommodate bipartitions with weak, positive covariances to other bipartitions in multiple communities.

In this example, the perspective offered by visualization and community detection methods reinforces the conflict that is also possible to see by manual inspection of the posterior probabilities on the MAP trees from different analyses. However, visualization and community detection tools give a richer view of the nature and strength of the conflict. In particular, they help us to recognize how the trees preferred by HKY are essentially intermediate between JC and GTR. For analyses with more taxa, manual inspection of summary trees (MAP or consensus) can be incredibly tedious and conflict can be difficult to summarize. The tools that we highlight here can be applied at much larger scales to focus attention automatically on relevant parts of trees where conflict exists (rather than where uncertainty is consistent across analyses) and give insights into the conflict's structure. They may also give the conflict a sense of direction by allowing us to see if each analysis occupies a completely distinct region of tree space or if we move in consistent directions as model assumptions change.

We focused on a relatively restricted set of models here for the purpose of illustrating clearly how visualization and community detection tools work and how they can be helpful in exploring model sensitivity, but the power of these approaches will become most apparent in analyses with larger trees and when comparing larger sets of models.

14.2.2 JOINT VERSUS INDEPENDENT INFERENCE OF GENE TREES

By modeling biological processes that cause gene trees to vary, we can jointly estimate topologies and branch lengths for all gene trees while also inferring the overarching species tree. The most common of these models, covered in detail in other chapters of this book, is the multispecies coalescent model (MSC). The MSC describes the variation across gene trees that we would expect to occur because of stochasticity in the coalescent process. While allowing gene trees to vary, the MSC still constrains their variation in important ways. For instance, gene trees should be more similar along those branches of the species tree that had small population sizes and relatively infrequent speciation events. Conversely, gene trees should vary more when ancestral population sizes were large and speciation events were rapid. In any case, sequences from all the genes in a joint analysis provide information that influences the inferred species tree, which in turn influences inference of each gene tree.

While we know that joint inference can alter each gene tree, we might like to have some sense for how strong this effect is and whether it acts in the same way across different genes. For instance, is the space of gene trees sampled during joint inference a subset of the space that is sampled when gene trees are inferred independently? Does joint inference change the precision of gene tree estimates or does it shift the overall distribution of sampled gene trees more substantially? Can we identify distinct regions in tree space occupied by gene trees resulting from joint versus independent inference? Nonlinear dimensionality reduction visualizations and community detection can help shed some light on answers to these questions, which we illustrate here with Bayesian analyses of 10 genes sampled from 23 primate species.

By comparing gene trees from independent inference to those from joint inference under the MSC, we can see several interesting patterns (figure 14.6). First, in this case the tree topologies sampled during joint inference generally do represent a subset of those

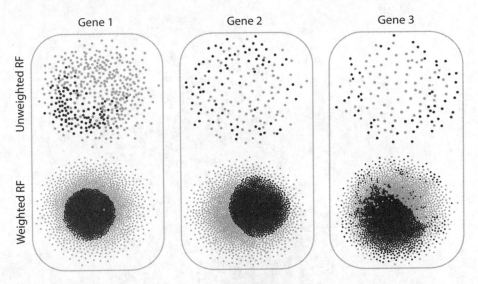

Figure 14.6. NLDR visualization of trees sampled during independent gene-tree inference (gray) or joint inference under the multispecies coalescent (MSC; black) for three genes using two distances (weighted and unweighted RF).

sampled during independent inference, and they do not seem to occupy a completely distinct region of tree space (figure 14.6, top row). This result should be comforting and suggests that the topological signal in each gene is roughly concordant with the topologies implied by the species tree. However, we can also see some interesting differences across genes. While the topologies sampled during joint inference generally occupy a subset of the space sampled during independent inference, the distribution of topologies changes (particularly for gene 1), indicating shifts in the posterior probability of different trees between the two analyses. Also, the tree space sampled by gene 1 includes a greater number of unique topologies, compared with genes 2 and 3, for both joint and independent inference. Both of these results suggest that gene 1 contains less topological information than genes 2 and 3. (Note that the overall size of the NLDR visualization is scaled to be equal for each plot, so the sizes of the plots themselves do not give us information about the expanse of tree space that they depict.)

Comparing projections of tree space generated using distances that do or do not take branch lengths into account also provides important information. The top row of figure 14.6 shows projected tree space based on unweighted RF distances (also known as symmetric differences; [87, 620]) for the three genes, where only the topology contributes to the distance. However, the bottom row shows projections based on weighted RF distances, where branch lengths are used, and these projections show much clearer distinctions between the trees sampled from joint and independent inference. The boundaries between the two are very clear for genes 1 and 2, although the two tree sets still overlap to some degree for gene 3. The increased distinction of the boundary in these plots suggests that the joint inference is having a substantial effect on estimated divergence times, possible more than on the topology.

A word of caution is also warranted here about the potential overinterpretation of NLDR visualizations. As discussed above, these visualizations frequently include some distortion of the true tree-to-tree distances, and these distortions can be resolved in

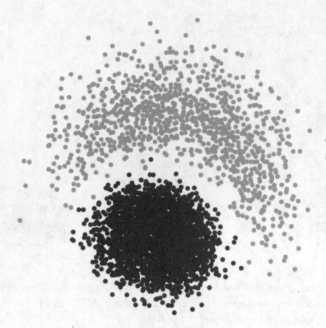

Figure 14.7. Alternate NLDR projection for Gene 1 from Fig. 14.6 using weighted RF distances. Kruskal-1 stress was used for this projection, while CCA stress was used in Fig. 14.6.

different ways, depending on the chosen NLDR method. While CCA stress [181] was optimized to generate all of the projections in figure 6, Kruskal-1 [382] stress was optimized to generate an alternative projection for gene 1 with weighted RF distances (figure 14.7). The CCA projection in figure 14.6 seems to suggest that the joint gene trees for gene 1 are nested inside the space of trees sampled from the independent analysis, but the Kruskal-1 projection suggests that these spaces are more distinct. In both cases, the space occupied by trees from the independent analysis remains larger than the space occupied by those from the joint analysis, and the distinction between the independent and joint tree sets is pronounced. A conservative practice with NLDR is to try several projections and see if patterns of interest persist. In cases in which one is interested in whether sets of trees are distinct in tree space, community detection with a topological affinity network has an advantage over NLDR since it uses the original tree-to-tree distances with no distortions from projection. Community detection on a topological affinity network can cleanly delineate the independent and joint trees as distinct communities for gene 1 (results not shown).

One significant practical challenge of using hierarchical models like the MSC in a Bayesian statistical framework is making sure that the Markov chain Monte Carlo analysis adequately samples tree space for each individual gene, as well as the species tree. Here, NLDR and community detection can also be helpful in checking for topological convergence. If replicate runs are sampling tree space in the same way, there should not be visually obvious differences in the distributions of trees from those runs and community detection methods on topological affinity networks should not be able to delineate the trees from different runs as separate communities. As an example of how these tools can help with assessing convergence, we separately shaded trees from replicate

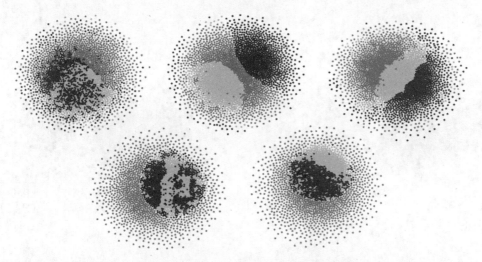

Figure 14.8. NLDR projections of gene-tree space for five different genes, based on three different analyses–independent gene tree analysis (medium grey), replicate one of a joint MSC analysis (black), and replicate two of a joint MSC analysis (light grey).

MSC analyses (figure 14.8), which were intentionally stopped before convergence was reached. For several genes, the replicate MSC analyses (shown in black and light grey) sampled the same general part of tree space but clearly did not yet mix well across this space. In other cases, the two replicates are sampling completely distinct regions of tree space (i.e., the black and light grey do not overlap; figure 14.8), perhaps because they have encountered local optima. Such diagnostics can be helpful in rapidly identifying loci with poor topological mixing.

14.2.3 UNDERSTANDING OF VARIATION ACROSS GENOMES

Modern phylogenomic studies involve the inference of phylogenies from every gene (or genomic region) across a set of completely sequenced genomes—a remarkable amount of information! This genome-wide perspective can shed light on the interplay between diverse biological processes that cause gene histories to differ from species histories. *Heliconius* butterflies are one well-studied group for which multiple biological processes and evolutionary forces (incomplete lineage sorting, horizontal gene flow, and selection) have combined to influence the true phylogenetic histories underlying different sections of genomes (i.e., the gene trees).

Edelman et al. [215] recently explored genome-wide patterns in gene tree variation for *Heliconius* and related groups. After filtering for quality and combining their data with other available genomes, they analyzed genome-wide alignments from 25 taxa in order to reconstruct the relationships among diverse *Heliconius* lineages. In doing so, they found strong evidence for the influence of gene flow in explaining the historical relationships among species. Gene flow and recombination will cause different sections of the genome to support different bifurcating phylogenetic trees, and [215] were able to map the distribution of support provided by 50 kb windows for different trees spanning the entire genome. They focused especially on the *erato-sara* clade, for which they had sampled six species, since many sections of the genome support a relatively small

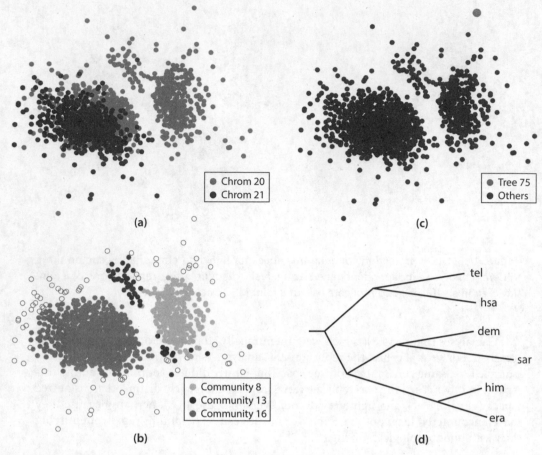

Figure 14.9. NLDR visualizations of trees from chromosomes 20 and 21 of the *erato-sara* clade. (A) Trees are colored by chromosome. (B) Trees are colored based on the results of community detection on a topology network with 24 communities. The three largest communities are shown in shades of grey, while trees from all other communities are shown as open circles. (C) Community detection on a topology network with two communities, where one community only contains a single tree (tree 75 from chromosome 20). (D) Phylogram of tree 75, with the outgroup removed.

number of topologies (two topologies, predominantly). To demonstrate how tree set visualization and community detection approaches could complement other analyses of these data, we focused on trees of the *erato-sara* clade inferred from 50 kb sliding windows across chromosomes 20 and 21.

Two- and three-dimensional visualizations of tree space based on NLDR projection with weighted RF distances show clear structuring in tree space for the trees sampled from these two chromosomes (figure 14.9). Labeling trees by chromosome allows us to see clearly that the two chromosomes do not share the same distribution of phylogenetic histories (figure 14.9a). Regions of chromosome 20 exhibit more variation in the trees they support, with trees from chromosome 21 being concentrated in a projected space roughly half the size of chromosome 20. This difference between chromosomes 20 and

21 (or, more broadly, between chromosome 21 and all others) was also noted by [215] based on tallies of different topologies across windows.

Community detection analyses based on topological networks give a more organized perspective on the variation across trees (figure 14.9b, c). Community detection (using the constant Potts model) first showed some natural structuring with two communities. One community contained only a single tree (inferred from the 75th 50 kb window on chromosome 20), while the other community contained all other trees (figure 14.9c). Further examination of tree 75 (figure 14.9d) shows that it has a topology distinct from the eight most common topologies across these genomes (figures 14.9d and 14.10d). As we increased the value of the constant Potts tuning parameter, other similar trees were added to the community with tree 75 (results not shown). These trees are uncommon in the genome and come from regions that may have been subject to rare hybridization events or influenced by unique evolutionary processes.

Three large communities emerge from the topological network as the tuning parameter is increased further (figure 14.9b). While no specific number of communities is strongly preferred, the structure of these large communities remains relatively stable. Visualizing these communities (figure 14.9b) shows both how they occupy central positions in tree space with high tree density and also how the use of networks based on original tree-to-tree distances is different than clustering in low-dimensional space. For instance, trees assigned to community 13 (figure 14.9b) are not nearest neighbors in the NLDR projection. Majority-rule consensus trees constructed from these three communities correspond to three of the most common topologies across the genome (community 8 = topology 1, community 13 = topology 5, and community 16 = topology 2; figure 14.10d), although the communities include some trees that are not identical to these topologies. Community 16 is the largest, which makes sense given the prevalence of topology 2 on chromosome 21. That none of these three communities corresponds directly to tree 3 (the third most common topology in the genome and second most common topology on these two chromosomes) suggests that trees with topology 3 may have relatively short internal branches and be close to other topologies in weighted RF tree space.

Bipartition networks show substantial structure and are dominated by two edges with large, negative weights (figure 14.10a). Natural community structure exists with two communities, although the exact composition of these communities varies depending on how stringently low- and high-frequency bipartitions are filtered (figure 14.10b, c). Less stringent filtering (removing only bipartitions present at frequencies above 0.99 or below 0.01) results in one community that includes several weakly connected bipartitions (figure 14.10b). Slightly more stringent filtering (removing bipartitions present at frequencies above 0.95 or below 0.05) results in two communities with two bipartitions each (figure 14.10c). Such filtering can be helpful because bipartitions present at either very low or very high frequencies often have weak covariances and are difficult to assign to any community with confidence. With either stringency, however, the two communities are separated by two large, negative edges.

The two large, negative edges correspond to the conflict between sets of bipartitions found in topology 1 and topology 3 (figure 14.10d). Topology 2, however, is composed of one bipartition from each community. This result highlights an important aspect of interpreting communities in bipartition networks. Bipartition communities may not precisely correspond to the most frequent topologies. For instance, the bipartition community in the bottom right of figure 14.10c contains bipartitions found in topology 1, while the bipartition community in the upper left contains bipartitions that are found in

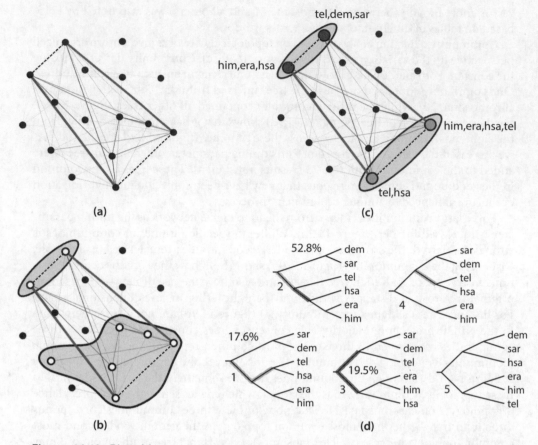

Figure 14.10. Bipartition covariance networks of the *erato-sara* clade, based on trees inferred from 50-kb windows along chromosomes 20 and 21. (A) The covariance network with nodes (black circles) representing different bipartitions. Thicker red lines represent strong negative covariances between bipartitions, while thicker blue lines represent strong positive covariances. (A color version of this figure is available through https://press.princeton.edu/ISBN/9780691207605.) (B) The covariance network with two communities (shaded areas) detected by community detection using bipartitions with frequencies ≥ 0.01 and ≤ 0.99. (C) The covariance network with two communities detected by community detection using bipartitions with frequencies ≥ 0.05 and ≤ 0.95. Bipartitions in these two communities are labeled with the taxa on one side of the induced split. (D) The five most common topologies observed in whole-genome analyses of the *erato-sara* clade, as identified by [215]. Note that the root of these trees is defined by the position of an outgroup that has been pruned. The frequencies of the first three topologies across chromosomes 20 and 21 are shown.

topology 3 (figure 14.10d). However, no bipartition community corresponds to topology 2, even though it is the most frequent topology across windows from chromosomes 20 and 21. At first, this result may seem troubling, but, in fact, it is informative. By mapping the bipartitions in different communities onto the most common topologies, the community detection results highlight parts of these topologies that conflict most strongly. The lack of large, positive covariances also tells us that the bipartitions in a community do not really "prefer" each other specifically. If only topologies 1 and 3 were present in the tree set, not topology 2, we would see strong, positive covariances within each of these communities. Further, that the bipartitions in these communities map to these topologies reinforces that they explain most of the conflict. (One bipartition also maps to topology 4, but this seems to be only because tree 4 shares that bipartition with topology 1). While these results from bipartition networks confirm what we could learn by manually examining the frequency of different topologies, manual inspection will be much more difficult for larger trees with more unique topologies.

In aggregate, these tools allow us to visualize the variation in phylogenetic signal across the genome in a way that is agnostic about the underlying evolutionary process, rapidly to identify "outlier" regions that may warrant particular attention, and to summarize the phylogenetic relationships that conflict most strongly across the genome.

14.2.4 PROSPECTS FOR FUTURE DEVELOPMENT AND APPLICATION

The combination of NLDR visualization and community detection methods show promise for a wide variety of applications in phylogenomics and species tree inference, and there is considerable room for future development. Since these tools are inherently exploratory, their most fruitful application may be in combination with software and databases that allow users to collect new data or test new hypotheses inspired by their initial explorations. The results from these analyses also need not be static. In the future, researchers could interactively explore integrated results. For example, one could slide the tuning parameter for community detection up and down and watch a network or NLDR plot to see how the community structure changes. If researchers are investigating genome-wide variation in gene trees, they could click on nodes in a topological network or points in an NLDR plot and immediately see summaries of the tree and its corresponding genomic region. They could also click on sets of bipartitions in a covariance network and have them highlighted in the corresponding topologies.

Mathematically, opportunities remain to explore approaches for community detection that incorporate properties found to be important for topological and bipartition networks in the context of phylogenetic studies (such as the ability to assign a particular node to more than one community simultaneously). Characterizing both network types for a range of biological data sets may also suggest new, tailored community detection models. Another opportunity could be to integrate additional information about the data that generated particular subsets of trees (e.g., characteristics of genes such as size, function, or variability) and use it to label nodes in topological networks. These labels could then be incorporated into the process of community detection.

Biologically, the application of these approaches (particularly community detection) in phylogenetics is new, and it will take time to understand what properties to expect of networks in different situations and how different community detection models may behave depending on the nature of the tree set. Bipartition covariance networks, in particular, seem to carry a lot of information, but the best way to characterize and interpret the structure of these networks is an active focus of research.

14.3 Appendix

Instructions for installing and running CloudForest, as well as workows and links to data for the examples presented in this chapter, are available at: https://treescaper .github.io/.

Bibliography

[1] Abbott, R., Albach, D., Ansell, S., Arntzen, J. W., Baird, S. J. E., Bierne, N., Boughman, J., et al. (2013). Hybridization and speciation. *Journal of Evolutionary Biology*, 26(2):229–246.

[2] Aberer, A. J., Krompass, D., and Stamatakis, A. (2013). Pruning rogue taxa improves phylogenetic accuracy: An efficient algorithm and webservice. *Systematic Biology*, 62(1):162–166.

[3] Adams, R. P., and Demeke, T. (1993). Systematic relationships in *Juniperus* based on random amplified polymorphic DNAs (RAPDs). *Taxon*, 42(3):553–571.

[4] Agrawal, A. (2017). Toward a predictive framework for convergent evolution: Integrating natural history, genetic mechanisms, and consequences for the diversity of life. *American Naturalist*, 190:S1–S12.

[5] Albalat, R., and Cañestro, C. (2016). Evolution by gene loss. *Nature Reviews Genetics*, 17(7): 379–391.

[6] Albert, V. A., Barbazuk, W., Depamphilis, C. W., Der, J. P., Leebens-Mack, J., Ma, H., and Palmer, J. D. (2013). The *Amborella* genome and the evolution of flowering plants. *Science*, 342(6165):1241089.

[7] Alexandrov, A., Ionescu, M. F., Schauser, K. E., and Scheiman, C. (1995). LogGP: Incorporating long messages into the LogP model—one step closer towards a realistic model for parallel computation. Technical Report, University of California at Santa Barbara, Santa Barbara, CA, USA.

[8] Ali, O. A., O'Rourke, S. M., Amish, S. J., Meek, M. H., Luikart, G., Jeffres, C., and Miller, M. R. (2016). RAD capture (rapture): Flexible and efficient sequence-based genotyping. *Genetics*, 202(2):389–400.

[9] Allman, E. S., Ané, C., and Rhodes, J. A. (2008). Identifiability of a Markovian model of molecular evolution with Gamma-distributed rates. *Advances in Applied Probability*, 40(1):229–249.

[10] Allman, E. S., Baños, H., and Rhodes, J. A. (2019). NANUQ: A method for inferring species networks from gene trees under the coalescent model. *Algorithms for Molecular Biology*, 14(1): 1–25.

[11] Allman, E. S., Degnan, J. H., and Rhodes, J. A. (2011a). Determining species tree topologies from clade probabilities under the coalescent. *Journal of Theoretical Biology*, 289(1):96–106.

[12] Allman, E. S., Degnan, J. H., and Rhodes, J. A. (2011b). Identifying the rooted species tree from the distribution of unrooted gene trees under the coalescent. *Journal of Mathematical Biology*, 62:833–862.

[13] Allman, E. S., Kubatko, L. S., and Rhodes, J. A. (2017a). Split scores: A tool to quantify phylogenetic signal in genome-scale data. *Systematic Biology*, 66(4):620–636.

[14] Allman, E. S., and Rhodes, J. A. (2003). Phylogenetic invariants for the general Markov model of sequence mutation. *Mathematical Biosciences*, 186(2):113–144.

[15] Allman, E. S., and Rhodes, J. A. (2007). Phylogenetic Invariants. In Gascuel, O., and Steel, M., editors, *Reconstructing Evolution: New Mathematical and Computational Advances*, pages 108–146. Oxford: Oxford University Press.

[16] Allman, E. S., and Rhodes, J. A. (2009). The identifiability of covarion models in phylogenetics. *IEEE/ACM Transactions on Computational Biology and Bioinformatics*, 6(1):76–88.

[17] Allman, E. S., Rhodes, J. A., and Sullivant, S. (2017b). Statistically consistent k-mer methods for phylogenetic tree reconstruction. *Journal of Computational Biology*, 24(2):153–171.

[18] Altenhoff, A., Boeckmann, B., Capella-Gutierrez, S., Dalquen, D., DeLuca, T., Forslund, K., Huerta-Cepas, J., et al. (2016). Standardized benchmarking in the quest for orthologs. *Nature Methods*, 13:425–430.

[19] Amdahl, G. M. (1967). Validity of the single processor approach to achieving large scale computing capabilities. In *Proceedings of the April 18–20, 1967, Spring Joint Computer Conference*, AFIPS '67 (Spring), pages 483–485. New York: Association for Computing Machinery.

[20] Amenta, N., and Klingner, J. (2002). Case study: Visualizing, sets of evolutionary trees. In *Proceedings of the IEEE Symposium on Information Visualization INFOVIS 2002*, page 71–74. https://doi.org/10.1109/INFVIS.2002.1173139.

[21] An, J., Zhu, L., Zhang, Y., and Tang, H. (2013). Efficient visible light photo-fenton-like degradation of organic pollutants using in situ surface-modified BiFeO3 as a catalyst. *Journal of Environmental Sciences*, 25(6):1213–1225.

[22] Andermann, T., Fernandes, A. M., Olsson, U., Töpel, M., Pfeil, B. E., Oxelman, B., Aleixo, A., Faircloth, B. C., and Antonelli, A. (2018). Allele phasing greatly improves the phylogenetic utility of ultraconserved elements. *Systematic Biology*. https://doi.org/10.1093/sysbio/syy039.

[23] Anderson, E. (1949). *Introgressive Hybridization*. New York: John Wiley.

[24] Anderson, E. (1953). Introgressive hybridization. *Biological Reviews of the Cambridge Philosophical Society*, 28:280–307.

[25] Anderson, E., Bai, Z., Bischof, C., Blackford, S., Demmel, J., Dongarra, J., Du Croz, J., et al. (1999). *LAPACK Users' Guide*. 3rd ed. Philadelphia: Society for Industrial and Applied Mathematics.

[26] Anderson, E., and Stebbins, G. L. (1954). Hybridization as an evolutionary stimulus. *Evolution*, 8:378–388.

[27] Ané, C., Larget, B., Baum, D. A., Smith, S. D., and Rokas, A. (2007). Bayesian estimation of concordance among gene trees. *Molecular Biology and Evolution*, 24:412–426.

[28] Angelis, K., and dos Reis, M. (2015). The impact of ancestral population size and incomplete lineage sorting on Bayesian estimation of species divergence times. *Current Zoology*, 61: 874–885.

[29] Anisimova, M., Bielawski, J., Dunn, K., and Yang, Z. (2007). Phylogenomic analysis of natural selection pressure in *Streptococcus* genomes. *BMC Evolutionary Biology*, 7(1):154.

[30] Arber, E., and Parkin, J. (1907). On the origin of angiosperms. *Botanical Journal of the Linnean Society*, 38(263):29–80.

[31] Arcila, D., Ortí, G., Vari, R., Armbruster, J. W., Stiassny, M. L., Ko, K. D., Sabaj, M. H., et al. (2017). Genome-wide interrogation advances resolution of recalcitrant groups in the tree of life. *Nature Ecology and Evolution*, 1:0020.

[32] Armstrong, J., Hickey, G., Diekhans, M., Fiddes, I. T., Novak, A. M., Deran, A., Fang, Q., et al. (2020). Progressive cactus is a multiple-genome aligner for the thousand-genome era. *Nature*, 587(7833):246–251.

[33] Arnold, B. C., and Strauss, D. (1991). Pseudolikelihood estimation: Some examples. *Sankhya: The Indian Journal of Statistics, Series B*, 53:233–243.

[34] Arnold, M., Hamrick, J., and Bennett, B. (1990). Allozyme variation in Louisiana irises—a test for introgression and hybrid speciation. *Heredity*, 65:297–306.

[35] Arvestad, L., Lagergren, J., and Sennblad, B. (2009). The gene evolution model and computing its associated probabilities. *Journal of the ACM*, 56(2):7.

[36] Avise, J., and Saunders, N. (1984). Hybridization and introgression among species of sunfish (*Lepomis*)—analysis by mitochondrial DNA and allozyme markers. *Genetics*, 108: 237–255.

[37] Avise, J. C., and Robinson, T. J. (2008). Hemiplasy: A new term in the lexicon of phylogenetics. *Systematic Biology*, 57(3):503–507.

[38] Avni, E., Cohen, R., and Snir, S. (2015). Weighted quartets phylogenetics. *Systematic Biology*, 64(2):233–242.

[39] Baele, G., Suchard, M., Rambaut, A., and Lemey, P. (2017). Emerging concepts of data integration in pathogen phylodynamics. *Systematic Biology*, 66(1):e47–e65.

[40] Baker, A., Haddrath, O., McPherson, J., and Cloutier, A. (2014). Genomic support for a moatinamou clade and adaptive morphological convergence in flightless ratites. *Molecular Biology and Evolution*, 31:1686–1696.

[41] Balbuena, J. A., Míguez-Lozano, R., and Blasco-Costa, I. (2013). PACo: a novel Procrustes application to cophylogenetic analysis. *PLOS ONE*, 8(4):e61048.

[42] Baldwin, B. G., and Markos, S. (1998). Phylogenetic utility of the external transcribed spacer (ETS) of 18S–26S rDNA: Congruence of ETS and ITS trees of *Calycadenia* (Compositae). *Molecular Phylogenetics and Evolution*, 10(3):449–463.

[43] Baldwin, B. G., Sanderson, M. J., Porter, J., Wojciechowski, M. F., Campbell, C. S., and Donoghue, M. J. (1995). The ITS region of nuclear ribosomal DNA: A valuable source of evidence on angiosperm phylogeny. *Annals of the Missouri Botanical Garden*, 82(2):247–277.

[44] Ballesteros, J. A., and Sharma, P. P. (2019). A critical appraisal of the placement of Xiphosura (Chelicerata) with account of known sources of phylogenetic error. *Systematic Biology*, 68(6):896–917.

[45] Bandelt, H.-J., and Dress, A. W. M. (1993). A relational approach to split decomposition. In Opitz, O., Lausen, B., Klar, R., Opitz, O., Lausen, B., and Klar, R., editors, *Information and Classification. Studies in Classification, Data Analysis and Knowledge Organization* pages 123–131. Berlin: Springer.

[46] Bansal, M., Burleigh, J., Eulenstein, O., and Fernández-Baca, D. (2010). Robinson-Foulds supertrees. *Algorithms for Molecular Biology*, 5:18.

[47] Barkman, T., and Simpson, B. (2002). Hybrid origin and parentage of *Dendrochilum acuiferum* (Orchidaceae) inferred in a phylogenetic context using nuclear and plastid DNA sequence data. *Systematic Botany*, 27:209–220.

[48] Barrowclough, G., Cracraft, J., Klicka, J., and Zink, R. (2016). How many kinds of birds are there and why does it matter? *PLOS ONE*, 11:e0166307.

[49] Barton, N. (2001). Speciation. *Trends in Ecology and Evolution*, 16:325.

[50] Bastide, P., Solís-Lemus, C., Kriebel, R., Sparks, K. W., and Ané, C. (2018). Phylogenetic comparative methods on phylogenetic networks with reticulations. *Systematic Biology*, 67(5): 800–820.

[51] Baudry, J.-P., Maugis, C., and Michel, B. (2012). Slope heuristics: Overview and implementation. *Statistics and Computing*, 22(2):455–470.

[52] Baum, D. A. (1998). Individuality and the existence of species through time. *Systematic Biology*, 47(4):641–653.

[53] Baum, D. A. (2007). Concordance trees, concordance factors, and the exploration of reticulate genealogy. *Taxon*, 56(2):417–426.

[54] Bayzid, M. S., Hunt, T., and Warnow, T. (2014). Disk covering methods improve phylogenomic analyses. *BMC Genomics*, 15(Suppl. 6):S7.

[55] Bayzid, M. S., Mirarab, S., Boussau, B., and Warnow, T. (2015). Weighted statistical binning: Enabling statistically consistent genome-scale phylogenetic analyses. *PLOS ONE*, 10(6): 30129183.

[56] Bayzid, M. S., and Warnow, T. (2013). Naive binning improves phylogenomic analyses. *Bioinformatics*, 29(18):2277–2284.

[57] Beaulieu, J. M., O'Meara, B. C., Crane, P., and Donoghue, M. J. (2015). Heterogeneous rates of molecular evolution and diversification could explain the Triassic age estimate for angiosperms. *Systematic Biology*, 64(5):869–878.

[58] Beckman, E. J., Benham, P. M., Cheviron, Z. A., and Witt, C. C. (2018). Detecting introgression despite phylogenetic uncertainty: The case of the South American siskins. *Molecular Ecology*, 27(22):4350–4367.

[59] Bedford, T., and Hartl, D. (2009). Optimization of gene expression by natural selection. *Proceedings of the National Academy of Sciences of the United States of America*, 106: 1133–1138.

[60] Bekele, W., Itaya, A., Boyle, B., Yan, W., Fetch, J. M., and Tinker, N. A. (2020). A targeted genotyping-by-sequencing tool (RAPTURE) for genomics-assisted breeding in oat. *Theoretical and Applied Genetics*, 133:653–664.

[61] Bell, C. D., Soltis, D. E., and Soltis, P. S. (2010). The age and diversification of the angiosperms re-revisited. *American Journal of Botany*, 97(8):1296–1303.

[62] Ben-Dor, A., Chor, B., Graur, D., Ophir, R., and Pelleg, D. (1998). Constructing phylogenies from quartets: Elucidation of Eutherian superordinal relationships. *Journal of Computational Biology*, 5(3):377–390.

[63] Benjamini, Y., and Hochberg, Y. (1995). Controlling the false discovery rate: A practical and powerful approach to multiple testing. *Journal of the Royal Statistical Society*, 57:289–300.

[64] Bensasson, D., Dicks, J., Ludwig, J. M., Bond, C. J., Elliston, A., Roberts, I. N., and James, S. A. (2018). Diverse lineages of *Candida albicans* live on old oaks. *Genetics*, 211(1):277–288.

[65] Bertrand, Y. J. K., Scheen, A.-C., Marcussen, T., Pfeil, B. E., de Sousa, F., and Oxelman, B. (2015). Assignment of homoeologues to parental genomes in allopolyploids for species tree inference, with an example from *Fumaria* (Papaveraceae). *Systematic Biology*, 64:448–471.

[66] Betancur-R, R., Naylor, G. J. P., and Ortí, G. (2014). Conserved genes, sampling error, and phylogenomic inference. *Systematic Biology*, 63(2):257–262.

[67] Bezanson, J., Karpinski, S., Shah, V. B., and Edelman, A. (2012). Julia: A Fast Dynamic Language for Technical Computing. https://doi.org/10.48550/arXiv.1209.5145.

[68] Bickley, S. R. B., and Logan, M. P. O. (2014). Regulatory modulation of the t-box gene tbx5 links development, evolution, and adaptation of the sternum. *Proceedings of the National Academy of Sciences of the United States of America*, 111:17917–17922.

[69] Billera, L., Holmes, S., and Vogtmann, K. (2001). Geometry of the space of phylogenetic trees. *Advances in Applied Mathematics*, 27:733–767.

[70] Bininda-Emonds, O. (2004a). *Phylogenetic Supertrees: Combining Information to Reveal the Tree of Life*. Computational Biology. Dordrecht: Kluwer Academic.

[71] Bininda-Emonds, O. R. P. (2004b). The evolution of supertrees. *Trends in Ecology and Evolution*, 19(6):315–322.

[72] Birge, L., and Massart, P. (2007). Minimal penalties for Gaussian model selection. *Probability Theory and Related Fields*, 138:33–73.

[73] Blair, C., and Ané, C. (2020). Phylogenetic trees and networks can serve as powerful and complementary approaches for analysis of genomic data. *Systematic Biology*, 69(3):593–601.

[74] Blanc, G., and Wolfe, K. H. (2004). Widespread paleopolyploidy in model plant species inferred from age distributions of duplicate genes. *The Plant Cell*, 16(7):1667–1678.

[75] Blanchette, M., Kent, W., Riemer, C., Elnitski, L., Smit, A., Roskin, K., Baertsch, R., et al. (2004). Aligning multiple genomic sequences with the threaded blockset aligner. *Genome Research*, 14:708–715.

[76] Blanco-Pastor, J., Vargas, P., and Pfeil, B. (2012). Coalescent simulations reveal hybridization and incomplete lineage sorting in Mediterranean *Linaria*. *PLOS ONE*, 7:39089.

[77] Blischak, P., Chifman, J., Wolfe, A. D., and Kubatko, L. S. (2018a). HyDe: A Python package for genome-scale hybridization detection. *Systematic Biology*, 67(5):821–829.

[78] Blischak, P. D., Latvis, M., Morales-Briones, D. F., Johnson, J. C., Di Stilio, V. S., Wolfe, A. D., and Tank, D. C. (2018b). Fluidigm2PURC: Automated processing and haplotype inference for double-barcoded PCR amplicons. *Applications in Plant Sciences*, 6:e1156.

[79] Blischak, P. D., Wenzel, A. J., and Wolfe, A. D. (2014). Gene prediction and annotation in *Penstemon* (Plantaginaceae): A workflow for marker development from low-coverage genome sequencing. *Applications in Plant Sciences*, 2:1400044.

[80] Blom, M. P. K., Bragg, J. G., Potter, S., and Moritz, C. (2017). Accounting for uncertainty in gene tree estimation: Summary-coalescent species tree inference in a challenging radiation of Australian lizards. *Systematic Biology*, 66(3):352–366.

[81] Bolger, A., Lohse, M., and Usadel, B. (2014). Trimmomatic: A flexible trimmer for Illumina sequence data. *Bioinformatics*, 30:2114–2120.

[82] Bolnick, D., Barrett, R., Oke, K., Rennison, D., and Stuart, Y. (2018). Non-parallel evolution. *Annual Review of Ecology, Evolution, and Systematics*, 49:303–330.

[83] Bonnard, C., Berry, V., and Lartillot, N. (2006). Multipolar consensus for phylogenetic trees. *Systematic Biology*, 55(5):837–843.

[84] Boucher, F., Casazza, G., Szövényi, P., and Conti, E. (2016). Sequence capture using RAD probes clarifies phylogenetic relationships and species boundaries in *Primula* sect. *Auricula*. *Molecular Phylogenetics and Evolution*, 104:60–72.

[85] Bouckaert, R., Heled, J., Kühnert, D., Vaughan, T., Wu, C.-H., Xie, D., Suchard, M. A., Rambaut, A., and Drummond, A. J. (2014). BEAST 2: software platform for Bayesian evolutionary analysis. *PLOS Computational Biology*, 10(4):e1003537.

[86] Bouckaert, R. R. (2010). DensiTree: Making sense of sets of phylogenetic trees. *Bioinformatics*, 26(10):1372–1373.

[87] Bourque, M. (1978). Arbres de steiner et reseaux dont varie l'emplacement de certain sommets. PhD diss., University of Montreal.

[88] Boussau, B., Szöllősi, G. J., Duret, L., Gouy, M., Tannier, E., and Daubin, V. (2013). Genome-scale coestimation of species and gene trees. *Genome Research*, 23(2):323–330.

[89] Boyle, E. A., Li, Y. I., and Pritchard, J. K. (2017). An expanded view of complex traits: From polygenic to omnigenic. *Cell*, 169(7):1177–1186.

[90] Bravo, G. A., Antonelli, A., Bacon, C. D., Bartoszck, K., Blom, M. P. K., Huynh, S., Jones, G., et al. (2019). Embracing heterogeneity: Coalescing the tree of life and the future of phylogenomics. *PeerJ*, 7:e6399.

[91] Brawand, D., Soumillon, M., Necsulea, A., Julien, P., Csárdi, G., Harrigan, P., Weier, M., et al. (2011). The evolution of gene expression levels in mammalian organs. *Nature*, 478(7369):343–348.

[92] Bremer, B., and Struwe, L. (1992). Phylogeny of the Rubiaceae and the Loganiaceae: Congruence or conflict between morphological and molecular data? *American Journal of Botany*, 79(10):1171–1184.

[93] Broderick, S. R., Stevens, M. R., Geary, B., Love, S. L., Jellen, E. N., Dockter, R. B., Daley, S. L., and Lindgren, D. T. (2011). A survey of *Penstemon*'s genome size. *Genome*, 54:160–173.

[94] Brooks, S. P., and Gelman, A. (1998). General methods for monitoring convergence of iterative simulations. *Journal of Computational and Graphical Statistics*, 7(4):434–455.

[95] Brown, J. M., and Thomson, R. C. (2016). Bayes factors unmask highly variable information content, bias, and extreme influence in phylogenomic analyses. *Systematic Biology*, 66(4):517–530.

[96] Brunsfeld, S. J., Miller, T. R., and Carstens, B. C. (2007). Insights into the biogeography of the Pacific Northwest of North America: Evidence from the phylogeography of *Salix melanopsis* (Salicaceae). *Systematic Botany*, 32:129–139.

[97] Brunsfeld, S. J., Sullivan, J., Soltis, D. S., and Soltis, P. S. (2001). Comparative phylogeography of northwestern North America: A synthesis. In Silvertown, J., and Antonovicks, J., editors, *Integrating Ecological and Evolutionary Processes in a Spatial Context*, pages 319–339. Oxford: Blackwell Science.

[98] Bryant, D., Bouckaert, R., Felsenstein, J., Rosenberg, N., and RoyChoudhury, A. (2012). Inferring species trees directly from biallelic genetic markers: bypassing gene trees in a full coalescent analysis. *Molecular Biology and Evolution*, 29(8):1917–1932.

[99] Bryant, D., and Steel, M. (2001). Constructing optimal trees from quartets. *Journal of Algorithms*, 38(1):237–259.

[100] Buckley, T. (2002). Model misspecification and probabilistic tests of topology: Evidence from empirical data sets. *Systematic Biology*, 51:509–523.

[101] Buckley, T., Cordeiro, M., Marshall, D., and Simon, C. (2006). Differentiating between hypotheses of lineage sorting and introgression in New Zealand alpine cicadas (Maoricicada Dugdale). *Systematic Biology*, 55:411–425.

[102] Buenrostro, J., Giresi, P., Zaba, L., Chang, H., and Greenleaf, W. (2013). Transposition of native chromatin for fast and sensitive epigenomic profiling of open chromatin, DNA-binding proteins and nucleosome position. *Nature Methods*, 10:1213–1218.

[103] Burbink, F. T., and Gehara, M. (2018). The biogeography of deep time reticulation. *Systematic Biology*, 67(5):743–744.

[104] Burggren, W., and Crews, D. (2014). Epigenetics in comparative biology: Why we should pay attention. *Integrative and Comparative Biology*, 54:7–20.

[105] Cai, R., and Ané, C. (2021). Assessing the fit of the multi-species network coalescent to multi-locus data. *Bioinformatics*, 37(5):634–641.

[106] Calamoneri, T., Donato, V. D., Mariottini, D., and Patrignani, M. (2020). Visualizing co-phylogenetic reconciliations. *Theoretical Computer Science*, 815:228–245.

[107] Cannon, S. B., McKain, M. R., Harkess, A., Nelson, M. N., Dash, S., Deyholos, M. K., and Peng, Y. (2014). Multiple polyploidy events in the early radiation of nodulating and nonnodulating legumes. *Molecular Biology and Evolution*, 32(1):193–210.

[108] Cao, Z., and Nakhleh, L. (2019). Empirical performance of tree-based inference of phylogenetic networks. In *Proceedings of the Workshop of Algorithms in Bioinformatics (WABI)*. Dagstuhl, Germany: Schloss Dagstuhl—Leibniz-Zentrum fuer Informatik.

[109] Carbone, L., Harris, R. A., Gnerre, S., Veeramah, K. R., Lorente-Galdos, B., Huddleston, J., Meyer, T. J., et al. (2014). Gibbon genome and the fast karyotype evolution of small apes. *Nature*, 513:195–201.

[110] Cardona, G., Llabrés, M., and Rosselló, F. (2010a). Two results on distances for phylogenetic networks. In *Advances in Bioinformatics*, pages 93–100. Berlin: Springer.

[111] Cardona, G., Llabrés, M., Rosselló, F., and Valiente, G. (2008a). A distance metric for a class of tree-sibling phylogenetic networks. *Bioinformatics*, 24(13):1481–1488.

[112] Cardona, G., Llabres, M., Rosselló, F., and Valiente, G. (2009a). Metrics for phylogenetic networks I: Generalizations of the Robinson-Foulds metric. *IEEE/ACM Transactions on Computational Biology and Bioinformatics*, 6(1):46–61.

[113] Cardona, G., Llabrés, M., Rosselló, F., and Valiente, G. (2009b). On Nakhleh's metric for reduced phylogenetic networks. *IEEE/ACM Transactions on Computational Biology and Bioinformatics*, 6(4):629–638.

[114] Cardona, G., Llabrés, M., Rosselló, F., and Valiente, G. (2010b). Comparison of galled trees. *IEEE/ACM Transactions on Computational Biology and Bioinformatics*, 8(2):410–427.

[115] Cardona, G., Llabrés, M., Rosselló, F., and Valiente, G. (2014). The comparison of tree-sibling time consistent phylogenetic networks is graph isomorphism-complete. *Scientific World Journal*, 2014: 254279.

[116] Cardona, G., Rosselló, F., and Valiente, G. (2008b). Extended Newick: it is time for a standard representation of phylogenetic networks. *BMC Bioinformatics*, 9:532.

[117] Cardona, G., Rosselló, F., and Valiente, G. (2009c). Comparison of tree-child phylogenetic networks. *IEEE/ACM Transactions on Computational Biology and Bioinformatics*, 6(4): 552–569.

[118] Carlson, J., Locke, A., Flickinger, M., Zawistowski, M., Levy, S., Myers, R., Boehnke, M., et al. (2018). Extremely rare variants reveal patterns of germline mutation rate heterogeneity in humans. *Nature Communications*, 9:3753.

[119] Carroll, S., Prud'homme, B., and Gompel, N. (2008). Regulating evolution. *Scientific American*, 298:60–67.

[120] Carstens, B. C., Brunsfeld, S. J., Demboski, J. R., Good, J. M., and Sullivan, J. (2005). Investigating the evolutionary history of the Pacific Northwest mesic forest ecosystem: Hypothesis testing within a comparative phylogeographic framework. *Evolution*, 59:1639–1652.

[121] Casola, C. (2018). From de novo to"de nono": The majority of novel protein-coding genes identified with phylostratigraphy are old genes or recent duplicates. *Genome Biology and Evolution*, 10:2906–2918.

[122] Castellanos, M. C., Wilson, P. S., Keller, S. J., Wolfe, A. D., and Thompson, J. D. (2006). Anther evolution: Pollen presentation strategies when pollinators differ. *American Naturalist*, 167:288–296.

[123] Castillo-Ramirez, S., Liu, L., Pearl, D., and Edwards, S. (2010). Bayesian estimation of species trees: A practical guide to optimal sampling and analysis. In Knowles, L. L., and Kubatko, L. S., editors, *Estimating Species Trees: Practical and Theoretical Aspects*, pages 15–33. Hoboken, NJ: Wiley-Blackwell.

[124] Cavender, J. A. (1989). Mechanized derivation of linear invariants. *Molecular Biology and Evolution*, 6(3):301–316.

[125] Cavender, J. A., and Felsenstein, J. (1987). Invariants of phylogenies in a simple case with discrete states. *Journal of Classification*, 4:57–71.

[126] Chan, R., Baldwin, B. G., and Ornduff, R. (2002). Cryptic goldfields: A molecular phylogenetic reinvestigation of *Lasthenia californica* sensu lato and close relatives (Compositae: Heliantheae sensu lato). *American Journal of Botany*, 89(7):1103–1112.

[127] Chan, Y., Marks, M., Jones, F., Villarreal, Jr., G., Shapiro, M., Brady, S., Southwick, A., et al. (2010). Adaptive evolution of pelvic reduction in sticklebacks by recurrent deletion of a Pitx1 enhancer. *Science*, 327:302–305.

[128] Chang, W. C., Burleigh, J. G., Baca, D. F., and Eulenstein, O. (2011). An ILP solution for the gene duplication problem. *BMC Bioinformatics*, 12(Suppl. 1):S14.

[129] Charleston, M. A. (1998). Jungles: A new solution to the host/parasite phylogeny reconciliation problem. *Mathematical Biosciences*, 149:191–223.

[130] Chase, M. W., Soltis, D. E., Olmstead, R. G., Morgan, D., Les, D. H., Mishler, B. D., and Duvall, M. R. (1993). Phylogenetics of seed plants: An analysis of nucleotide sequences from the plastid gene rbcL. *Annals of the Missouri Botanical Garden*, 80(3):528–580.

[131] Chaudhary, R., Boussau, B., Burleigh, J. G., and Fernández-Baca, D. (2014). Assessing approaches for inferring species trees from multi-copy genes. *Systematic Biology*, 64(2):325–339.

[132] Chaudhary, R., Burleigh, J. G., and Fernández-Baca, D. (2013). Inferring species trees from incongruent multi-copy gene trees using the Robinson-Foulds distance. *Algorithms for Molecular Biology*, 8:28.

[133] Chaudhary, R., Fernández-Baca, D., and Burleigh, J. G. (2015). MulRF: A software package for phylogenetic analysis using multi-copy gene trees. *Bioinformatics*, 31(3):432–433.

[134] Chen, F., and Li, W. (2001). Genomic divergences between humans and other hominoids and the effective population size of the common ancestor of humans and chimpanzees. *American Journal of Human Genetics*, 68:444–456.

[135] Chen, J., Swofford, R., Johnson, J., Cummings, B., Rogel, N., Lindblad-Toh, K., Haerty, W., Palma, F., and Regev, A. (2019). A quantitative framework for characterizing the evolutionary history of mammalian gene expression. *Genome Research*, 29:53–63.

[136] Chen, M.-Y., Liang, D., and Zhang, P. (2015). Selecting question-specific genes to reduce incongruence in phylogenomics: A case study of jawed vertebrate backbone phylogeny. *Systematic Biology*, 64(6):1104–1120.

[137] Cheng, S., Melkonian, M., Smith, S. A., Brockington, S., Archibald, J. M., Delaux, P.-M., and Li, F.-W. (2018). 10KP: A phylodiverse genome sequencing plan. *Gigascience*, 7(3):013.

[138] Chester, M., Gallagher, J. P., Symonds, V., Silva, A. V. C., Mavrodiev, E. V., Leitch, A. R., Soltis, P. S., and Soltis, D. E. (2012). Extensive chromosomal variation in a recently formed natural allopolyploid species, *Tragopogon miscellus* (Asteraceae). *Proceedings of the National Academy of Sciences of the United States of America*, 109(4):1176–1181.

[139] Chicco, D., Bi, H., Reimand, J., and Hoffman, M. (2019). Behst: Genomic set enrichment analysis enhanced through integration of chromatin long-range interactions. *bioRxiv*. https://doi.org/10.1101/168427.

[140] Chifman, J., and Kubatko, L. (2014). Quartet inference from SNP data under the coalescent model. *Bioinformatics*, 30(23):3317–3324.

[141] Chifman, J., and Kubatko, L. (2015). Identifiability of the unrooted species tree topology under the coalescent model with time-reversible substitution processes, site-specific rate variation, and invariable sites. *Journal of Theoretical Biology*, 374:35–47.

[142] Christin, P.-A., Edwards, E. J., Besnard, G., Boxall, S. F., Gregory, R., Kellogg, E. A., Hartwell, J., and Osborne, C. P. (2012). Adaptive evolution of C4 photosynthesis through recurrent lateral gene transfer. *Current Biology*, 22(5):445–449.

[143] Claramunt, S., and Cracraft, J. (2015). A new time tree reveals earth history's imprint on the evolution of modern birds. *Science Advances*, 1(11):e1501005.

[144] Clemente, J. C., Satou, K., and Valiente, G. (2007). Phylogenetic reconstruction from non-genomic data. *Bioinformatics*, 23(2):e110–e115.

[145] Cloutier, A., Sackton, T., Grayson, P., Clamp, M., Baker, A., and Edwards, S. (2019). Whole-genome analyses resolve the phylogeny of flightless birds (Palaeognathae) in the presence of an empirical anomaly zone. *Systematic Biology*, 68:937–955.

[146] Cloutier, A., Sackton, T., Grayson, P., Edwards, S., and Baker, A. (2018). First nuclear genome assembly of an extinct moa species, the little bush moa (anomalopteryx didiformis). *bioRxiv*. https://doi.org/10.1101/262816.

[147] Copes, L. E., Lucas, L. M., Thostenson, J. O., Hoekstra, H. E., and Boyer, D. M. (2016). A collection of non-human primate computed tomography scans housed in MorphoSource, a repository for 3D data. *Scientific Data*, 3:160001.

[148] Copetti, D., Búrquez, A., Bustamante, E., Charboneau, J. L. M., Childs, K. L., Eguiarte, L. E., Lee, S., et al. (2017). Extensive gene tree discordance and hemiplasy shaped the genomes of North American columnar cacti. *Proceedings of the National Academy of Sciences of the United States of America*, 114(45):12003–12008.

[149] Cracraft, J. (1974). Phylogeny and evolution of the ratite birds. *Ibis*, 116:494–521.

[150] Cracraft, J. (1988). The major clades of birds. In Benton, M. J., editor, *The Phylogeny and Classification of the Tetrapods. Volume 1: Amphibians, Reptiles, Birds*. Systematics Association Special Volume 35A:361.

[151] Cracraft, J., Donoghue, M., Dragoo, J., Hillis, D., and Yates, T., editors (2002). *Assembling the Tree of Life: Harnessing Life's History to Benefit Science and Society*. Washington, DC: National Science Foundation. Available at http://ucjeps.berkeley.edu/tol.pdf.

[152] Crane, P. R. (1985). Phylogenetic analysis of seed plants and the origin of angiosperms. *Annals of the Missouri Botanical Garden*, 72:716–793.

[153] Criscuolo, A., and Gascuel, O. (2008). Fast NJ-like algorithms to deal with incomplete distance matrices. *BMC Bioinformatics*, 9:166.

[154] Crombie, T. A., Zdraljevic, S., Cook, D. E., Tanny, R. E., Brady, S. C., Wang, Y., Evans, K. S., et al. (2019). Deep sampling of Hawaiian *Caenorhabditis elegans* reveals high genetic diversity and admixture with global populations. *eLife*, 8:e50465.

[155] Crosswhite, F. S. (1965). Hybridization of *Penstemon barbatus* (Scrophulariaceae) of section *Elmigera* with species of *Habroanthus*. *Southwestern Naturalist*, 10:234–237.

[156] Crotty, S. M., Minh, B. Q., Bean, N. G., Holland, B. R., Tuke, J., Jermiin, L. S., and Haeseler, A. V. (2019). GHOST: Recovering historical signal from heterotachously evolved sequence alignments. *Systematic Biology*, 69(2):249–264.

[157] Cui, R., Schumer, M., Kruesi, K., Walter, R., Andolfatto, P., and Rosenthal, G. G. (2013). Phylogenomics reveals extensive reticulate evolution in *Xiphophorus* fishes. *Evolution*, 67(8):2166–2179.

[158] Culler, D., Karp, R., Patterson, D., Sahay, A., Schauser, K. E., Santos, E., Subramonian, R., and von Eicken, T. (1993). LogP: Towards a realistic model of parallel computation. In *Proceedings of the Fourth ACM SIGPLAN Symposium on Principles and Practice of Parallel Programming, PPOPP '93*, pages 1–12, New York: Association for Computing Machinery.

[159] Cuénoud, P., Savolainen, V., Chatrou, L. W., Powell, M., Grayer, R. J., and Chase, M. W. (2002). Molecular phylogenetics of Caryophyllales based on nuclear 18S rDNA and plastid rbcL, atpB, and matK DNA sequences. *American Journal of Botany*, 89(1):132–144.

[160] Dasarathy, G., Nowak, R., and Roch, S. (2015). Data requirement for phylogenetic inference from multiple loci: A new distance method. *IEEE/ACM Transactions on Computational Biology and Bioinformatics*, 12(2):422–432.

[161] Datwyler, S. L., and Wolfe, A. D. (2004). Phylogenetic relationships and morphological evolution in *Penstemon* subg. *Dasanthera* (Veronicaceae). *Systematic Botany*, 29:165–176.

[162] Davidson, R., Vachaspati, P., Mirarab, S., and Warnow, T. (2015). Phylogenomic species tree estimation in the presence of incomplete lineage sorting and horizontal gene transfer. *BMC Genomics*, 16(Suppl. 10):S1.

[163] de Bakker, M. A., Fowler, D. A., Oude, J., Dondorp, E. M., Navas, M. A., Horbanczuk, J. O., Sire, J. Y., Szczerbinska, D., and Richardson, M. K. (2013). Digit loss in archosaur evolution and the interplay between selection and constraints. *Nature*, 500:445–448.

[164] De Maio, N., Schrempf, D., and Kosiol, C. (2015). PoMo: An allele frequency-based approach for species tree estimation. *Systematic Biology*, 64(6):1018–1031.

[165] De Oliveira Martins, L., Mallo, D., and Posada, D. (2016). A Bayesian supertree model for genome-wide species tree reconstruction. *Systematic Biology*, 65(3):397–416.

[166] De-Silva, D., Mota, L., Chazot, N., Mallarino, R., Silva-Brandao, K., Pinerez, L., and Freitas, A. (2017). North Andean origin and diversification of the largest ithomiine butterfly genus. *Scientific Reports*, 7:45966.

[167] de Vienne, D. M. (2018). Tanglegrams are misleading for visual evaluation of tree congruence. *Molecular Biology and Evolution*, 36(1):174–176.

[168] de Vienne, D. M., Refrégier, G., López-Villavicencio, M., Tellier, A., Hood, M. E., and Giraud, T. (2013). Cospeciation vs host-shift speciation: Methods for testing, evidence from natural associations and relation to coevolution. *New Phytologist*, 198(2):347–385.

[169] Dean, R., Harrison, P. W., Wright, A. E., Zimmer, F., and Mank, J. E. (2015). Positive selection underlies Faster-Z evolution of gene expression in birds. *Molecular Biology and Evolution*, 32(10):2646–2656.

[170] Deanna, R., Larter, M., Barboza, G., and Smith, S. (2019). Repeated evolution of a morphological novelty: A phylogenetic analysis of the inflated fruiting calyx in the Physalideae tribe (Solanaceae). *American Journal of Botany*, 106:270–279.

[171] Deanna, R., Smith, S., Sarkinen, T., and Chiarini, F. (2018). Patterns of chromosomal evolution in the florally diverse Andean clade Iochrominae (Solanaceae). *Perspectives in Plant Ecology Evolution and Systematics*, 35:31–43.

[172] DeGiorgio, M., and Degnan, J. H. (2013). Robustness to divergence time underestimation when inferring species trees from estimated gene trees. *Systematic Biology*, 63(1):66–82.

[173] Degnan, J. H. (2013). Anomalous unrooted gene trees. *Systematic Biology*, 62(4):574–590.

[174] Degnan, J. H. (2018). Modeling hybridization under the network multispecies coalescent. *Systematic Biology*, 67(5):786–799.

[175] Degnan, J. H., DeGiorgio, M., Bryant, D., and Rosenberg, N. A. (2009). Properties of consensus methods for inferring species trees from gene trees. *Systematic Biology*, 58(1):35–54.

[176] Degnan, J. H., and Rosenberg, N. A. (2006). Discordance of species trees with their most likely gene trees. *PLOS Genetics*, 2(5).

[177] Degnan, J. H., and Rosenberg, N. A. (2009). Gene tree discordance, phylogenetic inference and the multispecies coalescent. *Trends in Ecology and Evolution*, 24(6):332–340.

[178] Degnan, J. H., Rosenberg, N. A., and Stadler, T. (2012). A characterization of the set of species trees that produce anomalous ranked gene trees. *IEEE/ACM Transactions on Computational Biology and Bioinformatics*, 9(6):1558–1568.

[179] Degnan, J. H., and Salter, L. A. (2005). Gene tree distributions under the coalescent process. *Evolution*, 59(1):24–37.

[180] Delsuc, F., Brinkmann, H., and Philippe, H. (2005). Phylogenomics and the reconstruction of the tree of life. *Nature Reviews Genetics*, 6(5):361.

[181] Demartines, P., and Herault, J. (1997). Curvilinear component analysis: A self-organizing neural network for nonlinear mapping of data sets. *IEEE Trans Neural Networks*, 8(1):148–154.

[182] Demmel, J., and Yelick, K. (2013). *Communication Avoiding (CA) and Other Innovative Algorithms*, pages 243–250. Microsoft Corporation.

[183] Desper, R., and Gascuel, O. (2004). Theoretical foundation of the balanced minimum evolution method of phylogenetic inference and its relationship to weighted least-squares tree fitting. *Molecular Biology and Evolution*, 21(3):587–598.

[184] Dinh, V., Darling, A. E., and Matsen, F. A. (2018). Online Bayesian phylogenetic inference: Theoretical foundations via sequential Monte Carlo. *Systematic Biology*, 67(3):503–517.

[185] Dobzhansky, T. (1940). Speciation as a stage in evolutionary divergence. *American Naturalist*, 74:312–321.

[186] Dodt, W. G., Gallus, S., Phillips, M. J., and Nilsson, M. A. (2017). Resolving kangaroo phylogeny and overcoming retrotransposon ascertainment bias. *Scientific Reports*, 7(1):16811.

[187] Domazet-Loso, T., Carvunis, A., Alba, M., Sestak, M., Bakaric, R., Neme, R., and Tautz, D. (2017). No evidence for phylostratigraphic bias impacting inferences on patterns of gene emergence and evolution. *Molecular Biology and Evolution*, 34:843–856.

[188] Domazet-Loso, T., and Tautz, D. (2010). A phylogenetically based transcriptome age index mirrors ontogenetic divergence patterns. *Nature*, 468:815–818.

[189] Donoghue, M. J., and Doyle, J. A. (2000). Seed plant phylogeny: Demise of the anthophyte hypothesis? *Current Biology*, 10(3):106–109.

[190] Donoghue, M. J., Olmstead, R. G., Smith, J. F., and Palmer, J. D. (1992). Phylogenetic relationships of *Dipsacales* based on rbcL sequences. *Annals of the Missouri Botanical Garden*, 79:333–345.

[191] Doolittle, W. F. (2005). If the tree of life fell, would we recognize the sound. In Sapp, J., editor, Microbial Phylogeny and Evolution: Concepts and Controversies. Oxford: Oxford University Press

[192] Dornburg, A., Su, Z., and Townsend, J. P. (2018). Optimal rates for phylogenetic inference and experimental design in the era of genome-scale datasets. *Systematic Biology*, 68(1):145–156.

[193] Doronina, L., Churakov, G., Shi, J., Brosius, J., Baertsch, R., Clawson, H., and Schmitz, J. (2015). Exploring massive incomplete lineage sorting in arctoids (Laurasiatheria, Carnivora). *Molecular Biology and Evolution*, 32(12):3194–3204.

[194] dos Reis, M., Donoghue, P. C. J., and Yang, Z. (2016). Bayesian molecular clock dating of species divergences in the genomics era. *Nature Reviews Genetics*, 17:71–80.

[195] Doyle, J., and Doyle, J. (1987). A rapid DNA isolation procedure from small quantities of fresh leaf tissues. *Phytochemical Bulletin*, 19:11–15.

[196] Doyle, J. A. (1998). Molecules, morphology, fossils, and the relationship of angiosperms and Gnetales. *Molecular Phylogenetics and Evolution*, 9(3):448–462.

[197] Doyle, J. A., Donoghue, M. J., and Zimmer, E. A. (1994). Integration of morphological and ribosomal RNA data on the origin of angiosperms. *Annals of the Missouri Botanical Garden*, 81:419–450.

[198] Doyle, J. J. (1992). Gene trees and species trees: Molecular systematics as one-character taxonomy. *Systematic Botany*, 17(1):144–163.

[199] Doyle, J. J., and Doyle, J. L. (1990). Isolation of plant DNA from fresh tissue. *Focus*, 12(13):39–40.

[200] Doyon, J.-P. J., Ranwez, V., Daubin, V., and Berry, V. (2011). Models, algorithms and programs for phylogeny reconciliation. *Briefings in Bioinformatics*, 12(5):392–400.

[201] Drinkwater, B., and Charleston, M. A. (2014). An improved node mapping algorithm for the cophylogeny reconstruction problem. *Coevolution*, 2(1):1–17.

[202] Drosophila 12 Genomes Consortium (2007). Evolution of genes and genomes on the *Drosophila* phylogeny. *Nature*, 450(7167):203–218.

[203] Drummond, A. J., and Rambaut, A. (2007). BEAST: Bayesian evolutionary analysis by sampling trees. *BMC Evolutionary Biology*, 7(1):214.

[204] Dunn, C., Howison, M., and Zapata, F. (2013a). Agalma: An automated phylogenomics workflow. *BMC Bioinformatics*, 14:330.

[205] Dunn, C., Luo, X., and Wu, Z. (2013b). Phylogenetic analysis of gene expression. *Integrative and Comparative Biology*, 53:847–856.

[206] Dunn, C. W., Zapata, F., Munro, C., Siebert, S., and Hejnol, A. (2018). Pairwise comparisons across species are problematic when analyzing functional genomic data. *Proceedings of the National Academy of Sciences of the United States of America*, 115(3):E409–E417.

[207] Dunning, L. T., Olofsson, J. K., Parisod, C., Choudhury, R. R., Moreno-Villena, J. J., Yang, Y., and Dionora, J. (2019). Lateral transfers of large DNA fragments spread functional genes among grasses. *Proceedings of the National Academy of Sciences of the United States of America*, 116(10):4416–4425.

[208] Durand, E., Patterson, N., Reich, D., and Slatkin, M. (2011). Testing for ancient admixture between closely related populations. *Molecular Biology and Evolution*, 28:2239–2252.

[209] Eaton, D., and Ree, R. (2013). Inferring phylogeny and introgression using RADseq data: An example from flowering plants (Pedicularis: Orobanchaceae). *Systematic Biology*, 62:689–706.

[210] Eaton, D. A., Spriggs, E. L., Park, B., and Donoghue, M. J. (2017). Misconceptions on missing data in RAD-seq phylogenetics with a deep-scale example from flowering plants. *Systematic Biology*, 66(3):399–412.

[211] Eaton, D. A. R. (2014). PyRAD: Assembly of de novo RADseq loci for phylogenetic analyses. *Bioinformatics*, 30(13):1844–1849.

[212] Eaton, D. A. R. (2020). Toytree: A minimalist tree visualization and manipulation library for Python. *Methods in Ecology and Evolution*, 11(1):187–191.

[213] Ebersberger, I., Galgoczy, P., Taudien, S., Taenzer, S., Platzer, M., and von Haeseler, A. (2007). Mapping human genetic ancestry. *Molecular Biology and Evolution*, 24:2266–2276.

[214] Eckart, C., and Young, G. (1936). The approximation of one matrix by another of lower rank. *Psychometrika*, 1:211–218.

[215] Edelman, N., Frandsen, P., Miyagi, M., Clavijo, B., Davey, J., Dikow, R., García-Accinelli, G., et al. (2019). Genomic architecture and introgression shape a butterfly radiation. *Science*, 366:594–599.

[216] Edger, P. P., Heidel-Fischer, H. M., Bekaert, M., Rota, J., Glöckner, G., Platts, A. E., Heckel, D. G., et al. (2015). The butterfly plant arms-race escalated by gene and genome duplications. *Proceedings of the National Academy of Sciences of the United States of America*, 112(27):8362–8366.

[217] Edwards, S. (2009). Is a new and general theory of molecular systematics emerging? *Evolution*, 63(1):1–19.

[218] Edwards, S., and Beerli, P. (2000). Perspective: Gene divergence, population divergence, and the variance in coalescence time in phylogeographic studies. *Evolution*, 54(6):1839–1854.

[219] Edwards, S., Cloutier, A., and Baker, A. (2017). Conserved non-exonic elements: A novel class of marker for phylogenomics. *Systematic Biology*, 66:1028–1044.

[220] Edwards, S. V. (2016). Phylogenomic subsampling: A brief review. *Zoologica Scripta*, 45(S1):63–74.

[221] Edwards, S. V., Xi, Z., Janke, A., Faircloth, B. C., McCormack, J. E., Glenn, T. C., Zhong, B., et al. (2016). Implementing and testing the multispecies coalescent model: A valuable paradigm for phylogenomics. *Molecular Phylogenetics and Evolution*, 94:447–462.

[222] Efron, B., Halloran, E., and Holmes, S. (1996). Bootstrap confidence levels for phylogenetic trees. *Proceedings of the National Academy of Sciences of the United States of America*, 93:7085–7090.

[223] Eisen, J. A. (1998). Phylogenomics: Improving functional predictions for uncharacterized genes by evolutionary analysis. *Genome Research*, 8(3):163–167.

[224] El-Mabrouk, N., and Noutahi, E. (2019). Gene family evolution—an algorithmic framework. In Warnow, T., editor, *Bioinformatics and Phylogenetics*, pages 87–119. Cham, Switzerland: Springer.

[225] Ellegren, H. (2013). The evolutionary genomics of birds. *Annual Review of Ecology, Evolution, and Systematics*, 44:239–259.

[226] Elliott, K., Ricklefs, R., Gaston, A., Hatch, S., Speakman, J., and Davoren, G. (2013). High flight costs, but low dive costs, in auks support the biomechanical hypothesis for flightlessness in penguins. *Proceedings of the National Academy of Sciences of the United States of America*, 110:9380–9384.

[227] Elmer, K. R., Fan, S., Gunter, H. M., Jones, J. C., Boekhoff, S., Kuraku, S., and Meyer, A. (2010). Rapid evolution and selection inferred from the transcriptomes of sympatric crater lake cichlid fishes. *Molecular Ecology*, 19(S1):197–211.

[228] Elworth, R. A. L., Allen, C., Benedict, T., Dulworth, P., and Nakhleh, L. (2018). ALPHA: A toolkit for automated local phylogenomic analyses. *Bioinformatics*, 34(16):2848–2850.

[229] Elworth, R. L., Ogilvie, H. A., Zhu, J., and Nakhleh, L. (2019). Advances in computational methods for phylogenetic networks in the presence of hybridization. In Warnow, T., editor, *Bioinformatics and Phylogenetics*, pages 317–360. Cham, Switzerland: Springer.

[230] Emms, D., and Kelly, S. (2018). STAG: Species tree inference from all genes. *bioRxiv*. https://doi.org/10.1101/267914.

[231] Eschmeyer, W., Fricke, R., Fong, J., and Polack, D. (2010). Marine fish diversity: History of knowledge and discovery (Pisces). *Zootaxa*, 2525:19–50.

[232] Faircloth, B., McCormack, J., Crawford, N., Harvey, M., Brumfield, R., and Glenn, T. (2012). Ultraconserved elements anchor thousands of genetic markers spanning multiple evolutionary timescales. *Systematic Biology*, 61:717–726.

[233] Faisal, F. E., Meng, L., Crawford, J., and Milenković, T. (2015). The post-genomic era of biological network alignment. *Journal on Bioinformatics and Systems Biology*, 2015:3.

[234] Fang, B., Merilä, J., Matschiner, M., and Momigliano, P. (2020). Estimating uncertainty in divergence times among three-spined stickleback clades using the multispecies coalescent. *Molecular Phylogenetics and Evolution*, 142:106646.

[235] Farlie, P., Davidson, N., Baker, N., Raabus, M., Roeszler, K., Hirst, C., Major, A., et al. (2017). Co-option of the cardiac transcription factor nkx2.5 during development of the emu wing. *Nature Communications*, 8:132.

[236] Farris, J. S. (1970). Methods for computing Wagner trees. *Systematic Zoology*, 19(1):83.

[237] Farris, J. S. (1971). The hypothesis of nonspecificity and taxonomic congruence. *Annual Review of Ecology and Systematics*, 2(1):277–302.

[238] Feild, T. S., Arens, N. C., Doyle, J. A., Dawson, T. E., and Donoghue, M. J. (2004). Dark and disturbed: A new image of early angiosperm ecology. *Paleobiology*, 30(1):82–107.

[239] Felsenstein, J. (1978). Cases in which parsimony or compatibility methods will be positively misleading. *Systematic Zoology*, 27:401–410.

[240] Felsenstein, J. (1985a). Confidence limits on phylogenies: An approach using the bootstrap. *Evolution*, 39(4):783–791.

[241] Felsenstein, J. (1985b). Phylogenies and the comparative method. *American Naturalist*, 125: 1–15.

[242] Felsenstein, J. (2004). *Inferring Phylogenies*. Sunderland, MA: Sinauer Associates.

[243] Flouri, T., Jiao, X., Rannala, B., and Yang, Z. (2018). Species tree inference with BPP using genomic sequences and the multispecies coalescent. *Molecular Biology and Evolution*, 35(10):2585–2593.

[244] Flouri, T., Rannala, B., and Yang, Z. (2020). A tutorial on the use of BPP for species tree estimation and species delimitation. In Scornavacca, C., Delsuc, F., and Galtier, N., editors, *Phylogenetics in the Genomic Era*. Licensed under CC BY-NC-ND 4.0.

[245] Floyd, S. K., and Friedman, W. (2001). Developmental evolution of endosperm in basal angiosperms: Evidence from *Amborella* (Amborellaceae), *Nuphar* (Nymphaeaceae), and *Illicium* (Illiciaceae). *Plant Systematics and Evolution*, 228(3–4):153–169.

[246] Folk, R. A., Mandel, J. R., and Freudenstein, J. V. (2016). Ancestral gene flow and parallel organellar genome capture result in extreme phylogenomic discord in a lineage of angiosperms. *Systematic Biology*, 66:320–337.

[247] Folk, R. A., Visger, C. J., Soltis, P. S., Soltis, D. E., and Guralnick, R. P. (2018). Geographic range dynamics drove ancient hybridization in a lineage of angiosperms. *American Naturalist*, 192(2):171–187.

[248] Fontaine, M. C., Pease, J. B., Steele, A., Waterhouse, R. M., Neafsey, D. E., Sharakhov, I. V., Jiang, X., et al. (2015). Extensive introgression in a malaria vector species complex revealed by phylogenomics. *Science*, 347(6217):1258524.

[249] Foote, A. D., Liu, Y., Thomas, G. W., Vinař, T., Alföldi, J., Deng, J., Dugan, S., et al. (2015). Convergent evolution of the genomes of marine mammals. *Nature Genetics*, 47(3):272.

[250] Forst, C. V., Flamm, C., Hofacker, I. L., and Stadler, P. F. (2006). Algebraic comparison of metabolic networks, phylogenetic inference, and metabolic innovation. *BMC Bioinformatics*, 7(1):67.

[251] Forst, C. V., and Schulten, K. (2001). Phylogenetic analysis of metabolic pathways. *Journal of Molecular Evolution*, 52(6):471–489.

[252] Fortunato, S. (2010). Community detection in graphs. *Physics Reports*, 486:75–174.

[253] Foster, P. G., and Hickey, D. A. (1999). Compositional bias may affect both DNA-based and protein-based phylogenetic reconstructions. *Journal of Molecular Evolution*, 48(3):284–290.

[254] Friedman, W. E., and Ryerson, K. C. (2009). Reconstructing the ancestral female gametophyte of angiosperms: Insights from *Amborella* and other ancient lineages of flowering plants. *American Journal of Botany*, 96(1):129–143.

[255] Frigo, M., Leiserson, C. E., Prokop, H., and Ramachandran, S. (1999). Cache-oblivious algorithms. In *Proceedings of the 40th Annual Symposium on Foundations of Computer Science*, FOCS '99, page 285. Washington, DC: IEEE Computer Society.

[256] Gaither, J., and Kubatko, L. (2016). Hypothesis tests for phylogenetic quartets, with applications to coalescent-based species tree inference. *Journal of Theoretical Biology*, 408:179–186.

[257] García, N., Folk, R. A., Meerow, A. W., Chamala, S., Gitzendanner, M. A., de Oliveira, R. S., Soltis, D. E., and Soltis, P. S. (2017). Deep reticulation and incomplete lineage sorting obscure the diploid phylogeny of rain-lillies and allies (Amaryllidaceae tribe Hippeastreae). *Molecular Phylogenetics and Evolution*, 111:231–247.

[258] Garrigan, D., Kingan, S. B., Geneva, A. J., Andolfatto, P., Clark, A. G., Thornton, K. R., and Presgraves, D. C. (2012). Genome sequencing reveals complex speciation in the *Drosophila simulans* clade. *Genome Research*, 22(8):1499–1511.

[259] Gates, D., Pilson, D., and Smith, S. (2018). Filtering of target sequence capture individuals facilitates species tree construction in the plant subtribe Iochrominae (Solanaceae). *Molecular Phylogenetics and Evolution*, 123:26–34.

[260] Gates, D., Strickler, S., Mueller, L., Olson, B., and Smith, S. (2016). Diversification of R2R3-MYB transcription factors in the tomato family Solanaceae. *Journal of Molecular Evolution*, 83:26–37.

[261] Gatesy, J., DeSalle, R., and Wahlberg, N. (2007). How many genes should a systematist sample? Conflicting insights from a phylogenomic matrix characterized by replicated incongruence. *Systematic Biology*, 56(2):355–363.

[262] Gatesy, J., Sloan, D. B., Warren, J. M., Baker, R. H., Simmons, M. P., and Springer, M. S. (2019). Partitioned coalescence support reveals biases in species-tree methods and detects gene trees that determine phylogenomic conflicts. *Molecular Phylogenetics and Evolution*, 139:106539.

[263] Gatesy, J., and Springer, M. (2013). Concatenation versus coalescence versus "concatalescence." *Proceedings of the National Academy of Sciences of the United States of America*, 110(13):E1179–E1179.

[264] Gatesy, J., and Springer, M. (2014). Phylogenetic analysis at deep timescales: Unreliable gene trees, bypassed hidden support, and the coalescence/concatalescence conundrum. *Molecular Phylogenetics and Evolution*, 80:231–266.

[265] Gauthier, O., and Lapointe, F.-J. (2007). Hybrids and phylogenetics revisited: A statistical test of hybridization using quartets. *Systematic Botany*, 32(1):8–15.

[266] Gehrke, A., Schneider, I., Calle-Mustienes, E., Tena, J., Gomez-Marin, C., Chandran, M., Nakamura, T., Braasch, I., Postlethwait, J., and Gomez-Skarmeta, J. (2015). Deep conservation of wrist and digit enhancers in fish. *Proceedings of the National Academy of Sciences of the United States of America*, 112:803–808.

[267] Ghiselin, M. T. (1997). *Metaphysics and the Origin of Species*. Albany: State University of New York Press.

[268] Giarla, T. C., and Esselstyn, J. A. (2015). The challenges of resolving a rapid, recent radiation: Empirical and simulated phylogenomics of Philippine shrews. *Systematic Biology*, 64(5):727–740.

[269] Girvan, M., and Newman, M. (2002). Community structure in social and biological networks. *Proceedings of the National Academy of Sciences of the United States of America*, 99:7821–7826.

[270] Gong, L., and Flegal, J. M. (2016). A practical sequential stopping rule for high-dimensional Markov chain Monte Carlo. *Journal of Computational and Graphical Statistics*, 25(3):684–700.

[271] Goremykin, V. V., Hirsch-Ernst, K. I., Wölfl, S., and Hellwig, F. H. (2003). Analysis of the *Amborella trichopoda* chloroplast genome sequence suggests that *Amborella* is not a basal angiosperm. *Molecular Biology and Evolution*, 20(9):1499–1505.

[272] Grant, P., Grant, B., Markert, J., Keller, L., and Petren, K. (2004). Convergent evolution of Darwin's finches caused by introgressive hybridization and selection. *Evolution*, 58:1588–1599.

[273] Grant, V. (1971). *Plant Speciation*. New York: Columbia University Press.

[274] Grant, V., and Grant, K. (1971). Natural hybridization between Cholla cactus species *Opuntia spinosior* and *Opuntia versicolor*. In *Proceedings of the National Academy of Sciences of the United States of America*, 68:1993–1995.

[275] Grayson, D., and Stillman, M. (2021). Macaulay2, a software system for research in algebraic geometry, version 1.19.1. Available at http://www.math.uiuc.edu/Macaulay2/.

[276] Grealy, A., Phillips, M., Miller, G., Gilbert, M., Rouillard, J.-M., Lambert, D., Bunce, M., and Haile, J. (2017). Eggshell palaeogenomics: Palaeognath evolutionary history revealed through ancient nuclear and mitochondrial DNA from Madagascan elephant bird (*Aepyornis* sp.) eggshell. *Molecular Phylogenetics and Evolution*, 109:151–163.

[277] Green, R. E., Krause, J., Briggs, A. W., Maricic, T., Stenzel, U., Kircher, M., Patterson, N., et al. (2010). A draft sequence of the Neandertal genome. *Science*, 328(5979):710–722.

[278] Gregg, W. T., Ather, S. H., and Hahn, M. W. (2017). Gene-tree reconciliation with mul-trees to resolve polyploidy events. *Systematic Biology*, 66(6):1007–1018.

[279] Gropp, W., Olson, L. N., and Samfass, P. (2016). Modeling MPI communication performance on SMP nodes: Is it time to retire the ping pong test. In *Proceedings of the 23rd European MPI Users' Group Meeting*, EuroMPI 2016, pages 41–50, New York: Association for Computing Machinery.

[280] Gruenstaeudl, M., Reid, N. M., Wheeler, G. L., and Carstens, B. C. (2015). Posterior predictive checks of coalescent models: P2C2M, an R package. *Molecular Ecology Resources*, 16:193–205.

[281] Grünewald, S., Spillner, A., Bastkowski, S., Bögershausen, A., Moulton, V., Grünewald, S., and Bögershausen, A. (2013). SuperQ: Computing supernetworks from quartets. *IEEE/ACM Transactions on Computational Biology and Bioinformatics*, 10(1):151–60.

[282] Guang, A., Zapata, F., Howison, M., Lawrence, C., and Dunn, C. (2016). An integrated perspective on phylogenetic workflows. *Trends in Ecology and Evolution*, 31:116–126.

[283] Guerrero, R. F., and Hahn, M. W. (2018). Quantifying the risk of hemiplasy in phylogenetic inference. *Proceedings of the National Academy of Sciences of the United States of America*, 115(50):12787–12792.

[284] Guindon, S., Dufayard, J.-F., Lefort, V., Anisimova, M., Hordijk, W., and Gascuel, O. (2010). New algorithms and methods to estimate maximum-likelihood phylogenies: Assessing the performance of PhyML 3.0. *Systematic Biology*, 59(3):307–321.

[285] Guindon, S., and Gascuel, O. (2003). A simple, fast, and accurate algorithm to estimate large phylogenies by maximum likelihood. *Systematic Biology*, 52(5):696–704.

[286] Haak, D. C., Ballenger, B. A., and Moyle, L. C. (2014). No evidence for phylogenetic constraint on natural defense evolution among wild tomatoes. *Ecology*, 95(6):1633–1641.

[287] Hackett, S. J., Kimball, R. T., Reddy, S., Bowie, R. C., Braun, E. L., Braun, M. J., Chojnowski, J. L., et al. (2008). A phylogenomic study of birds reveals their evolutionary history. *Science*, 320(5884):1763–1768.

[288] Haddrath, O., and Baker, A. (2012). Multiple nuclear genes and retroposons support vicariance and dispersal of the palaeognaths, and an early cretaceous origin of modern birds. *Proceedings of the Royal Society B: Biological Sciences*, 279:4617–4625.

[289] Hahn, M. W. (2019). *Molecular Population Genetics*. Sunderland, MA: Sinauer Associates.

[290] Hahn, M. W., and Nakhleh, L. (2016). Irrational exuberance for resolved species trees. *Evolution*, 70(1):7–17.

[291] Harris, S. A. (1995). Systematics and randomly amplified polymorphic DNA in the genus *Leucaena* (Leguminosae, Mimosoideae. *Plant Systematics and Evolution*, 197(1–4):195–208.

[292] Harshman, J., Braun, E., Braun, M., Huddleston, C., Bowie, R., Chojnowski, J., Hackett, S., et al. (2008). Phylogenomic evidence for multiple losses of flight in ratite birds. *Proceedings of the National Academy of Sciences of the United States of America*, 105:13462–13467.

[293] Harvey, M. G., Judy, C. D., Seeholzer, G. F., Maley, J. M., Graves, G. R., and Brumfield, R. T. (2015). Similarity thresholds used in DNA sequence assembly from short reads can reduce the comparability of population histories across species. *PeerJ*, 3:e895.

[294] He, C., Liang, D., and Zhang, P. (2020). Asymmetric distribution of gene trees can arise under purifying selection if differences in population size exist. *Molecular Biology and Evolution*, 37(3):881–892.

[295] Heath, M. T. (2015). A tale of two laws. *International Journal of High Performance Computing Applications*, 29(3):320–330.

[296] Hedges, S., and Kumar, S. (2009). *The Timetree of Life*. New York: Oxford University Press.

[297] Hejase, H., and Liu, K. (2016). Mapping the genomic architecture of adaptive traits with interspecific introgressive origin: A coalescent-based approach. *BMC Genomics*, 17:8.

[298] Heled, J., and Drummond, A. J. (2010). Bayesian inference of species trees from multilocus data. *Molecular Biology and Evolution*, 27(3):570–580.

[299] Hellmuth, M., Hernandez-Rosales, M., Long, Y., and Stadler, P. F. (2017). Inferring phylogenetic trees from the knowledge of rare evolutionary events. *Journal of Mathematical Biology*, 76(7):1623–1653.

[300] Heubl, G., Bringmann, G., and Meimberg, H. (2006). Molecular phylogeny and character evolution of carnivorous plant families in Caryophyllales—revisited. *Plant Biology*, 8(06): 821–830.

[301] Hibbins, M. S., and Hahn, M. W. (2019). The timing and direction of introgression under the multispecies network coalescent. *Genetics*, 211(3):1059–1073.

[302] Hiller, M., Schaar, B., Indjeian, V., Kingsley, D., Hagey, L., and Bejerano, G. (2012). A "forward genomics" approach links genotype to phenotype using independent phenotypic losses among related species. *Cell Reports*, 2:817–823.

[303] Hillis, D. (1998). Taxonomic sampling, phylogenetic accuracy, and investigator bias. *Systematic Biology*, 47:3–8.

[304] Hillis, D., Heath, T., and St. John, K. (2005). Analysis and visualization of tree space. *Systematic Biology*, 54:471–482.

[305] Hillis, D. M. (2019). Species delimitation in herpetology. *Journal of Herpetology*, 53:3–12.

[306] Hobolth, A., Andersen, L. N., and Mailund, T. (2011). On computing the coalescence time density in an isolation-with-migration model with few samples. *Genetics*, 187(4):1241–1243.

[307] Hobolth, A., Christensen, O. F., Mailund, T., and Schierup, M. H. (2007). Genomic relationships and speciation times of human, chimpanzee, and gorilla inferred from a coalescent hidden Markov model. *PLOS Genetics*, 3(2):e7.

[308] Hoefler, T., Lumsdaine, A., and Rehm, W. (2007). Implementation and performance analysis of non-blocking collective. operations for MPI. In *SC '07: Proceedings of the 2007 ACM/IEEE Conference on Supercomputing*, pages 1–10. New York: Association for Computing Machinery

[309] Holland, B., Conner, G., Huber, K., and Moulton, V. (2007). Imputing supertrees and supernetworks from quartets. *Systematic Biology*, 56(1):57–67.

[310] Holland, B., and Moulton, V. (2003). Consensus networks: A method for visualising incompatibilities in collections of trees. In *International Workshop on Algorithms in Bioinformatics*, page 165–176. Berlin: Springer.

[311] Holland, B. R., Huber, K. T., Moulton, V., and Lockhart, P. J. (2004). Using consensus networks to visualize contradictory evidence for species phylogeny. *Molecular Biology and Evolution*, 21(7):1459–1461.

[312] Holmes, S. (2003). Statistics for phylogenetic trees. *Theoretical Population Biology*, 63(1):17–32.

[313] Hon, C., and Carninci, P. (2020). Expanded ENCODE delivers invaluable genomic encyclopedia. *Nature*, 583:685–686.

[314] Hosner, P., Faircloth, B., Glenn, T., Braun, E., and Kimball, R. (2016). Avoiding missing data biases in phylogenomic inference: An empirical study in the landfowl (Aves: Galliformes). *Molecular Biology and Evolution*, 33(4):1110–1125.

[315] Hossain, A., Blackburne, B., Shah, A., and Whelan, S. (2015). Evidence of statistical inconsistency of phylogenetic methods in the presence of multiple sequence alignment uncertainty. *Genome Biology and Evolution*, 7:2102–2116.

[316] Howison, M., Zapata, F., Edwards, E., and Dunn, C. (2014). Bayesian genome assembly and assessment by Markov chain Monte Carlo sampling. *PLOS ONE*, 9:e99497.

[317] Hu, Z., Sackton, T., Edwards, S., and Liu, J. (2019). Bayesian detection of convergent rate changes of conserved noncoding elements on phylogenetic trees. *Molecular Biology and Evolution*, 36:1086–1100.

[318] Huang, H., and Knowles, L. L. (2016). Unforeseen consequences of excluding missing data from next-generation sequences: Simulation study of RAD sequences. *Systematic Biology*, 65(3): 357–365.

[319] Huang, H., Sukumaran, J., Smith, S. A., and Knowles, L. L. (2017). Cause of gene tree discord? Distinguishing incomplete lineage sorting and lateral gene transfer in phylogenetics. *PeerJ Preprints*, 5:e3489v1.

[320] Huang, J.-P., and Knowles, L. L. (2015). The species versus subspecies conundrum: Quantitative delimitation from integrating multiple data types within a single Bayesian approach in Hercules beetles. *Systematic Biology*, 65(4):685–699.

[321] Huang, S.-S. C., and Fraenkel, E. (2009). Integrating proteomic, transcriptional, and interactome data reveals hidden components of signaling and regulatory networks. *Science Signaling*, 2(81):ra40.

[322] Huang, W., Zhou, G., Marchand, M., Ash, J. R., Morris, D., Van Dooren, P., Brown, J. M., Gallivan, K. A., and Wilgenbusch, J. C. (2016). Treescaper: Visualizing and extracting phylogenetic signal from sets of trees. *Molecular Biology and Evolution*, 33:3314–3316.

[323] Huber, K., and Moulton, V. (2006). Phylogenetic networks from multi-labelled trees. *Journal of Mathematical Biology*, 72:699–725.

[324] Huber, K. T., Linz, S., Moulton, V., and Wu, T. (2015). Spaces of phylogenetic networks from generalized nearest-neighbor interchange operations. *Journal of Mathematical Biology*, 72: 699–725.

[325] Huber, K. T., Moulton, V., Steel, M., and Wu, T. (2016). Folding and unfolding phylogenetic trees and networks. *Journal of Mathematical Biology*, 73(6–7):1761–1780.

[326] Huber, K. T., Spillner, A., Suchecki, R., and Moulton, V. (2011). Metrics on multilabeled trees: interrelationships and diameter bounds. *IEEE/ACM Transactions on Computational Biology and Bioinformatics* (TCBB), 8(4):1029–1040.

[327] Hubisz, M., Pollard, K., and Siepel, A. (2011). Phast and rphast: Phylogenetic analysis with space/time models. *Briefings in Bioinformatics*, 12:41–51.

[328] Hudson, R. R. (1983). Testing the constant-rate neutral allele model with protein sequence data. *Evolution*, 37(1):203–217.

[329] Hudson, R. R. (2002). Generating samples under a Wright–Fisher neutral model of genetic variation. *Bioinformatics*, 18:337–338.

[330] Hudson, R. R., and Kaplan, N. L. (1988). The coalescent process in models with selection and recombination. *Genetics*, 120(3):831–840.

[331] Huelsenbeck, J., and Rannala, B. (2004). Frequentist properties of Bayesian posterior probabilities of phylogenetic trees under simple and complex substitution models. *Systematic Biology*, 53:904–913.

[332] Huelsenbeck, J. P., and Bollback, J. P. (2001). Empirical and hierarchical Bayesian estimation of ancestral states. *Systematic Biology*, 50(3):351–366.

[333] Huelsenbeck, J. P., and Hillis, D. M. (1993). Success of phylogenetic methods in the four-taxon case. *Systematic Biology*, 42:247–264.

[334] Huelsenbeck, J. P., and Ronquist, F. (2001). MRBAYES: Bayesian inference of phylogenetic trees. *Bioinformatics*, 17(8):754–755.

[335] Hunter, J. D. (2007). Matplotlib: A 2D graphics environment. *Computing in Science and Engineering*, 9(3):90–95.

[336] Huson, D. H., Albrecht, B., Patz, S., and Steel, M. (2019). Anti-consensus: Detecting trees that have an evolutionary signal that is lost in consensus. *bioRxiv*. https://doi.org/10.1101/706416.

[337] Huson, D. H., and Bryant, D. (2005). Application of phylogenetic networks in evolutionary studies. *Molecular Biology and Evolution*, 23(2):254–267.

[338] Huson, D. H., Richter, D. C., Rausch, C., Dezulian, T., Franz, M., and Rupp, R. (2007). Dendroscope: An interactive viewer for large phylogenetic trees. *BMC Bioinformatics*, 8(1):460.

[339] Huson, D. H., Rupp, R., and Scornavacca, C. (2010). *Phylogenetic Networks: Concepts, Algorithms and Applications*. Cambridge: Cambridge University Press.

[340] Huson, D. H., and Scornavacca, C. (2012). Dendroscope 3: An interactive tool for rooted phylogenetic trees and networks. *Systematic Biology*, 61(6):1061–1067.

[341] Höhna, S., Landis, M., Heath, T., Boussau, B., Lartillot, N., Moore, B., Huelsenbeck, J., and Ronquist, F. (2016). RevBayes: Bayesian phylogenetic inference using graphical models and an interactive model-specification language. *Systematic Biology*, 65:726–736.

[342] Ihmels, J., Bergmann, S., Gerami-Nejad, M., Yanai, I., McClellan, M., Berman, J., and Barkai, N. (2005). Rewiring of the yeast transcriptional network through the evolution of motif usage. *Science*, 309(5736):938–940.

[343] Jarvis, E., Mirarab, S., Aberer, A. J., Li, B., Houde, P., Li, C., Ho, S., et al. (2014). Whole-genome analyses resolve early branches in the tree of life of modern birds. *Science*, 346(6215):1320–1331.

[344] Jeffroy, O., Brinkmann, H., Delsuc, F., and Philippe, H. (2006). Phylogenomics: The beginning of incongruence? *Trends in Genetics*, 22(4):225–231.

[345] Jewett, E., and Rosenberg, N. A. (2012). iGLASS: An improvement to the GLASS method for estimating species trees from gene trees. *Journal of Computational Biology*, 19(3):293–315.

[346] Jhwueng, D.-C., and O'Meara, B. (2015). Trait evolution on phylogenetic networks. *bioRxiv*. https://doi.org/10.1101/023986.

[347] Jiang, T., Kearney, P., and Li, M. (2001). A polynomial time approximation scheme for inferring evolutionary trees from quartet topologies and its application. *SIAM Journal on Computing*, 30(6):1942–1961.

[348] Johnson, M. G., Gardner, E. M., Liu, Y., Medina, R., Bernard Goffinet, A. S., Zerega, N. J., and Wickett, N. J. (2016). Hybpiper: Extracting coding sequence and introns for phylogenetics from high-throughput sequencing reads using target enrichment. *Applications in Plant Sciences*, 4:7.

[349] Joly, S., McLenachan, P. A., and Lockhart, P. J. (2009). A statistical approach for distinguishing hybridization and incomplete lineage sorting. *American Naturalist*, 174(2):54- 70.

[350] Jombart, T., Kendall, M., Almagro-Garcia, J., and Colijn, C. (2017). Treespace: Statistical exploration of landscapes of phylogenetic trees. *Molecular Ecology Resources*, 17:1385–1392.

[351] Jones, G., Sagitov, S., and Oxelman, B. (2013). Statistical inference of allopolyploid species networks in the presence of incomplete lineage sorting. *Systematic Biology*, 62(3):467–478.

[352] Joshi, N. A., and Fash, J. N. (2011). Sickle: Sliding-window, adaptive, quality-based trimming tool for FASTQ files (version 1.33). Available at https://github.com/najoshi/sickle.

[353] Kamneva, O. K., Syring, J., Liston, A., and Rosenberg, N. A. (2017). Evaluating allopolyploid origins in strawberries (*Fragaria*) using haplotypes generated from target sequence capture. *BMC Evolutionary Biology*, 17:180.

[354] Kane, D., and Tao, T. (2017). A bound on partitioning clusters. *Electronic Journal of Combinatorics*, 24:P2.31.

[355] Karimi, N., Grover, C. E., Gallagher, J. P., Wendel, J. F., Ané, C., and Baum, D. A. (2019). Reticulate evolution helps explain apparent homoplasy in floral biology and pollination in baobabs (Adansonia; Bombacoideae; Malvaceae). *Systematic Biology*, 69(3):462–478.

[356] Katoh, S. (2013). MAFFT multiple sequence alignment software version 7: Improvements in performance and usability. *Molecular Biology and Evolution*, 30:772–780.

[357] Kaur, K. M., Malé, P.-J. G., Spence, E., Gomez, C., and Frederickson, M. E. (2019). Using text-mined trait data to test for cooperate-and-radiate co-evolution between ants and plants. *PLOS Computational Biology*, 15(10):e1007323.

[358] Keck, D. D. (1945). Studies in *Penstemon*–XIII: A cyto-taxonomic account of the section *Spermunculus*. *American Midland Naturalist*, 33:128–206.

[359] Kelchner, S. A., and Thomas, M. A. (2007). Model use in phylogenetics: Nine key questions. *Trends in Ecology and Evolution*, 22(2):87–94.

[360] Kelleher, J., Etheridge, A. M., and McVean, G. (2016). Efficient coalescent simulation and genealogical analysis for large sample sizes. *PLOS Computational Biology*, 12(5):e1004842.

[361] Kellogg, E. A., Appels, R., and Mason-Gamer, R. J. (1996). When genes tell different stories: The diploid genera of Triticeae (Gramineae). *Systematic Botany*, 21(3):321–347.

[362] Kent, W. (2002). Blat–the blast-like alignment tool. *Genome Research*, 12:656–664.

[363] Kingman, J. F. C. (1982a). Exchangeability and the evolution of large populations. In G. Koch and F. Spizzichino, editors, *Exchangeability in Probability and Statistics*, pages 97–112. Amsterdam: North-Holland.

[364] Kingman, J. F. C. (1982b). On the genealogy of large populations. *Journal of Applied Probability*, 19:27–43.

[365] Kingman, J. F. C. (1982c). The coalescent. *Stochastic Processes and Applications*, 13:235–248.

[366] Kirkpatrick, M., and Barrett, B. (2015). Chromosome inversions, adaptive cassettes and the evolution of species' ranges. *Molecular Ecology*, 24:2046–2055.

[367] Knaack, S., Thompson, D., and Roy, S. (2016). Reconstruction and analysis of the evolution of modular transcriptional regulatory programs using arboretum. *Methods in Molecular Biology*, 1361:375–389.

[368] Knowles, L. (2009). Estimating species trees: Methods of phylogenetic analysis when there is incongruence across genes. *Systematic Biology*, 58(5):463–467.

[369] Knowles, L., Huang, H., Sukumaran, J., and Smith, S. A. (2018a). A matter of phylogenetic scale: Distinguishing incomplete lineage sorting from lateral gene transfer as the cause of gene tree discord in recent versus deep diversification histories. *American Journal of Botany*, 105(3): 376–384.

[370] Knowles, L., and Kubatko, L. (2010). Estimating species trees: An introduction to concepts and models. In Knowles, L., and Kubatko, L., editors, *Estimating Species Trees: Practical and Theoretical Aspects*. New York: Wiley-Blackwell.

[371] Knowles, L. L., Huang, H., Sukumaran, J., and Smith, S. A. (2018b). A matter of phylogenetic scale: Distinguishing incomplete lineage sorting from lateral gene transfer as the cause of gene tree discord in recent versus deep diversification histories. *American Journal of Botany*, 105(3):376–384.

[372] Knowles, L. L., and Kubatko, L. S. (2011). *Estimating Species Trees: Practical and Theoretical Aspects*. Hoboken, NJ: John Wiley and Sons.

[373] Kobert, K., Salichos, L., Rokas, A., and Stamatakis, A. (2016). Computing the internode certainty and related measures from partial gene trees. *Molecular Biology and Evolution*, 33(6):1606–1617.

[374] Koenen, E. J., Ojeda, D. I., Steeves, R., Migliore, J., Bakker, F. T., Wieringa, J. J., and Kidner, C. (2019). Large-scale genomic sequence data resolve the deepest divergences in the legume phylogeny and support a near-simultaneous evolutionary origin of all six subfamilies. *New Phytologist*, 225(3):1355–1369.

[375] Kolaczkowski, B., and Thornton, J. (2004). Performance of maximum parsimony and likelihood phylogenetics when evolution is heterogeneous. *Nature*, 431:980–984.

[376] Kopp, A., and True, J. R. (2002). Phylogeny of the Oriental *Drosophila melanogaster* species group: A multilocus reconstruction. *Systematic Biology*, 51(5):786–805.

[377] Kosakovsky Pond, S. L., and Muse, S. V. (2005). HyPhy: Hypothesis testing using phylogenies. In Nielsen, R., editor, *Statistical Methods in Molecular Evolution*, pages 125–181. New York: Springer.

[378] Kosiol, C., Vinař, T., da Fonseca, R. R., Hubisz, M. J., Bustamante, C. D., Nielsen, R., and Siepel, A. (2008). Patterns of positive selection in six mammalian genomes. *PLOS Genetics*, 4(8):e1000144.

[379] Kozlov, A., Aberer, A., and Stamatakis, A. (2015). Examl version 3: A tool for phylogenomic analyses on supercomputers. *Bioinformatics*, 31(15):2577–2579.

[380] Kozlov, A. M., Darriba, D., Flouri, T., Morel, B., and Stamatakis, A. (2019a). RAxML-NG: A fast, scalable and user-friendly tool for maximum likelihood phylogenetic inference. *Bioinformatics*, 35(21):4453–4455.

[381] Kozlov, A. M., Morel, B., and Stamatakis, A. (2019b). ParGenes: A tool for massively parallel model selection and phylogenetic tree inference on thousands of genes. *Bioinformatics*, 35(10):1771–1773.

[382] Kruskal, J. (1964). Multidimensional scaling by optimizing goodness of fit to a nonmetric hypothesis. *Psychometrika*, 29:1–27.

[383] Kubatko, L. (2019). The multispecies coalescent. In Balding, D. J., Moltke, I., and Marioni, J., editors, *Handbook of Statistical Genomics, 4th ed., volume 1*, pages 219–245. Hoboken, NJ: John Wiley and Sons.

[384] Kubatko, L., and Chifman, J. (2020). Identifiability of speciation times under the multispecies coalescent. *bioRxiv*. https://doi.org/10.1101/2020.11.24.396424.

[385] Kubatko, L., and Degnan, J. (2007). Inconsistency of phylogenetic estimates from concatenated data under coalescence. *Systematic Biology*, 56:17–24.

[386] Kubatko, L., Shah, P., Herbei, R., and Gilchrist, M. A. (2016). A codon model of nucleotide substitution with selection on synonymous codon usage. *Molecular Phylogenetics and Evolution*, 94:290–297.

[387] Kubatko, L. S. (2009). Identifying hybridization events in the presence of coalescence via model selection. *Systematic Biology*, 58(5):478–488.

[388] Kubatko, L. S., Carstens, B. C., and Knowles, L. L. (2009). STEM: Species tree estimation using maximum likelihood for gene trees under coalescence. *Bioinformatics*, 25(7):971–973.

[389] Kubatko, L. S., and Chifman, J. (2019). An invariants-based method for efficient identification of hybrid speciation from large-scale genomic data. *BMC Evolutionary Biology*, 19:112.

[390] Kubatko, L. S., Gibbs, H. L., and Bloomquist, E. (2011). Inferring species-level phylogenies using multi-locus data for a recent radiation of *Sistrurus* rattlesnakes. *Systematic Biology*, 60(4):393–409.

[391] Kuhner, M., and McGill, J. (2014). Correcting for sequencing error in maximum likelihood phylogeny inference. *G3-Genes Genomes Genetics*, 4:2544–2551.

[392] Kuhner, M. K., and Felsenstein, J. (1994). A simulation comparison of phylogeny algorithms under equal and unequal evolutionary rates. *Molecular Biology and Evolution*, 11(3):459–468.

[393] Künster, A., Wolf, J. B. W., Backström, N., Whitney, O., Balakrishnan, C. N., Day, L., Edwards, S. V., et al. (2010). Comparative genomics based on massive parallel transcriptome sequencing reveals patterns of substitution and selection across 10 bird species. *Molecular Ecology*, 19(S1):266–276.

[394] Küpper, C., Stocks, M., Risse, J. E., dos Remedios, N., Farrell, L. L., McRae, S. B., Morgan, T. C., et al. (2015). A supergene determines highly divergent male reproductive morphs in the ruff. *Nature Genetics*, 48(1):79–83.

[395] Kuritzin, A., Kischka, T., Schmitz, J., and Churakov, G. (2016). Incomplete lineage sorting and hybridization statistics for large-scale retroposon insertion data. *PLOS Computational Biology*, 12(3):e1004812.

[396] Lafond, M., and Scornavacca, C. (2019). On the weighted quartet consensus problem. *Theoretical Computer Science*, 769:1–17.

[397] Lagomarsino, L. P., Condamine, F. L., Antonelli, A., Mulch, A., and Davis, C. C. (2016). The abiotic and biotic drivers of rapid diversification in Andean bellflowers (Campanulaceae). *New Phytologist*, 210(4):1430–1442.

[398] Lake, J. A. (1987). A rate independent technique for analysis of nucleic acid sequences: Evolutionary parsimony. *Molecular Biology and Evolution*, 4(2):167–191.

[399] Lamarck, J.-B. (1796). *Encyclopédie méthodique. Botanique*. Volume 4. Paris, Liège: Panckoucke; Plomteux, 1783–1808, Paris.

[400] Lamichhaney, S., Berglund, J., Almén, M., Maqbool, K., Grabherr, M., Martinez-Barrio, A., Promerová, M., et al. (2015). Evolution of Darwin's finches and their beaks revealed by genome sequencing. *Nature*, 518:371–375.

[401] Lamichhaney, S., Card, D., Grayson, P., Tonini, J., Bravo, G., Näpflin, K., Termignoni-Garcia, F., et al. (2019). Integrating natural history collections and comparative genomics to study the genetic architecture of convergent evolution. *Philosophical Transaction of the Royal Society B: Biological Sciences*, 374:20180248.

[402] Lamichhaney, S., Han, F., Berglund, J., Wang, C., Almen, M. S., Webster, M. T., Grant, B. R., Grant, P. R., and Andersson, L. (2016). A beak size locus in Darwin's finches facilitated character displacement during a drought. *Science*, 352(6284):470–474.

[403] Lande, R., and Arnold, S. (1983). The measurement of selection on correlated characters. *Evolution*, 37:1210–1226.

[404] Lanier, H. C., and Knowles, L. L. (2015). Applying species-tree analyses to deep phylogenetic histories: Challenges and potential suggested from a survey of empirical phylogenetic studies. *Molecular Phylogenetics and Evolution*, 83:191–199.

[405] Larget, B., Kotha, S., Dewey, C., and Ané, C. (2010). BUCKy: Gene tree/species tree reconciliation with the Bayesian concordance analysis. *Bioinformatics*, 26(22):2910–2911.

[406] Larget, B., and Simon, D. L. (1999). Markov chain Monte Carlo algorithms for the Bayesian analysis of phylogenetic trees. *Molecular Biology and Evolution*, 16(6):750–759.

[407] Larson, D. A., Walker, J. F., Vargas, O. M., and Smith, S. A. (2020). A consensus phylogenomic approach highlights paleopolyploid and rapid radiation in the history of Ericales. *American Journal of Botany*, 107(5):773–789.

[408] Lawrence, T. J., and Datwyler, S. L. (2016). Testing the hypothesis of allopolyploidy in the origin of *Penstemon azureus* (Plantaginaceae). *Frontiers in Ecology and Evolution*, 4:60.

[409] Lawson, D. J., van Dorp, L., and Falush, D. (2018). A tutorial on how not to over-interpret STRUCTURE and ADMIXTURE bar plots. *Nature Communications*, 9:3258.

[410] Lawson, D. J., Hellenthal, G., Myers, S., and Falush, D. (2012). Inference of population structure using dense haplotype data. *PLOS Genetics*, 8(1):e1002453.

[411] Le, T., Sy, A., Molloy, E. K., Zhang, Q., Rao, S., and T., W. (2019). Using INC within divide-and-conquer phylogeny estimation. In Holmes, I., Martín-Vide, C., Vega-Rodríguez, M., editors, *Algorithms for Computational Biology. AlCoB 2019*. Lecture Notes in Computer Science, volume 11488. Cham, Switzerland: Springer.

[412] Le Duc, D., Renaud, G., Krishnan, A., Almén, M. S., Huynen, L., Prohaska, S. J., Ongyerth, M., et al. (2015). Kiwi genome provides insights into evolution of a nocturnal lifestyle. *Genome Biology*, 16(1):147.

[413] Leaché, A. D., Banbury, B. L., Felsenstein, J., De Oca, A. N.-M., and Stamatakis, A. (2015). Short tree, long tree, right tree, wrong tree: New acquisition bias corrections for inferring SNP phylogenies. *Systematic Biology*, 64(6):1032–1047.

[414] Leaché, A. D., and Oaks, J. R. (2017). The utility of single nucleotide polymorphism (SNP) data in phylogenetics. *Annual Review of Ecology, Evolution, and Systematics*, 48(1):69–84.

[415] Leavitt, S. D., Grewe, F., Widhelm, T., Muggia, L., Wray, B., and Lumbsch, H. T. (2016). Resolving evolutionary relationships in lichen-forming fungi using diverse phylogenomic datasets and analytical approaches. *Scientific Reports*, 6:22262.

[416] Lee, J., and Verleysen, M. (2007). *Nonlinear Dimensionality Reduction*. New York: Springer Science + Business Media

[417] Lefort, V., Desper, R., and Gascuel, O. (2015). FastME 2.0: A comprehensive, accurate, and fast distance-based phylogeny inference program. *Molecular Biology and Evolution*, 32(10):2798–2800.

[418] Legried, B., Molloy, E. K., Warnow, T., and Roch, S. (2021). Polynomial-time statistical estimation of species trees under gene duplication and loss. *Journal of Computational Biology*, 28(5):452–468.

[419] Lei, B. R., and Olival, K. J. (2014). Contrasting patterns in mammal-bacteria coevolution: *Bartonella* and *Leptospira* in bats and rodents. *PLOS Neglected Tropical Diseases*, 8(3):e2738.

[420] Lemmon, A., Emme, S., and Lemmon, E. (2012). Anchored hybrid enrichment for massively high-throughput pylogenomics. *Systematic Biology*, 61:727–744.

[421] Lemmon, A., and Moriarty, E. (2004). The importance of proper model assumption in Bayesian phylogenetics. *Systematic Biology*, 53:265–277.

[422] Lemmon, E. M., and Lemmon, A. R. (2013). High-throughput genomic data in systematics and phylogenetics. *Annual Review of Ecology, Evolution, and Systematics*, 44:99–121.

[423] Lemoine, F., Domelevo Entfellner, J.-B., Wilkinson, E., Correia, D., Dávila Felipe, M., De Oliveira, T., and Gascuel, O. (2018). Renewing Felsenstein's phylogenetic bootstrap in the era of big data. *Nature*, 556(7702):452–456.

[424] Levin, D. A. (1973). The role of trichomes in plant defense. *Quarterly Review of Biology*, 48:3–15.

[425] Lewin, H. A., Robinson, G. E., Kress, W. J., Baker, W. J., Coddington, J., Crandall, K. A., Durbin, R., et al. (2018). Earth biogenome project: Sequencing life for the future of life. *Proceedings of the National Academy of Sciences of the United States of America*, 115(17):4325–4333.

[426] Lewis, P., Chen, M.-H., Kuo, L., Lewis, L. A., Fučíková, K., Neupane, S., Wang, Y.-B., and She, D. (2016). Estimating Bayesian phylogenetic information content. *Systematic Biology*, 65(6):1009–1023.

[427] Lewis, P. O. (2001). A likelihood approach to estimating phylogeny from discrete morphological character data. *Systematic Biology*, 50(6):913–925.

[428] Lewis, Z. R., and Dunn, C. W. (2018). Genome evolution: We are not so special. *eLife*, 7:e38726.

[429] Li, H., Handsaker, B., Wysoker, A., Fennell, T., Ruan, J., Homer, N., and Marth, G. (2009). The sequence alignment/map format and SAMtools. *Bioinformatics*, 25:2078–2079.

[430] Liang, C., Musser, J., Cloutier, A., Prum, R., and Wagner, G. (2018). Pervasive correlated evolution in gene expression shapes cell and tissue type transcriptomes. *Genome Biology and Evolution*, 10:538–552.

[431] Linck, E., and Battey, C. J. (2019). Minor allele frequency thresholds strongly affect population structure inference with genomic data sets. *Molecular Ecology Resources*, 19(3):639–647.

[432] Liu, K., Linder, C. R., and Warnow, T. (2011). RAxML and FastTree: Comparing two methods for large-scale maximum likelihood phylogeny estimation. *PLOS ONE*, 6(11):e27731.

[433] Liu, K., Raghavan, S., Nelesen, S., Linder, C. R., and Warnow, T. (2009a). Rapid and accurate large-scale coestimation of sequence alignments and phylogenetic trees. *Science*, 324(5934):1561–1564.

[434] Liu, K. J., Dai, J., Truong, K., Song, Y., Kohn, M. H., and Nakhleh, L. (2014a). An HMM-based comparative genomic framework for detecting introgression in eukaryotes. *PLOS Computational Biology*, 10(6):e1003649.

[435] Liu, L. (2008). BEST: Bayesian estimation of species trees under the coalescent model. *Bioinformatics*, 24(21):2542–2543.

[436] Liu, L., and Pearl, D. K. (2007). Species trees from gene trees: reconstructing Bayesian posterior distributions of a species phylogeny using estimated gene tree distributions. *Systematic Biology*, 56(3):504–514.

[437] Liu, L., Xi, Z., and Davis, C. (2014b). Coalescent methods are robust to the simultaneous effects of long branches and incomplete lineage sorting. *Molecular Biology and Evolution*, 32: 791–805.

[438] Liu, L., Xi, Z., Wu, S., Davis, C. C., and Edwards, S. V. (2015). Estimating phylogenetic trees from genome-scale data. *Annals of the New York Academy of Sciences*, 1360(1):36–53.

[439] Liu, L., and Yu, L. (2011). Estimating species trees from unrooted gene trees. *Systematic Biology*, 60(5):661–667.

[440] Liu, L., Yu, L., and Edwards, S. (2010). A maximum pseudo-likelihood approach for estimating species trees under the coalescent model. *BMC Evolutionary Biology*, 10:302.

[441] Liu, L., Yu, L., Kubatko, L., Pearl, D. K., and Edwards, S. V. (2009b). Coalescent methods for estimating multilocus phylogenetic trees. *Molecular Phylogenetics and Evolution*, 53:320–328.

[442] Liu, L., Yu, L., Pearl, D. K., and Edwards, S. V. (2009c). Estimating species phylogenies using coalescence times among sequences. *Systematic Biology*, 58(5):468–477.

[443] Liu, L., Zhang, J., Rheindt, F., Lei, F., Qu, Y., Wang, Y., Zhang, Y., et al. (2017). Genomic evidence reveals a radiation of placental mammals uninterrupted by the KPg boundary. *Proceedings of the National Academy of Sciences of the United States of America*, 114:E7282–E7290.

[444] Lockhart, P. J., Steel, M. A., Hendy, M. D., and Penny, D. (1994). Recovering evolutionary trees under a more realistic model of sequence evolution. *Molecular Biology and Evolution*, 11:605–612.

[445] Long, C., and Kubatko, L. (2018). The effect of gene flow on coalescent-based species-tree inference. *Systematic Biology*, 67(5):770–785.

[446] Long, C., and Kubatko, L. (2019). Identifiability and reconstructibility of species phylogenies under a modified coalescent. *Bulletin of Mathematical Biology*, 81:408–430.

[447] Long, C., and Kubatko, L. (2021). Hypothesis testing with rank conditions in phylogenetics. *Frontiers in Genetics*, 12:664357.

[448] Longo, S., Faircloth, B., Meyer, A., Westneat, M., Alfaro, M., and Wainwright, P. (2017). Phylogenomic analysis of a rapid radiation of misfit fishes (Syngnathiformes) using ultraconserved elements. *Molecular Phylogenetics and Evolution*, 113:33–48.

[449] Lopez, P., Casane, D., and Philippe, H. (2002). Heterotachy, an important process of protein evolution. *Molecular Biology and Evolution*, 19:1–7.

[450] Losos, J. (2011). Convergence, adaptation, and constraint. *Evolution*, 65:1827–1840.

[451] Lott, M., Spillner, A., Huber, K. T., and Moulton, V. (2009a). PADRE: A package for analyzing and displaying reticulate evolution. *Bioinformatics*, 25:1199–1200.

[452] Lott, M., Spillner, A., Huber, K. T., Petri, A., Oxelman, B., and Moulton, V. (2009b). Inferring polyploid phylogenies from multiply-labeled gene trees. *BMC Evolutionary Biology*, 9(1):216.

[453] Ma, B., Xin, L., and Zhang, K. (2008). A new quartet approach for reconstructing phylogenetic trees: Quartet joining method. *Journal of Combinatorial Optimization*, 16(3):293–306.

[454] Mabry, M. E., Brose, J. M., Blischak, P. D., Sutherland, B., Dismukes, W. T., Bottoms, C. A., and Edger, P. P. (2020). Phylogeny and multiple independent whole-genome duplication events in the Brassicales. *American Journal of Botany*, 107(8):1148–1164.

[455] Maddison, W. P. (1997). Gene trees in species trees. *Systematic Biology*, 46:523–536.

[456] Maddison, W. P., and Knowles, L. L. (2006). Inferring phylogeny despite incomplete lineage sorting. *Systematic Biology*, 55(1):21–30.

[457] Maddison, W. P., and Maddison, D. R. (2004). Mesquite: A modular system for evolutionary analysis. Version 1.01. http://mesquiteproject.org.

[458] Maddison, W. P., and Maddison., D. R. (2018). Mesquite: A modular system for evolutionary analysis. Version 3.51 http://www.mesquiteproject.org.

[459] Magoč, T., and Salzberg, S. L. (2011). FLASH: Fast length adjustment of short reads to improve genome assemblies. *Bioinformatics*, 27:2957–2963.

[460] Mai, U., and Mirarab, S. (2018). TreeShrink: Fast and accurate detection of outlier long branches in collections of phylogenetic trees. *BMC Genomics*, 19(S5):272.

[461] Mallet, J. (2005). Hybridization as an invasion of the genome. *TREE*, 20(5):229–237.

[462] Mallet, J. (2007). Hybrid speciation. *Nature*, 446:279–283.

[463] Mallet, J., Besansky, N., and Hahn, M. (2016). How reticulated are species? *Bioessays*, 38:140–149.

[464] Mallo, D., and Posada, D. (2016). Multilocus inference of species trees and DNA barcoding. *Philosophical Transactions of the Royal Society B: Biological Sciences*, 371:20150335.

[465] Manthey, J. D., Campillo, L. C., Burns, K. J., and Moyle, R. G. (2016). Comparison of target-capture and restriction-site associated DNA sequencing for phylogenomics: A test in cardinalid tanagers (Aves, genus: *Piranga*). *Systematic Biology*, 65(4):640–650.

[466] Marcovitz, A., Jia, R., and Bejerano, G. (2016). "Reverse genomics" predicts function of human conserved noncoding elements. *Molecular Biology and Evolution*, 33:1358–1369.

[467] Marcussen, T., Jakobsen, K. S., Danihelka, J., Ballard, H. E., Blaxland, K., Brysting, A. K., and Oxelman, B. (2012). Inferring species networks from gene trees in high-polyploid North American and Hawaiian violets (*Viola*, Violaceae). *Systematic Biology*, 61:107–126.

[468] Marin, J., and Hedges, S. B. (2018). Undersampling genomes has biased time and rate estimates throughout the tree of life. *Molecular Biology and Evolution*, 35:2077–2084.

[469] Markin, A., and Eulenstein, O. (2021). Quartet-based inference is statistically consistent under the unified duplication-loss-coalescence model. *Bioinformatics*, 37:4064–4074.

[470] Martin, C. H., Höhna, S., Crawford, J. E., Turner, B. J., Richards, E. J., and Simons, L. H. (2017). The complex effects of demographic history on the estimation of substitution rate: Concatenated gene analysis results in no more than twofold overestimation. *Proceedings of the Royal Society B: Biological Sciences*, 284(1860):20170537.

[471] Martin, S. H., Davey, J. W., and Jiggins, C. D. (2014). Evaluating the use of ABBA˘BABA statistics to locate introgressed loci. *Molecular Biology and Evolution*, 32(1):244–257.

[472] Martin, S. H., Davey, J. W., Salazar, C., and Jiggins, C. D. (2019). Recombination rate variation shapes barriers to introgression across butterfly genomes. *PLOS Biology*, 17(2):e2006288.

[473] Martin, S. H., and Van Belleghem, S. M. (2017). Exploring evolutionary relationships across the genome using topology weighting. *Genetics*, 206(1):429–438.

[474] Martínez-Aquino, A. (2016). Phylogenetic framework for coevolutionary studies: A compass for exploring jungles of tangled trees. *Current Zoology*, 62(4):393–403.

[475] Massatti, R., Reznicek, A. A., and Knowles, L. L. (2016). Utilizing RADseq data for phylogenetic analysis of challenging taxonomic groups: A case study in *Carex* sect. Racemosae. *American Journal of Botany*, 103(2):337–347.

[476] Mathews, S., and Donoghue, M. J. (1999). The root of angiosperm phylogeny inferred from duplicate phytochrome genes. *Science*, 286(5441):947–950.

[477] Matsen, F. A., Billey, S. C., Kas, A., and Konvalinka, M. (2018). Tanglegrams: A reduction tool for mathematical phylogenetics. *IEEE/ACM Transactions on Computational Biology and Bioinformatics*, 15(1):343–349.

[478] Mattick, J. (2004). RNA regulation: A new genetics? *Nature Reviews Genetics*, 5:316–323.

[479] Mattick, J., and Gagen, M. (2001). The evolution of controlled multitasked gene networks: The role of introns and other noncoding RNAs in the development of complex organisms. *Molecular Biology and Evolution*, 18:1611–1630.

[480] Mayr, E. (1942a). Grounds for a statistical method in regional animal geography. *Geographical Review*, 32:694–695.

[481] Mayr, E. (1942b). Zoogeographical studies of the Tsugitaka Mountains of Formosa. *Geographical Review*, 32:693–694.

[482] Mayrose, I., Doron-Faigenbom, A., Bacharach, E., and Pupko, T. (2007). Towards realistic codon models: Among site variability and dependency of synonymous and non-synonymous rates. *Bioinformatics*, 23(13):i319–i327.

[483] McCormack, J., Hird, S., Zellmer, A., Carstens, B., and Brumfield, R. (2013a). Applications of next-generation sequencing to phylogeography and phylogenetics. *Molecular Phylogenetics and Evolution*, 66:526–538.

[484] McCormack, J. E., Harvey, M. G., Faircloth, B. C., Crawford, N. G., Glenn, T. C., and Brumfield, R. T. (2013b). A phylogeny of birds based on over 1,500 loci collected by target enrichment and high-throughput sequencing. *PLOS One*, 8(1):e54848.

[485] McDade, L. (1990). Hybrids and phylogenetic systematics I. Patterns of character expression in hybrids and their implications for cladistic analysis. *Evolution*, 44:1685–1700.

[486] McDade, L. (1992). Hybrids and phylogenetic systematics II. The impact of hybrids on cladistic analysis. *Evolution*, 46:1329–1346.

[487] McInerney, J. O., Pisani, D., Bapteste, E., and O'Connell, M. J. (2011). The public goods hypothesis for the evolution of life on earth. *Biology Direct*, 6(1):41.

[488] McKain, M. R., Johnson, M. G., Uribe-Convers, S., Eaton, D., and Yang, Y. (2018). Practical considerations for plant phylogenomics. *Applications in Plant Sciences*, 6(3).

[489] McKenzie, P. F., and Eaton, D. A. R. (2020). ipcoal: An interactive Python package for simulating and analyzing genealogies and sequences on a species tree or network. *Bioinformatics*, 36:4193–4196.

[490] McVean, G. A., and Cardin, N. J. (2005). Approximating the coalescent with recombination. *Philosophical Transactions of the Royal Society B: Biological Sciences*, 360(1459):1387–1393.

[491] Meadows, J. R., and Lindblad-Toh, K. (2017). Dissecting evolution and disease using comparative vertebrate genomics. *Nature Reviews Genetics*, 18(10):624.

[492] Mehta, R. S., Bryant, D., and Rosenberg, N. A. (2016). The probability of monophyly of a sample of gene lineages on a species tree. *Proceedings of the National Academy of Sciences of the United States of America*, 113(29):8002–8009.

[493] Meiklejohn, K. A., Faircloth, B. C., Glenn, T. C., Kimball, R. T., and Braun, E. L. (2016). Analysis of a rapid evolutionary radiation using ultraconserved elements: Evidence for a bias in some multispecies coalescent methods. *Systematic Biology*, 65(4):612–627.

[494] Mendes, F. K., Fuentes-González, J. A., Schraiber, J. G., and Hahn, M. W. (2018). A multispecies coalescent model for quantitative traits. *eLife*, 7:e36482.

[495] Mendes, F. K., and Hahn, M. W. (2016). Gene tree discordance causes apparent substitution rate variation. *Systematic Biology*, 65(4):711–721.

[496] Mendes, F. K., Hahn, Y., and Hahn, M. W. (2016). Gene tree discordance can generate patterns of diminishing convergence over time. *Molecular Biology and Evolution*, 33(12):3299–3307.

[497] Mendes, F. K., Livera, A. P., and Hahn, M. W. (2019). The perils of intralocus recombination for inferences of molecular convergence. *Proceedings of the Royal Society B: Biological Sciences*, 374(1777):20180244.

[498] Meng, C., and Kubatko, L. S. (2009). Detecting hybrid speciation in the presence of incomplete lineage sorting using gene tree incongruence: A model. *Theoretical Population Biology*, 75(1): 35–45.

[499] Menozzi, P., Piazza, A., and Cavalli-Sforza, L. (1978). Synthetic maps of human gene frequencies in Europeans. *Science*, 201(4358):786–792.

[500] Mickevich, M. F. (1978). Taxonomic congruence. *Systematic Zoology*, 27(2):143–158.

[501] Miller, W., Rosenbloom, K., Hardison, R., Hou, M., Taylor, J., Raney, B., Burhans, R., et al. (2007). 28-way vertebrate alignment and conservation track in the UCSC genome browser. *Genome Research*, 17:1797–1808.

[502] Mirarab, S. (2017). Phylogenomics: Constrained gene tree inference. *Nature Ecology and Evolution*, 1(2):0056.

[503] Mirarab, S., Bayzid, M. S., Boussau, B., and Warnow, T. (2014a). Statistical binning enables an accurate coalescent-based estimation of the avian tree. *Science*, 346(6215):1250463.

[504] Mirarab, S., Bayzid, M. S., and Warnow, T. (2016). Evaluating summary methods for multi-locus species tree estimation in the presence of incomplete lineage sorting. *Systematic Biology*, 65(3).

[505] Mirarab, S., Nguyen, N., Guo, S., Wang, L.-S., Kim, J., and Warnow, T. (2015). PASTA: Ultra-large multiple sequence alignment for nucleotide and amino-acid sequences. *Journal of Computational Biology*, 22:377–386.

[506] Mirarab, S., Reaz, R., Bayzid, M. S., Zimmermann, T., Swenson, M. S., and Warnow, T. (2014b). ASTRAL: Genome-scale coalescent-based species tree estimation. *Bioinformatics*, 30(17):i541–i548.

[507] Mirarab, S., and Warnow, T. (2015). ASTRAL-II: Coalescent-based species tree estimation with many hundreds of taxa and thousands of genes. *Bioinformatics*, 31:i44–i52.

[508] Mitchell, J. D., Allman, E. S., and Rhodes, J. A. (2019). Hypothesis testing near singularities and boundaries. *Electronic Journal of Statistics*, 13(1):2150.

[509] Mitchell, K., Llamas, B., Soubrier, J., Rawlence, N., Worthy, T., Wood, J., Lee, M., and Cooper, A. (2014). Ancient DNA reveals elephant birds and kiwi are sister taxa and clarifies ratite bird evolution. *Science*, 344:898–900.

[510] Miyamoto, M. M. (1985). Consensus cladograms and general classifications. *Cladistics*, 1(2):186–189.

[511] Molloy, E. K., and Warnow, T. (2018). To include or not to include: The impact of gene filtering on species tree estimation methods. *Systematic Biology*, 67(2):285–303.

[512] Molloy, E. K., and Warnow, T. (2019a). Statistically consistent divide-and-conquer pipelines for phylogeny estimation using NJMerge. *Algorithms for Molecular Biology*, 14:14.

[513] Molloy, E. K., and Warnow, T. (2019b). TreeMerge: A new method for improving the scalability of species tree estimation methods. *Bioinformatics*, 35(14):i417–i426.

[514] Molloy, E. K., and Warnow, T. (2020). FastMulRFS: Fast and accurate species tree estimation under generic gene duplication and loss models. *Bioinformatics*, 36(Suppl. 1):i57–i65.

[515] Morales-Briones, D. F., Liston, A., and Tank, D. C. (2018). Phylogenomic analyses reveal a deep history of hybridization and polyploidy in the Neotropical genus *Lachemilla* (Rosaceae). *New Phytologist*, 218(4):1668–1684.

[516] Moran, G., Bell, J., and Matheson, A. (1980). The genetic structure and levels of inbreeding in a *Pinus radiata* D. Don seed orchard. *Silvae Genetica*, 29:190–193.

[517] Morin, M. M., and Moret, B. M. E. (2006). NetGen: Generating phylogenetic networks with diploid hybrids. *Bioinformatics*, 22(15):1921–1923.

[518] Morris, J. L., Puttick, M. N., Clark, J. W., Edwards, D., Kenrick, P., Pressel, S., Wellman, C. H., et al. (2018). The timescale of early land plant evolution. *Proceedings of the National Academy of Sciences of the United States of America*, 115(10):2274–2283.

[519] Morrison, D. (2014). Phylogenetic networks: A review of methods to display evolutionary history. *Annual Research and Review in Biology*, 4(10):1518–1543.

[520] Morrison, D. A. (2011). *An Introduction to Phylogenetic Networks*. RJR Productions.

[521] Morrison, D. A. (2018). Multiple sequence alignment is not a solved problem. *arXiv*. https://doi.org/10.48550/arXiv.1808.07717.

[522] Mossel, E., and Roch, S. (2011). Incomplete lineage sorting: Consistent phylogeny estimation from multiple loci. *IEEE/ACM Transactions on Computational Biology and Bioinformatics*, 7(1):166–171.

[523] Moyers, B., and Zhang, J. (2015). Phylostratigraphic bias creates spurious patterns of genome evolution. *Molecular Biology and Evolution*, 32:258–267.

[524] Moyers, B., and Zhang, J. (2017). Further simulations and analyses demonstrate open problems of phylostratigraphy. *Genome Biology and Evolution*, 9:1519–1527.

[525] Moyers, B., and Zhang, J. (2018). Toward reducing phylostratigraphic errors and biases. *Genome Biology and Evolution*, 10:2037–2048.

[526] Muchhala, N., Johnsen, S., and Smith, S. (2014). Competition for hummingbird pollination shapes flower color variation in Andean Solanaceae. *Evolution*, 68:2275–2286.

[527] Myers, R., Stamatoyannopoulos, J., Snyder, M., Dunham, I., Hardison, R., Bernstein, B., Gingeras, T., et al. (2011). A user's guide to the encyclopedia of DNA elements (ENCODE). *PLOS Biology*, 9(4):e1001046.

[528] Nakhleh, L. (2010a). Evolutionary phylogenetic networks: Models and issues. In Heath, L., and Ramakrishnan, N., editors, *Problem Solving Handbook in Computational Biology and Bioinformatics*, pages 125–158. Boston: Springer.

[529] Nakhleh, L. (2010b). A metric on the space of reduced phylogenetic networks. *IEEE/ACM Transactions on Computational Biology and Bioinformatics*, 7(2):218–222.

[530] Nakhleh, L. (2013). Computational approaches to species phylogeny inference and gene tree reconciliation. *Trends in Ecology and Evolution*, 28:719–728.

[531] Naser-Khdour, S., Minh, B. Q., Zhang, W., Stone, E. A., and Lanfear, R. (2019). The prevalence and impact of model violations in phylogenetic analysis. *Genome Biology and Evolution*, 11(12):3341–3352.

[532] Nelesen, S., Liu, K., Wang, L.-S., Linder, C. R., and Warnow, T. (2012). DACTAL: Divide-and-conquer trees (almost) without alignments. *Bioinformatics*, 28(12):i274–i282.

[533] Nesbitt, S., and Clarke, J. (2016). The anatomy and taxonomy of the exquisitely preserved Green River Formation (early Eocene) lithornithids (Aves) and the relationships of Lithornithidae. *Bulletin of the American Museum of Natural History, 46*.

[534] Nevado, B., Atchison, G. W., Hughes, C. E., and Filatov, D. A. (2016). Widespread adaptive evolution during repeated evolutionary radiations in New World lupins. *Nature Communications*, 7(1):12384.

[535] Newman, M. (2010). *Networks: An Introduction*. Oxford: Oxford University Press.

[536] Nguyen, L., Schmidt, H., Haeseler, A., and Minh, B. (2015a). IQ-TREE: A fast and effective stochastic algorithm for estimating maximum-likelihood phylogenies. *Molecular Biology and Evolution*, 32:268–274.

[537] Nguyen, N., Mirarab, S., Kumar, K., and Warnow, T. (2015b). Ultra-large alignments using phylogeny-aware profiles. *Genome Biology*, 16(1):124.

[538] Nichols, R. (2001). Gene trees and species trees are not the same. *Trends in Ecology and Evolution*, 16:358–364.

[539] Nielsen, R. (1998). Maximum likelihood estimation of population divergence times and population phylogenies under the infinite sites model. *Theoretical Population Biology*, 53:143–151.

[540] Nielsen, R. (2005). Molecular signatures of natural selection. *Annual Review of Genetics*, 39:197–218.

[541] Nold, R. (1999). *Penstemons*. Portland, OR: Timber Press.

[542] Novembre, J., and Stephens, M. (2008). Interpreting principal component analyses of spatial population genetic variation. *Nature Genetics*, 40(5):646–649.

[543] Nute, M., Chou, J., Molloy, E., and Warnow, T. (2018). The performance of coalescent-based species tree estimation methods under models of missing data. *BMC Genomics*, 19 (Suppl. 5):286.

[544] Oakley, T. H., Gu, Z., Abouheif, E., Patel, N. H., and Li, W.-H. (2005). Comparative methods for the analysis of gene-expression evolution: An example using yeast functional genomic data. *Molecular Biology and Evolution*, 22(1):40–50.

[545] Oaks, J. R. (2019). Full Bayesian comparative phylogeography from genomic data. *Systematic Biology*, 68(3):371–395.

[546] Oaks, J. R., Siler, C. D., and Brown, R. M. (2019). The comparative biogeography of philippine geckos challenges predictions from a paradigm of climate-driven vicariant diversification across an island archipelago. *Evolution*, 73(6):1151–1167.

[547] Oberprieler, C., Wagner, F., Tomasello, S., and Konowalik, K. (2017). A permutation approach for inferring species networks from gene trees in polyploid complexes by minimising deep coalescences. *Methods in Ecology and Evolution*, 8(7):835–849.

[548] Ochman, H., Lawrence, J. G., and Groisman, E. A. (2000). Lateral gene transfer and the nature of bacterial innovation. *Nature*, 405(6784):299.

[549] Ogilvie, H. A., Bouckaert, R. R., and Drummond, A. J. (2017). StarBEAST2 brings faster species tree inference and accurate estimates of substitution rates. *Molecular Biology and Evolution*, 34(8):2101–2114.

[550] Ogilvie, H. A., Heled, J., Xie, D., and Drummond, A. J. (2016). Computational performance and statistical accuracy of *BEAST and comparisons with other methods. *Systematic Biology*, 65(3):381–396.

[551] Olave, M., Solà, E., and Knowles, L. L. (2014). Upstream analyses create problems with DNA-based species delimitation. *Systematic Biology*, 63(2):263–271.

[552] O'Leary, M. A., Bloch, J. I., Flynn, J. J., Gaudin, T. J., Giallombardo, A., Giannini, N. P., Goldberg, et al. (2013). The placental mammal ancestor and the post-K-Pg radiation of placentals. *Science*, 339(6120):662–667.

[553] O'Meara, B. C. (2012). Evolutionary inferences from phylogenies: A review of methods. *Annual Review of Ecology, Evolution, and Systematics*, 43(1):267–285.

[554] One Thousand Plant Transcriptomes Initiative (2019). One thousand plant transcriptomes and the phylogenomics of green plants. *Nature*, 574(7780):679.

[555] Ovadia, Y., Fielder, D., Conow, C., and Libeskind-Hadas, R. (2011). The cophylogeny reconstruction problem is NP-complete. *Journal of Computational Biology*, 18(1):59–65.

[556] Oxelman, B., Brysting, A. K., Jones, G. R., Marcussen, T., Oberprieler, C., and Pfeil, B. E. (2017). Phylogenetics of allopolyploids. *Annual Review of Ecology, Evolution, and Systematics*, 48: 543–557.

[557] Page, R. D. M. (2003). *Tangled Trees: Phylogeny, Cospeciation, and Coevolution*. Chicago: University of Chicago Press. .

[558] Pagel, M. (1999). Inferring the historical patterns of biological evolution. *Nature*, 401(6756): 877–884.

[559] Pamilo, P., and Nei, M. (1988). Relationships between gene trees and species trees. *Molecular Biology and Evolution*, 5:568–583.

[560] Pardi, F., and Scornavacca, C. (2015). Reconstructible phylogenetic networks: Do not distinguish the indistinguishable. *PLOS Computational Biology*, 11(4):e1004135.

[561] Parkinson, C. L., Adams, K. L., and Palmer, J. D. (1999). Multigene analyses identify the three earliest lineages of extant flowering plants. *Current Biology*, 9(24):1485–1491.

[562] Patel, S. (2013). Error in phylogenetic estimation for bushes in the tree of life. *Journal of Phylogenetics and Evolutionary Biology*, 01.

[563] Paten, B., Earl, D., Nguyen, N., Diekhans, M., Zerbino, D., and Haussler, D. (2011). Cactus: Algorithms for genome multiple sequence alignment. *Genome Research*, 21:1512–1528.

[564] Patterson, N., Moorjani, P., Luo, Y., Mallick, S., Rohland, N., Zhan, Y., and Genschoreck, T. (2012). Ancient admixture in human history. *Genetics*, 192:1065–1093.

[565] Patterson, N., Richter, D. J., Gnerre, S., Lander, E. S., and Reich, D. (2006). Genetic evidence for complex speciation of humans and chimpanzees. *Nature*, 441:1103–1108.

[566] Pease, J. B. (2018). Why phylogenomic uncertainty enhances introgression analyses. *Molecular Ecology*, 27(22):4347–4349.

[567] Pease, J. B., Brown, J. W., Walker, J. F., Hinchliff, C. E., and Smith, S. A. (2018). Quartet sampling distinguishes lack of support from conflicting support in the green plant tree of life. *American Journal of Botany*, 105(3):385–403.

[568] Pease, J. B., Haak, D. C., Hahn, M. W., and Moyle, L. C. (2016). Phylogenomics reveals three sources of adaptive variation during a rapid radiation. *PLOS Biology*, 14(2):1002379.

[569] Pease, J. B., and Hahn, M. W. (2015). Detection and polarization of introgression in a five-taxon phylogeny. *Systematic Biology*, 64(4):651–662.

[570] Pease, J. B., and Rosenzweig, B. K. (2018). Encoding data using biological principles: The multi-sample variant format for phylogenomics and population genomics. *IEEE/ACM Transactions on Computational Biology and Bioinformatics*, 15(4):1231–1238.

[571] Pei, J., and Wu, Y. (2017). STELLS2: Fast and accurate coalescent-based maximum likelihood inference of species trees from gene tree topologies. *Bioinformatics*, 33(12):1789–1797.

[572] Peng, J., Swofford, D., and Kubatko, L. (2021). Estimation of speciation times under the multispecies coalesent. *bioRXiv*. https://doi.org/10.1101/681023.

[573] Pfeiffer, F., Gröber, C., Blank, M., Händler, K., Beyer, M., Schultze, J. L., and Mayer, G. (2018). Systematic evaluation of error rates and causes in short samples in next-generation sequencing. *Scientific Reports*, 8(1).

[574] Philippe, H., de Vienne, D. M., Ranwez, V., Roure, B., Baurain, D., and Delsuc, F. (2017). Pitfalls in supermatrix phylogenomics. *European Journal of Taxonomy*, 283.

[575] Phillips, M. J., Delsuc, F., and Penny, D. (2004). Genome-scale phylogeny and the detection of systematic biases. *Molecular Biology and Evolution*, 21(7):1455–1458.

[576] Phillips, M. J., Gibb, G. C., Crimp, E. A., and Penny, D. (2010). Tinamous and moa flock together: Mitochondrial genome sequence analysis reveals independent losses of flight among ratites. *Systematic Biology*, 59(1):90–107.

[577] Piaggio-Talice, R., Burleigh, J. G., and Eulenstein, O. (2004). Quartet supertrees. In Bininda-Emonds, O., editor, *Phylogenetic Supertrees*, volume 4, pages 173–191. Dordrecht: Springer.

[578] Pick, K., Hervé Philippe, F., Erpenbeck, D., Jackson, D., Wrede, P., and Wiens, M. (2010). Improved phylogenomic taxon sampling noticeably affects nonbilaterian relationships. *Molecular Biology and Evolution*, 27(9):1983–1987.

[579] Pickrell, J. K., and Pritchard, J. K. (2012). Inference of population splits and mixtures from genome-wide allele frequency data. *PLOS Genetics*, 8(11):e1002967.

[580] Pickrell, J. K., and Reich, D. (2014). Toward a new history and geography of human genes informed by ancient DNA. *Trends in Genetics*, 30(9):377–389.

[581] Poe, S., and Chubb, A. (2004). Birds in a bush: Five genes indicate explosive evolution of avian orders. *Evolution*, 58:404–415.

[582] Pollard, D., Iyer, V., Moses, A., and Eisen, M. (2006). Widespread discordance of gene trees with species trees in *Drosophila*: Evidence for incomplete lineage sorting. *PLOS Genetics*, 2:e173.

[583] Pollard, K., Hubisz, M., Rosenbloom, K., and Siepel, A. (2010). Detection of nonneutral substitution rates on mammalian phylogenies. *Genome Research*, 20:110–121.

[584] Posada, D., and Crandall, K. A. (1998). MODELTEST: Testing the model of DNA substitution. *Bioinformatics*, 14:817–818.

[585] Potapov, V., and Ong, J. L. (2017). Examining sources of error in PCR by single-molecule sequencing. *PLOS ONE*, 12(1):e0169774.

[586] Price, M. N., Dehal, P. S., and Arkin, A. P. (2010). FastTree 2—approximately maximum-likelihood trees for large alignments. *PLOS ONE*, 5(3):e9490.

[587] Pritchard, J. K., Stephens, M., and Donnelly, P. (2000a). Inference of population structure using multilocus genotype data. *Genetics*, 155(2):945–959.

[588] Pritchard, J. K., Stephens, M., Rosenberg, N. A., and Donnelly, P. (2000b). Association mapping in structured populations. *American Journal of Human Genetics*, 67(1):170–181.

[589] Prum, R., Berv, J., Dornburg, A., Field, D., Townsend, J., Lemmon, E., and Lemmon, A. (2015). A comprehensive phylogeny of birds (Aves) using targeted next-generation DNA sequencing. *Nature*, 526:569–573.

[590] Puttick, M. N., Morris, J. L., Williams, T. A., Cox, C. J., Edwards, D., Kenrick, P., and Pressel, S. (2018). The interrelationships of land plants and the nature of the ancestral embryophyte. *Current Biology*, 28(5):733–745.

[591] Qiu, Y.-L., Li, L., Wang, B., Chen, Z., Knoop, V., Groth-Malonek, M., and Dombrovska, O. (2006). The deepest divergences in land plants inferred from phylogenomic evidence. *Proceedings of the National Academy of Sciences of the United States of America*, 103(42):15511–15516.

[592] Qiu, Y.-L., Yangrae Cho, J. C., and Palmer, J. D. (1998). The gain of three mitochondrial introns identifies liverworts as the earliest land plants. *Nature*, 394(6694):671.

[593] Queiroz, A. d., and Gatesy, J. (2007). The supermatrix approach to systematics. *Trends in Ecology and Evolution*, 22(1):34–41.

[594] Queiroz, A. d., Donoghue, M. J., and Kim, J. (1995). Separate versus combined analysis of phylogenetic evidence. *Annual Review of Ecology and Systematics*, 26(1):657–681.

[595] Rabiee, M., Sayyari, E., and Mirarab, S. (2019). Multi-allele species reconstruction using ASTRAL. *Molecular Phylogenetics and Evolution*, 130:286–296.

[596] Rabosky, D. L., Chang, J., Title, P. O., Cowman, P. F., Sallan, L., Friedman, M., Kaschner, K., et al. (2018). An inverse latitudinal gradient in speciation rate for marine fishes. *Nature*, 559(7714):392.

[597] Ragan, M. A. (1992). Phylogenetic inference based on matrix representation of trees. *Molecular Phylogenetics and Evolution*, 1(1):53–58.

[598] Rambaut, A., and Grassly, N. C. (1997). Seq-Gen: An application for the Monte Carlo simulation of DNA sequence evolution along phylogenetic trees. *Bioinformatics*, 13(3):235–238.

[599] Ramsden, C., Holmes, E. C., and Charleston, M. A. (2008). Hantavirus evolution in relation to its rodent and insectivore hosts: No evidence for codivergence. *Molecular Biology and Evolution*, 26(1):143–153.

[600] Ran, J.-H., Shen, T.-T., Wang, M.-M., and Wang, X.-Q. (2018). Phylogenomics resolves the deep phylogeny of seed plants and indicates partial convergent or homoplastic evolution between Gnetales and angiosperms. *Proceedings of the Royal Society B: Biological Sciences*, 285(1881):20181012.

[601] Rannala, B., and Yang, Z. (2003). Likelihood and Bayes estimation of ancestral population sizes in hominoids using data from multiple loci. *Genetics*, 164:1645–1656.

[602] Rannala, B., and Yang, Z. (2017). Efficient Bayesian species tree inference under the multispecies coalescent. *Systematic Biology*, 66:823–842.

[603] Ranwez, V., and Gascuel, O. (2001). Quartet-based phylogenetic inference: Improvements and limits. *Molecular Biology and Evolution*, 18(6):1103–1116.

[604] Rasmussen, M., and Kellis, M. (2012). Unified modeling of gene duplication, loss, and coalescence using a locus tree. *Genome Research*, 22(4):755–765.

[605] Ratan, A. (2009). Assembly algorithms for next-generation sequence data. PhD diss., Pennsylvania State University.

[606] Ray, D. A., Xing, J., Salem, A. H., and Batzer, M. A. (2006). Sines of a nearly perfect character. *Systematic Biology*, 55(6):928–35.

[607] Ray, J. (1690). *Synopsis methodica stirpium Britannicarum*. London: Samuel Smith.

[608] Reaz, R., Bayzid, M. S., and Rahman, M. S. (2014). Accurate phylogenetic tree reconstruction from quartets: A heuristic approach. *PLOS ONE*, 9(8):e104008.

[609] Reddy, S., Kimball, R. T., Pandey, A., Hosner, P. A., Braun, M. J., Hackett, S. J., Han, K.-L., et al. (2017). Why do phylogenomic data sets yield conflicting trees? Data type influences the avian tree of life more than taxon sampling. *Systematic Biology*, 66(5):857–879.

[610] Redelings, B. (2021). Bali-phy version 3: Model-based co-estimation of alignment and phylogeny. *Bioinformatics*, 37:3032–3034.

[611] Ree, R. H., and Hipp, A. L. (2015). Inferring phylogenetic history from restriction site associated DNA (RADseq). In Hörandl, E., and Appelhans, M. S., editors, *Next-Generation Sequencing in Plant Systematics*, pages 181–204. Bratislava, Slovakia: International Association for Plant Taxonomy.

[612] Reich, D., Thangaraj, K., Patterson, N., Price, A. L., and Singh, L. (2009). Reconstructing Indian population history. *Nature*, 461(7263):489–494.

[613] Reid, N., Demboski, J., and Sullivan, J. (2012). Phylogeny estimation of the radiation of western North American chipmunks (*Tamias*) in the face of introgression using reproductive protein genes. *Systematic Biology*, 61:44–62.

[614] Reid, N. M., Hird, S. M., Brown, J. M., Pelletier, T. A., McVay, J. D., Satler, J. D., and Carstens, B. C. (2014). Poor fit to the multispecies coalescent is widely detectable in empirical data. *Systematic Biology*, 63:322–333.

[615] Rettig, J., Wilson, H. D., and Manhart, J. R. (1992). Phylogeny of the Caryophyllales-gene sequence data. *Taxon*, 41(2):201–209.

[616] Rhodes, J. A., Baños, H., Mitchell, J. D., and Allman, E. S. (2021). MSCquartets 1.0: Quartet methods for species trees and networks under the multispecies coalescent model in R. *Bioinformatics*, 37(12):1766–1768.

[617] Richards, E. J., Brown, J. M., Barley, A. J., Chong, R. A., and Thomson, R. C. (2018). Variation across mitochondrial gene trees provides evidence for systematic error: How much gene tree variation is biological? *Systematic Biology*, 67(5):847–860.

[618] Rieseberg, L. (1995). The role of hybridization in evolution—Old wine in new skins. *American Journal of Botany*, 82:944–953.

[619] Robinson, D., and Foulds, L. (1979). Comparison of weighted labelled trees. In Horadam, A. F., and Wallis, W. D., *Combinatorial Mathematics VI*. Lecture Notes in Mathematics 748, page 119–126. Berlin: Springer.

[620] Robinson, D., and Foulds, L. (1981). Comparison of phylogenetic trees. *Mathematical Biosciences*, 53(1–2):131–147.

[621] Roch, S. (2006). A short proof that phylogenetic tree reconstruction by maximum likelihood is hard. *IEEE/ACM Transactions on Computational Biology and Bioinformatics*, 3(1):92–94.

[622] Roch, S., Nute, M., and Warnow, T. (2018). Long-branch attraction in species tree estimation: Inconsistency of partitioned likelihood and topology-based summary methods. *Systematic Biology*, 68(2):281–297.

[623] Roch, S., and Snir, S. (2013). Recovering the treelike trend of evolution despite extensive lateral genetic transfer: A probabilistic analysis. *Journal of Computational Biology*, 20(2):93–112.

[624] Roch, S., and Steel, M. (2015). Likelihood-based tree reconstruction on a concatenation of aligned sequence data sets can be statistically inconsistent. *Theoretical Population Biology*, 100:56–62.

[625] Rochette, N. C., Rivera-Colón, A. G., and Catchen, J. M. (2019). Stacks 2: Analytical methods for paired-end sequencing improve RADseq-based population genomics. *Molecular Ecology*, 28(21):4737–4754.

[626] Rodman, J. E., Oliver, M. K., Nakamura, R. R., McClammer, Jr., J. U., and Bledsoe, A. H. (1984). A taxonomic analysis and revised classification of Centrospermae. *Systematic Botany*, 9(3):297–323.

[627] Rogers, J. (2018). Adding resolution and dimensionality to comparative genomics: Moving from reference genomes to clade genomics. *Genome Biology*, 19(1):115.

[628] Rohlfs, R., and Nielsen, R. (2015). Phylogenetic ANOVA: The expression variance and evolution model for quantitative trait evolution. *Systematic Biology*, 64:695–708.

[629] Rokas, A., and Carroll, S. B. (2005). More genes or more taxa? The relative contribution of gene number and taxon number to phylogenetic accuracy. *Molecular Biology and Evolution*, 22(5):1337–1344.

[630] Rokas, A., Williams, B., King, N., and Carroll, S. (2003). Genome-scale approaches to resolving incongruence in molecular phylogenies. *Nature*, 425:798–804.

[631] Romero, I. G., Ruvinsky, I., and Gilad, Y. (2012). Comparative studies of gene expression and the evolution of gene regulation. *Nature Reviews Genetics*, 13(7):505–516.

[632] Ronquist, F. (1997). Phylogenetic approaches in coevolution and biogeography. *Zoologica Scripta*, 26(4):313–322.

[633] Rosenberg, N. A. (2002). The probability of topological concordance of gene trees and species trees. *Theoretical Population Biology*, 61:225–247.

[634] Rosenberg, N. A. (2003). The shapes of neutral gene genealogies in two species: Probabilities of monophyly, paraphyly, and polyphyly in a coalescent model. *Evolution*, 57(7):1465–1477.

[635] Rosenblum, E., Rompler, H., Schoneberg, T., and Hoekstra, H. (2010). Molecular and functional basis of phenotypic convergence in white lizards at white sands. *Proceedings of the National Academy of Sciences of the United States of America*, 107:2113–2117.

[636] Rosenzweig, B. K., Pease, J. B., Besansky, N. J., and Hahn, M. W. (2016). Powerful methods for detecting introgressed regions from population genomic data. *Molecular Ecology*, 25(11):2387–2397.

[637] Rothfels, R. C., Li, F.-W., and Pryer, K. M. (2017). Next-generation polyploid phylogenetics: Rapid resolution of hybrid polyploid complexes using PacBio single-molecule sequencing. *New Phytologist*, 213:413–429.

[638] Roure, B., Baurain, D., and Philippe, H. (2012). Impact of missing data on phylogenies inferred from empirical phylogenomic data sets. *Molecular Biology and Evolution*, 30:197–214.

[639] Roy, S., Wapinski, I., Pfiffner, J., French, C., Socha, A., Konieczka, J., Habib, N., et al. (2013). Arboretum: Reconstruction and analysis of the evolutionary history of condition-specific transcriptional modules. *Genome Research*, 23:1039–1050.

[640] Rubin, B., Ree, R., and Moreau, C. (2012). Inferring phylogenies from RAD sequence data. *PLOS One*, 7:33394.

[641] Ruprecht, C., Vaid, N., Proost, S., Persson, S., and Mutwil, M. (2017). Beyond genomics: Studying evolution with gene coexpression networks. *Trends in Plant Science*, 22(4):298–307.

[642] Ryan, J. F., Pang, K., Schnitzler, C. E., Nguyen, A.-D., Moreland, R. T., Simmons, D. K., Koch, B. J., et al. (2013). The genome of the ctenophore *Mnemiopsis leidyi* and its implications for cell type evolution. *Science*, 342(6164):1242592.

[643] Sackton, T., Grayson, P., Cloutier, A., Hu, Z., Liu, J., Wheeler, N., Gardner, P., et al. (2019). Convergent regulatory evolution and loss of flight in paleognathous birds. *Science*, 364:74–78.

[644] Saglam, I. K., Baumsteiger, J., and Miller, M. R. (2017). Failure to differentiate between divergence of species and their genes can result in over-estimation of mutation rates in recently diverged species. *Proceedings of the Royal Society B: Biological Sciences*, 284(1860):20170021.

[645] Saitou, N., and Nei, M. (1987). The neighbor-joining method: A new method for reconstructing phylogenetic trees. *Molecular Biology and Evolution*, 4:406–425.

[646] Salichos, L., and Rokas, A. (2013). Inferring ancient divergences requires genes with strong phylogenetic signals. *Nature*, 497(7449):327.

[647] Salichos, L., Stamatakis, A., and Rokas, A. (2014). Novel information theory-based measures for quantifying incongruence among phylogenetic trees. *Molecular Biology and Evolution*, 31(5):1261–1271.

[648] Sanderson, M. J. (2002). Estimating absolute rates of molecular evolution and divergence times: A penalized likelihood approach. *Molecular Biology and Evolution*, 19(1):101–109.

[649] Sanderson, M. J., and Driskell, A. C. (2003). The challenge of constructing large phylogenetic trees. *Trends in Plant Science*, 8(8):374–379.

[650] Sanderson, M. J., McMahon, M. M., and Steel, M. (2010). Phylogenomics with incomplete taxon coverage: The limits to inference. *BMC Evolutionary Biology*, 10(1):155.

[651] Sanderson, M. J., McMahon, M. M., and Steel, M. (2011). Terraces in phylogenetic tree space. *Science*, 333(6041):448–450.

[652] Sarver, B. A. J., Herrera, N. D., Sneddon, D., Hunter, S. S., Settles, M. L., Kronenberg, Z., Demboski, J. R., Good, J. M., and Sullivan, J. (2021). Diversification, introgression, and rampant cytonuclear discordance in Rocky Mountains chipmunks (Sciuridae: *Tamias*). *Systematic Biology*, 70(5):908–921.

[653] Satler, J. D., Herre, E. A., Jandér, K. C., Eaton, D. A. R., Machado, C. A., Heath, T. A., and Nason, J. D. (2019). Inferring processes of coevolutionary diversification in a community of Panamanian strangler figs and associated pollinating wasps. *Evolution*, 73(11):2295–2311.

[654] Sato, S., Tabata, S., Hirakawa, H., Asamizu, E., Shirasawa, K., Isobe, S., and Kaneko, T. (2012). The tomato genome sequence provides insights into fleshy fruit evolution. *Nature*, 485:635–641.

[655] Sauquet, H., Balthazar, M., Magallón, S., Doyle, J. A., Endress, P. K., Bailes, E. J., and Morais, E. B. (2017). The ancestral flower of angiosperms and its early diversification. *Nature Communications*, 8:16047.

[656] Sayyari, E., and Mirarab, S. (2016a). Anchoring quartet-based phylogenetic distances and applications to species tree reconstruction. *BMC Genomics*, 17(S10):101–113.

[657] Sayyari, E., and Mirarab, S. (2016b). Fast coalescent-based computation of local branch support from quartet frequencies. *Molecular Biology and Evolution*, 33:1654–1668.

[658] Sayyari, E., and Mirarab, S. (2018). Testing for polytomies in phylogenetic species trees using quartet frequencies. *Genes*, 9(3):132.

[659] Sayyari, E., Whitfield, J. B., and Mirarab, S. (2017). Fragmentary gene sequences negatively impact gene tree and species tree reconstruction. *Molecular Biology and Evolution*, 34(12):3279–3291.

[660] Sayyari, E., Whitfield, J. B., and Mirarab, S. (2018). DiscoVista: Interpretable visualizations of gene tree discordance. *Molecular Phylogenetics and Evolution*, 122:110–115.

[661] Schierup, M., and Hein, J. (2000). Consequences of recombination on traditional phylogenetic analysis. *Genetics*, 156:879–891.

[662] Schrempf, D., Minh, B., De Maio, N., von Haeseler, A., and Kosiol, C. (2016). Reversible polymorphism-aware phylogenetic models and their application to tree inference. *Journal of Theoretical Biology*, 407:362–370.

[663] Schrider, D. R., Hourmozdi, J. N., and Hahn, M. W. (2011). Pervasive multinucleotide mutational events in eukaryotes. *Current Biology*, 21(12):1051–1054.

[664] Schrider, D. R., and Kern, A. D. (2018). Supervised machine learning for population genetics: A new paradigm. *Trends in Genetics*, 34(4):301–312.

[665] Sedio, B. E. (2017). Recent breakthroughs in metabolomics promise to reveal the cryptic chemical traits that mediate plant community composition, character evolution and lineage diversification. *New Phytologist*, 214(3):952–958.

[666] Seki, R., Li, C., Fang, Q., Hayashi, S., Egawa, S., Hu, J., Xu, L., et al. (2017). Functional roles of Aves class-specific cis-regulatory elements on macroevolution of bird-specific features. *Nature Communications*, 8:14229.

[667] Semple, C., and Steel, M. (2003). *Phylogenetics*. Oxford: Oxford University Press.

[668] Seo, T.-K. (2008). Calculating bootstrap probabilities of phylogeny using multilocus sequence data. *Molecular Biology and Evolution*, 25(5):960–971.

[669] Sharma, V., Hecker, N., Walther, F., Stuckas, H., and Hiller, M. (2020). Convergent losses of TLR5 suggest altered extracellular flagellin detection in four mammalian lineages. *Molecular Biology and Evolution*, 37:1847–1854.

[670] Shaw, A. (2000). Phylogeny of the Sphagnopsida based on chloroplast and nuclear DNA sequences. *Bryologist*, 103(2):277–306.

[671] Shaw, J., Lickey, E. B., Beck, J. T., Farmer, S. B., Liu, W., Miller, J., Siripun, K. C., et al. (2005). The tortoise and the hare II: Relative utility of 21 noncoding chloroplast DNA sequences for phylogenetic analysis. *American Journal of Botany*, 92(1):142–166.

[672] Shaw, J., Lickey, E. B., Schilling, E. E., and Small, R. L. (2007). Comparison of whole chloroplast genome sequences to choose noncoding regions for phylogenetic studies in angiosperms: The tortoise and the hare III. *American Journal of Botany*, 94(3):275–288.

[673] Sheehan, H., Feng, T., Walker-Hale, N., Lopez-Nieves, S., Pucker, B., Guo, R., and Yim, W. C. (2020). Evolution of l-DOPA 4, 5-dioxygenase activity allows for recurrent specialisation to betalain pigmentation in Caryophyllales. *New Phytologist*, 227(3):914–929.

[674] Shekhar, S., Roch, S., and Mirarab, S. (2017). Species tree estimation using ASTRAL: How many genes are enough? *IEEE/ACM Transactions on Computational Biology and Bioinformatics*, 15(5):1738–1747.

[675] Shen, X.-X., Hittinger, C. T., and Rokas, A. (2017). Contentious relationships in phylogenomic studies can be driven by a handful of genes. *Nature Ecology and Evolution*, 1(5): 0126.

[676] Shi, C.-M., and Yang, Z. (2018). Coalescent-based analyses of genomic sequence data provide a robust resolution of phylogenetic relationships among major groups of gibbons. *Molecular Biology and Evolution*, 35:159–179.

[677] Shimodaira, H., and Hasegawa, M. (1999). Multiple comparisons of log-likelihoods with applications to phylogenetic inference. *Molecular Biology and Evolution*, 16:1114–1116.

[678] Shultz, A., and Sackton, T. (2019). Immune genes are hotspots of shared positive selection across birds and mammals. *eLife*, 8:e41815.

[679] Siepel, A., Bejerano, G., Pedersen, J., Hinrichs, A., Hou, M., Rosenbloom, K., Clawson, H., et al. (2005). Evolutionarily conserved elements in vertebrate, insect, worm, and yeast genomes. *Genome Research*, 15:1034–1050.

[680] Siepel, A., and Haussler, D. (2004). Phylogenetic estimation of context-dependent substitution rates by maximum likelihood. *Molecular Biology and Evolution*, 21:468–488.

[681] Simmons, M. P., and Gatesy, J. (2015). Coalescence vs. concatenation: Sophisticated analyses vs. first principles applied to rooting the angiosperms. *Molecular Phylogenetics and Evolution*, 91:98–122.

[682] Simmons, M. P., Sloan, D. B., and Gatesy, J. (2016). The effects of subsampling gene trees on coalescent methods applied to ancient divergences. *Molecular Phylogenetics and Evolution*, 97(January):76–89.

[683] Simmons, M. P., Sloan, D. B., Springer, M. S., and Gatesy, J. (2019). Gene-wise resampling outperforms site-wise resampling in phylogenetic coalescence analyses. *Molecular Phylogenetics and Evolution*, 131(November 2018):80–92.

[684] Simpson, G. G. (1942). The beginnings of vertebrate paleontology in North America. *Proceedings of the American Philosophical Society*, 86(1):130–188.

[685] Simpson, G. G. (1943). Criteria for genera, species, and subspecies in zoology and paleozoology. *Annals of the New York Academy of Sciences*, 44(2):145–178.

[686] Slatkin, M., and Pollack, J. L. (2005). The concordance of gene trees and species trees at two linked loci. *Genetics*, 172(3):1979–1984.

[687] Sloan, P. R. (1972). John Locke, John Ray, and the problem of the natural system. *Journal of the History of Biology*, 5(1):1–53.

[688] Smirnov, V., and Warnow, T. (2020). Unblended disjoint tree merging using GTM improves species tree estimation. *BMC Genomics*, 21(2):1–17.

[689] Smith, J., Braun, E., and Kimball, R. (2013). Ratite nonmonophyly: Independent evidence from 40 novel loci. *Systematic Biology*, 62:35–49.

[690] Smith, N., and Clarke, J. (2014). Osteological histology of the pan-alcidae (aves, charadriiformes): correlates of wing-propelled diving and flightlessness. *Anatomical Record (Hoboken)*, 297:188–199.

[691] Smith, S., and Baum, D. (2006). Phylogenetics of the florally diverse Andean clade Iochrominae (Solanaceae). *American Journal of Botany*, 93:1140–1153.

[692] Smith, S., and Baum, D. (2007). Systematics of Iochrominae (Solanaceae): Patterns in floral diversity and interspecific crossability. *Acta Horticulturae*, 745:241–254.

[693] Smith, S., Hall, S., Izquierdo, P., and Baum, D. (2008). Comparative pollination biology of sympatric and allopatric Andean *Iochroma* (Solanaceae). *Annals of the Missouri Botanical Garden*, 95:600–617.

[694] Smith, S., and Leiva, S. (2011). A new species of *Iochroma* (Solanaceae) from Ecuador. *Novon*, 21:491–495.

[695] Smith, S., and Pease, J. (2017). Heterogeneous molecular processes among the causes of how sequence similarity scores can fail to recapitulate phylogeny. *Briefings in Bioinformatics*, 18:451–457.

[696] Smith, S. A., Brown, J. W., and Walker, J. F. (2018a). So many genes, so little time: A practical approach to divergence-time estimation in the genomic era. *PloS ONE*, 13(5):0197433.

[697] Smith, S. A., Brown, J. W., Yang, Y., Bruenn, R., Drummond, C. P., Brockington, S. F., Walker, J. F., et al. (2018b). Disparity, diversity, and duplications in the Caryophyllales. *New Phytologist*, 217(2):836–854.

[698] Smith, S. A., and Donoghue, M. J. (2008). Rates of molecular evolution are linked to life history in flowering plants. *Science*, 322(5898):86–89.

[699] Smith, S. A., and Dunn, C. (2008). Phyutility: A phyloinformatics utility for trees, alignments, and molecular data. *Bioinformatics*, 24:715–716.

[700] Smith, S. A., Moore, M. J., Brown, J. W., and Yang, Y. (2015). Analysis of phylogenomic datasets reveals conflict, concordance, and gene duplications with examples from animals and plants. *BMC Evolutionary Biology*, 15(1):150.

[701] Smith, S. A., and O'Meara, B. C. (2012). treePL: Divergence time estimation using penalized likelihood for large phylogenies. *Bioinformatics*, 28(20):2689–2690.

[702] Smith, S. A., Walker-Hale, N., Walker, J. F., and Brown, J. W. (2019). Phylogenetic conflicts, combinability, and deep phylogenomics in plants. *Systematic Biology*, 69(3):579–592.

[703] Smith, S. D., Pennell, M. W., Dunn, C. W., and Edwards, S. V. (2020). Phylogenetics is the new genetics (for most of biodiversity). *Trends in Ecology and Evolution*, 35(5):415–425.

[704] Sneath, P. H. A. (1975). Cladistic representation of reticulate evolution. *Systematic Zoology*, 24(3):360–368.

[705] Sneath, P. H. A., and Sokal, R. R. (1973). *Numerical Taxonomy. The Principles and Practice of Numerical Classification*. San Francisco: W. H. Freeman.

[706] Snir, S., and Rao, S. (2010). Quartets MaxCut: A divide and conquer quartets algorithm. *IEEE/ACM Transactions on Computational Biology and Bioinformatics*, 7(4):704–718.

[707] Snir, S., and Rao, S. (2012). Quartet MaxCut: A fast algorithm for amalgamating quartet trees. *Molecular Phylogenetics and Evolution*, 62(1):1–8.

[708] Sokal, R. R., and Camin, J. H. (1965). The two taxonomies: Areas of agreement and conflict. *Systematic Biology*, 14(3):176.

[709] Sokal, R. R., Camin, J. H., Rohlf, F. J., and Sneath, P. H. A. (1965). Numerical taxonomy: Some points of view. *Systematic Zoology*, 14(3):237.

[710] Solís-Lemus, C., and Ané, C. (2016). Inferring phylogenetic networks with maximum pseudo-likelihood under incomplete lineage sorting. *PLOS Genetics*, 12(3):e1005896.

[711] Solís-Lemus, C., Bastide, P., and Ané, C. (2017). PhyloNetworks: A package for phylogenetic networks. *Molecular Biology and Evolution*, 34(12):3292–3298.

[712] Solís-Lemus, C., Knowles, L. L., and Ané, C. (2015). Bayesian species delimitation combining multiple genes and traits in a unified framework. *Evolution*, 69(2):492–507.

[713] Solís-Lemus, C., Yang, M., and Ané, C. (2016). Inconsistency of species-tree methods under gene flow. *Systematic Biology*, 65(5):843–851.

[714] Soltis, D. E., and Soltis, P. S. (1999). Polyploidy: Recurrent formation and genome evolution. *Trends in Ecology and Evolution*, 14(9):348–352.

[715] Soltis, D. E., Soltis, P. S., Chase, M. W., Mort, M. E., Albach, D. C., Zanis, M., and Savolainen, V. (2000). Angiosperm phylogeny inferred from 18S rDNA, rbcL, and atpB sequences. *Botanical Journal of the Linnean Society*, 133(4):381–461.

[716] Springer, M., and Gatesy, J. (2016). The gene tree delusion. *Molecular Phylogenetics and Evolution*, 94:1–33.

[717] Springer, M., Molloy, E., Sloan, D., Simmons, M., and Gatesy, J. (2019). ILS-aware analysis of low-homoplasy retroelement insertions: Inference of species trees and introgression using quartets. *Journal of Heredity*, 111(2):147–168.

[718] Springer, M. S., and Gatesy, J. (2018). On the importance of homology in the age of phylogenomics. *Systematics and Biodiversity*, 16(3):210–228.

[719] Stadler, T., and Steel, M. (2012). Distribution of branch lengths and phylogenetic diversity under homogeneous speciation models. *Journal of Theoretical Biology*, 297:33–40.

[720] Stamatakis, A. (2014). RAxML version 8: A tool for phylogenetic analysis and post-analysis of large phylogenies. *Bioinformatics*, 30(9):1312–1313.

[721] Stamatakis, A., Hoover, P., and Rougemont, J. (2008). A rapid bootstrap algorithm for the RAxML web servers. *Systematic Biology*, 57:758–771.

[722] Stamos, D. N. (2005). Pre-Darwinian taxonomy and essentialism—a reply to Mary Winsor. *Biology and Philosophy*, 20(1):79–96.

[723] Stebbins, G. (1959). The synthetic approach to problems of organic evolution. *Cold Spring Harbor Symposia on Quantitative Biology*, 24:305–311.

[724] Stebbins, G. L. (1950). *Variation and Evolution in Plants*. New York: Columbia University Press.

[725] Steel, M. (1992). The complexity of reconstructing trees from qualitative characters and subtrees. *Journal of Classification*, 9:91–116.

[726] Steel, M., and Penny, D. (2000). Parsimony, likelihood, and the role of models in molecular phylogenetics. *Molecular Biology and Evolution*, 17(6):839–850.

[727] Stenz, N., Larget, B., Baum, D. A., and Ané, C. (2015). Exploring tree-like and non-tree-like patterns using genome sequences: An example using the inbreeding plant species *Arabidopsis thaliana* (L.) Heynh. *Systematic Biology*, 64(5):809–823.

[728] Stiller, J., and Zhang, G. (2019). Comparative phylogenomics, a stepping stone for bird biodiversity studies. *Diversity*, 11(7):115.

[729] Straw, R. M. (1955). Hybridization, homogamy, and sympatric speciation. *Evolution*, 9:441–444.

[730] Straw, R. M. (1966). A redefinition of *Penstemon* (Scrophulariaceae). *Brittonia*, 18:80–95.

[731] Streicher, J. W., Schulte, J. A., and Wiens, J. J. (2016). How should genes and taxa be sampled for phylogenomic analyses with missing data? An empirical study in Iguanian lizards. *Systematic Biology*, 65(1):128–145.

[732] Strickler, D. (1997). *Northwest Penstemons*. Detroit: Flower Press.

[733] Strimmer, K., Goldman, N., and von Haeseler, A. (1997). Bayesian probabilities and quartet puzzling. *Molecular Biology and Evolution*, 14(2):210–211.

[734] Strimmer, K., and von Haeseler, A. (1997). Likelihood-mapping: A simple method to visualize phylogenetic content of a sequence alignment. *Proceedings of the National Academy of Sciences of the United States of America*, 94(13):6815–6819.

[735] Stull, G., Walker-Hale, N., Walker, J. F., and Brown (2013). A targeted enrichment strategy for massively parallel sequencing of angiosperm plastid genomes. *Applications in Plant Sciences*, 1(2):1200497.

[736] Sukumaran, J., and Holder, M. T. (2010). DendroPy: A Python library for phylogenetic computing. *Bioinformatics*, 26(12):1569–1571.

[737] Sukumaran, J., and Knowles, L. L. (2017). Multispecies coalescent delimits structure, not species. *Proceedings of the National Academy of Sciences of the United States of America*, 114:1607–1612.

[738] Swenson, M., Suri, R., Linder, C., and Warnow, T. (2012). SuperFine: Fast and accurate supertree estimation. *Systematic Biology*, 61(2):214–227.

[739] Swenson, M. S., Suri, R., Linder, C. R., and Warnow, T. (2011). An experimental study of Quartets MaxCut and other supertree methods. *Algorithms for Molecular Biology*, 6(1):7.

[740] Swenson, U., and Manns, U. (2003). Phylogeny of *Pericallis* (Asteraceae): A total evidence approach reappraising the double origin of woodiness. *Taxon*, 52(3):533–548.

[741] Swofford, D. L. (1996). *PAUP*: Phylogenetic Analysis Using Parsimony (and Other Methods)*, version 4.0. Sunderland, MA: Sinauer.

[742] Szöllősi, G. J., Tannier, E., Daubin, V., and Boussau, B. (2014). The inference of gene trees with species trees. *Systematic Biology*, 64(1):e42–e62.

[743] Tajima, F. (1983). Evolutionary relationship of DNA sequences in finite populations. *Genetics*, 105:437–460.

[744] Takahata, N. (1989). Gene genealogy in three related populations: Consistency probability between gene and population trees. *Genetics*, 122:957–966.

[745] Takahata, N., and Nei, M. (1985). Gene genealogy and variance of interpopulational nucleotide differences. *Genetics*, 110:325–344.

[746] Tavaré, S. (1984). Line-of-descent and genealogical processes, and their applications in population genetics models. *Theoretical Population Biology*, 26:119–164.

[747] Tavaré, S. (1986). Some probabilistic and statistical problems in the analysis of DNA sequences. In Miura, R. M., editor, *Some Mathematical Questions in Biology: DNA Sequence Analysis*, pages 57–86. Lectures on Mathematics in the Life Sciences, volume 17, pages 57–86. Providence, RI: American Mathematical Society.

[748] Than, C., and Nakhleh, L. (2009). Species tree inference by minimizing deep coalescences. *PLOS Computational Biology*, 5(9):e1000501.

[749] Than, C., Ruths, D., and Nakhleh, L. (2008). PhyloNet: A software package for analyzing and reconstructing reticulate evolutionary relationships. *BMC Bioinformatics*, 9(1):322.

[750] Than, C. V., and Rosenberg, N. A. (2011). Consistency properties of species tree inference by minimizing deep coalescences. *Journal of Computational Biology*, 18(1):1–15.

[751] The Heliconius Genome Consortium (2012). Butterfly genome reveals promiscuous exchange of mimicry adaptations among species. *Nature*, 487:94–98.

[752] Theys, K., Lemey, P., Vandamme, A.-M., and Baele, G. (2019). Advances in visualization tools for phylogenomic and phylodynamic studies of viral diseases. *Frontiers in Public Health*, 7:208.

[753] Tian, Y., and Kubatko, L. (2017). Rooting phylogenetic trees under the coalescent model using site pattern probabilities. *BMC Evolutionary Biology*, 17:263.

[754] Tidwell, H., and Nakhleh, L. (2019). Integrated likelihood for phylogenomics under a no-common-mechanism model. *BMC Genomics 21:219*.

[755] Toll-Riera, M., Laurie, S., and Alba, M. M. (2010). Lineage-specific variation in intensity of natural selection in mammals. *Molecular Biology and Evolution*, 28(1):383–398.

[756] Traag, V., Dooren, P., and Nesterov, Y. (2011). Narrow scope for resolution-limit-free community detection. *Physical Review E*, 84:016114.

[757] Tuffley, C., and Steel, M. (1997). Links between maximum likelihood and maximum parsimony under a simple model of site substitution. *Bulletin of Mathematical Biology*, 59(3):581–607.

[758] Uetz, P., and Stylianou, A. (2018). The original descriptions of reptiles and their subspecies. *Zootaxa*, 4375:257–264.

[759] Uribe-Convers, S., Settles, M. L., and Tank, D. C. (2016). A phylogenomic approach based on PCR enrichment and high throughput sequencing: Resolving diversity within the South American species of *Bartsia* L. (Orobanchaceae). *PLOS ONE*, 11:e0148203.

[760] Uricchio, L. H., Kitano, H. C., Gusev, A., and Zaitlen, N. A. (2019). An evolutionary compass for detecting signals of polygenic selection and mutational bias. *Evolution Letters*, 3(1):69–79.

[761] Vachaspati, P., and Warnow, T. (2015). ASTRID: Accurate species trees from internode distances. *BMC Genomics*, 16(Suppl. 10):S3.

[762] Vachaspati, P., and Warnow, T. (2018). SVDquest: Improving SVDquartets species tree estimation using exact optimization within a constrained search space. *Molecular Phylogenetics and Evolution*, 124:122–136.

[763] Varin, C., Reid, N., and Firth, D. (2011). An overview of composite likelihood methods. *Statistica Sinica*, 21:5–42.

[764] Venkat, A., Hahn, M. W., and Thornton, J. W. (2018). Multinucleotide mutations cause false inferences of lineage-specific positive selection. *Nature Ecology and Evolution*, 2(8):1280–1288.

[765] Villar, D., Berthelot, C., Aldridge, S., Rayner, T., Lukk, M., Pignatelli, M., Park, T., et al. (2015). Enhancer evolution across 20 mammalian species. *Cell*, 160:554–566.

[766] Villiers, M., Pirie, M., Hughes, M., Moller, M., Edwards, T., and Bellstedt, D. (2013). An approach to identify putative hybrids in the 'coalescent stochasticity zone,' as exemplified in the African plant genus *Streptocarpus* (Gesneriaceae). *New Phytologist*, 198:284–300.

[767] Vinogradov, A. E. (2010). Systemic factors dominate mammal protein evolution. *Proceedings of the Royal Society B: Biological Sciences*, 277(1686):1403–1408.

[768] Wagner, R., Toups, B., Deng, Z., Gallivan, K., Brown, J., and Wilgenbusch, J. (2021). Investigating the genomic distribution of phylogenetic signal with cloudforest. In *Practice and Experience in Advanced Research Computing (PEARC '21)*, pages 1–4. Boston: Association for Computing Machinery.

[769] Wagner, Jr., W. H. (1954). Reticulate evolution in the aspleniums Aspleniums. *Evolution*, 8(2):103–118.

[770] Wakeley, J. (2009). *Coalescent Theory: An Introduction*. Greenwood Village, CO: Roberts and Company.

[771] Walker, J. F., Brown, J. W., and Smith, S. A. (2018a). Analyzing contentious relationships and outlier genes in phylogenomics. *Systematic Biology*, 5:916–924.

[772] Walker, J. F., Walker-Hale, N., Vargas, O. M., Larson, D. A., and Stull, G. W. (2019). Characterizing gene tree conflict in plastome-inferred phylogenies. *PeerJ*, 7:7747.

[773] Walker, J. F., Yang, Y., Feng, T., Timoneda, A., Mikenas, J., Hutchison, V., and Edwards, C. (2018b). From cacti to carnivores: Improved phylotranscriptomic sampling and hierarchical homology inference provide further insight into the evolution of Caryophyllales. *American Journal of Botany*, 3:446–462.

[774] Walker, J. F., Yang, Y., Moore, M. J., Mikenas, J., Timoneda, A., Brockington, S. F., and Smith, S. A. (2017). Widespread paleopolyploidy, gene tree conflict, and recalcitrant relationships among the carnivorous Caryophyllales. *American Journal of Botany*, 104(6):858–867.

[775] Wang, J., and Guo, M. (2018). A review of metrics measuring dissimilarity for rooted phylogenetic networks. *Briefings in Bioinformatics*, 20(6):1972–1980.

[776] Wang, J., Street, N. R., Park, E.-J., Liu, J., and Ingvarsson, P. K. (2020). Evidence for widespread selection in shaping the genomic landscape during speciation of *Populus*. *Molecular Ecology* 29(6):1120–1136.

[777] Wang, L.-S., Warnow, T., Moret, B., Jansen, R., and Raubeson, L. (2006). Distance-based genome rearrangement phylogeny. *Journal of Molecular Evolution*, 63(4):473–483.

[778] Wang, N., Yang, Y., Moore, M. J., Brockington, S. F., Walker, J. F., Brown, J. W., and Liang, B. (2018). Evolution of Portulacineae marked by gene tree conflict and gene family expansion associated with adaptation to harsh environments. *Molecular Biology and Evolution*, 36(1):112–126.

[779] Wang, X., Ye, X., Zhao, L., Li, D., Guo, Z., and Zhuang, H. (2017). Genome-wide RAD sequencing data provide unprecedented resolution of the phylogeny of temperate bamboos (Poaceae: Bambusoideae). *Scientific Reports*, 7:11546.

[780] Warnow, T. (2015). Concatenation analyses in the presence of incomplete lineage sorting. *PLOS Currents*. https://doi.org/10.1371%2Fcurrents.tol.8d41ac0f13d1abedf4c4a59f5d17b1f7.

[781] Warnow, T. (2017). *Computational Phylogenetics: An Introduction to Designing Methods for Phylogeny Estimation*. Cambridge: Cambridge University Press.

[782] Warren, D., Geneva, A., and Lanfear, R. (2017). RWTY (R We There Yet): An R package for examining convergence of Bayesian phylogenetic analyses. *Molecular Biology and Evolution*, 34:1016–1020.

[783] Wascher, M., and Kubatko, L. (2021). Consistency of SVDQuartets and maximum likelihood for coalescent-based species tree estimation. *Systematic Biology*, 70(1):33–48.

[784] Watterson, G. A. (1984). Lines of descent and the coalescent. *Theoretical Population Biology*, 26:77–92.

[785] Wegner, K. M., Piel, D., Bruto, M., John, U., Mao, Z., Alunno-Bruscia, M., Petton, B., and Roux, F. L. (2019). Molecular targets for coevolutionary interactions between Pacific oyster larvae and their sympatric *Vibrios*. *Frontiers in Microbiology*, 10:2067.

[786] Wehe, A., Bansal, M., Burleigh, J., and Eulenstein, O. (2008). DupTree: A program for large-scale phylogenetic analyses using gene tree parsimony. *Bioinformatics*, 24(13):1540–1541.

[787] Wen, D., and Nakhleh, L. (2018). Co-estimating reticulate phylogenies and gene trees from multi-locus sequence data. *Systematic Biology*, 67(3):439–457.

[788] Wen, D., Yu, Y., Hahn, M., and Nakhleh, L. (2016a). Reticulate evolutionary history and extensive introgression in mosquito species revealed by phylogenetic network analysis. *Molecular Ecology*, 25:2361–2372.

[789] Wen, D., Yu, Y., and Nakhleh, L. (2016b). Bayesian inference of reticulate phylogenies under the multispecies network coalescent. *PLOS Genetics*, 12(5):e1006006.

[790] Wen, D., Yu, Y., Zhu, J., and Nakhleh, L. (2018). Inferring phylogenetic networks using PhyloNet. *Systematic Biology*, 67(4):735–740.

[791] Wessinger, C. A., Freeman, C. C., Mort, M. E., Rausher, M. D., and Hileman, L. C. (2016). Multiplexed shotgun genotyping resolves species relationships within the North American genus *Penstemon*. *American Journal of Botany*, 103:912–922.

[792] Westesson, O., and Holmes, I. (2009). Accurate detection of recombinant breakpoints in whole-genome alignments. *PLOS Computational Biology*, 5(3):e1000318.

[793] Westover, K. M., Rusinko, J. P., Hoin, J., and Neal, M. (2013). Rogue taxa phenomenon: A biological companion to simulation analysis. *Molecular Phylogenetics and Evolution*, 69(1):1–3.

[794] Whidden, C., and Matsen, IV, F. (2015). Quantifying MCMC exploration of phylogenetic tree space. *Systematic Biology*, 64:472–491.

[795] White, J. R., Nagarajan, N., and Pop, M. (2009). Statistical methods for detecting differentially abundant features in clinical metagenomic samples. *PLOS Computational Biology*, 5(4): e1000352.

[796] Whitfield, J., and Lockhart, P. (2007). Deciphering ancient rapid radiations. *Trends in Ecology and Evolution*, 22:258–265.

[797] Whitfield, J. B., and Kjer, K. M. (2008). Ancient rapid radiations of insects: Challenges for phylogenetic analysis. *Annual Review of Entomology*, 53:449–472.

[798] Wickett, N., Mirarab, S., Nguyen, N., Warnow, T., Carpenter, E., Matasci, N., Ayyampalayam, S., et al. (2014). Phylotranscriptomic analysis of the origin and early diversification of land plants. *Proceedings of the National Academy of Sciences of the United States of America*, 111(45):E4859–E4868.

[799] Wiens, J. J. (1998). Combining data sets with different phylogenetic histories. *Systematic Biology*, 47(4):568–581.

[800] Wiens, J. J. (2003). Incomplete taxa, incomplete characters, and phylogenetic accuracy: Is there a missing data problem? *Journal of Vertebrate Paleontology*, 23:297–310.

[801] Wiens, J. J. (2006). Missing data and the design of phylogenetic analyses. *Journal of Biomedical Informatics*, 39(1):34–42.

[802] Wilgenbusch, J., Huang, W., and Gallivan, K. (2017). Visualizing phylogenetic tree landscapes. *BMC Bioinformatics*, 18(85).

[803] Willson, J., Roddur, M. S., Liu, B., Zaharias, P., and Warnow, T. (2021). DISCO: species tree inference using multi-copy gene family tree decomposition. *Systematic Biology*. https://doi.org/10.1093/sysbio/syab070.

[804] Wilson, P. S., Wolfe, A. D., Armbruster, W. S., and Thompson, J. D. (2007). Constrained lability in floral evolution: Counting convergent origins of hummingbird pollination in *Penstemon* and *Keckiella*. *New Phytologist*, 176:883–890.

[805] Winsor, M. P. (2003). Non-essentialist methods in pre-Darwinian taxonomy. *Biology and Philosophy*, 18(3):387–400.

[806] Winsor, M. P. (2006a). The creation of the essentialism story: An exercise in metahistory. *History and Philosophy of the Life Sciences*, 28(3):149–174.

[807] Winsor, M. P. (2006b). Linnaeus's biology was not essentialist. *Annals of the Missouri Botanical Garden*, 93(1):2–7.

[808] Wolfe, A. D. (2005). ISSR techniques for evolutionary biology. *Methods in Enzymology*, 395:134–144.

[809] Wolfe, A. D., Datwyler, S. L., and Randle, C. P. (2002). A phylogenetic and biogeographic analysis of the Cheloneae (Scrophulariaceae) based on ITS and matK sequence data. *Systematic Botany*, 27:138–148.

[810] Wolfe, A. D., Randle, C. P., Datwyler, S. L., Morawetz, J. J., Arguedas, N., and Diaz, J. (2006). Phylogeny, taxonomic affinities, and biogeography of *Penstemon* (Plantaginaceae) based on ITS and cpDNA sequence data. *American Journal of Botany*, 93:1699–1713.

[811] Wolfe, A. D., Xiang, Q.-Y., and Kephart, S. R. (1998a). Assessing hybridization in natural populations of *Penstemon* (Scrophulariaceae) using hypervariable intersimple sequence repeat (ISSR) bands. *Molecular Ecology*, 7:1107–1125.

[812] Wolfe, A. D., Xiang, Q.-Y., and Kephart, S. R. (1998b). Diploid hybrid speciation in *Penstemon* (Scrophulariaceae). *Proceedings of the National Academy of Sciences of the United States of America*, 95:5112–5115.

[813] Wu, C.-I. (1991). Inferences of species phylogeny in relation to segregation of ancient polymorphisms. *Genetics*, 127:429–435.

[814] Wu, F., Mueller, L., Crouzillat, D., Petiard, V., and Tanksley, S. (2006). Combining bioinformatics and phylogenetics to identify large sets of single-copy orthologous genes (COSII) for comparative, evolutionary and systematic studies: A test case in the euasterid plant clade. *Genetics*, 174:1407–1420.

[815] Wu, M., Kostyun, J. L., Hahn, M. W., and Moyle, L. C. (2018). Dissecting the basis of novel trait evolution in a radiation with widespread phylogenetic discordance. *Molecular Ecology*, 27(16):3301–3316.

[816] Wu, Y. (2012). Coalescent-based species tree inference from gene tree topologies under incomplete lineage sorting by maximum likelihood. *Evolution: International Journal of Organic Evolution*, 66(3):763–775.

[817] Wu, Y.-C., Rasmussen, M. D., Bansal, M. S., and Kellis, M. (2013). TreeFix: Statistically informed gene tree error correction using species trees. *Systematic Biology*, 62(1):110–120.

[818] Xi, Z., Liu, L., and Davis, C. (2015). Genes with minimal phylogenetic information are problematic for coalescent analyses when gene tree estimation is biased. *Molecular Phylogenetics and Evolution*, 92:63–71.

[819] Xi, Z., Liu, L., and Davis, C. C. (2016). The impact of missing data on species tree estimation. *Molecular Biology and Evolution*, 33(3):838–860.

[820] Xi, Z., Liu, L., Rest, J. S., and Davis, C. C. (2014). Coalescent versus concatenation methods and the placement of *Amborella* as sister to water lilies. *Systematic Biology*, 63(6):919–932.

[821] Xiao, S., Xie, D., Cao, X., Yu, P., Xing, X., Chen, C., Musselman, M., et al. (2012). Comparative epigenomic annotation of regulatory DNA. *Cell*, 149:1381–1392.

[822] Xin, L., Ma, B., and Zhang, K. (2007). A new quartet approach for reconstructing phylogenetic trees: Quartet joining method. In *Proceedings, Computing and Combinatorics (COCOON) 2007*. Lecture Notes in Computer Science, volume 4598, pages 40–50. Berlin: Springer-Verlag.

[823] Xu, B., and Yang, Z. (2016). Challenges in species tree estimation under the multispecies coalescent model. *Genetics*, 204:1353–1368.

[824] Xu, C., and Jackson, S. A. (2019). Machine learning and complex biological data. *Genome Biology*, 20(1):76.

[825] Yang, J., Grünewald, S., Xu, Y., and Wan, X.-F. (2014). Quartet-based methods to reconstruct phylogenetic networks. *BMC Systems Biology*, 8:21.

[826] Yang, Y., Moore, M. J., Brockington, S. F., Mikenas, J., Olivieri, J., Walker, J. F., and Smith, S. A. (2018). Improved transcriptome sampling pinpoints 26 ancient and more recent polyploidy events in Caryophyllales, including two allopolyploidy events. *New Phytologist*, 217(2):855–870.

[827] Yang, Y., Moore, M. J., Brockington, S. F., Soltis, D. E., Wong, G. K.-S., Carpenter, E. J., Zhang, Y., et al. (2015). Dissecting molecular evolution in the highly diverse plant clade Caryophyllales using transcriptome sequencing. *Molecular Biology and Evolution*, 32(8):2001–2014.

[828] Yang, Y., Moore, M. J., Brockington, S. F., Timoneda, A., Feng, T., Marx, H. E., Walker, J. F., and Smith, S. A. (2017). An efficient field and laboratory workflow for plant phylotranscriptomic projects. *Applications in Plant Sciences*, 5(3):1600128.

[829] Yang, Y., and Smith, S. A. (2014). Orthology inference in nonmodel organisms using transcriptomes and low-coverage genomes: Improving accuracy and matrix occupancy for phylogenomics. *Molecular Biology and Evolution*, 31(11):3081–3092.

[830] Yang, Z. (2007). PAML 4: A program package for phylogenetic analysis by maximum likelihood. *Molecular Biology and Evolution*, 24:1586–1591.

[831] Yang, Z. (2015). The BPP program for species tree estimation and species delimitation. *Current Zoology*, 61(5):854–865.

[832] Yang, Z., and Bielawski, J. P. (2000). Statistical methods for detecting molecular adaptation. *Trends in Ecology and Evolution*, 15(12):496–503.

[833] Yang, Z., and Nielsen, R. (2008). Mutation-selection models of codon substitution and their use to estimate selective strengths on codon usage. *Molecular Biology and Evolution*, 25(3):568–579.

[834] Yang, Z., and Rannala, B. (2010). Bayesian species delimitation using multilocus sequence data. *Proceedings of the National Academy of Sciences of the United States of America*, 107(20):9264–9269.

[835] Yang, Z., and Rannala, B. (2014). Unguided species delimitation using DNA sequence data from multiple loci. *Molecular Biology and Evolution*, 31(12):3125–3135.

[836] Yang, Z., and Zhu, T. (2018). Bayesian selection of misspecified models is overconfident and may cause spurious posterior probabilities for phylogenetic trees. *Proceedings of the National Academy of Sciences of the United States of America*, 115(8):1854–1859.

[837] Yeates, D. K., Zwick, A., and Mikheyev, A. S. (2016). Museums are biobanks: Unlocking the genetic potential of the three billion specimens in the world's biological collections. *Current Opinion in Insect Science*, 18:83–88.

[838] Yin, J., Zhang, C., and Mirarab, S. (2019). ASTRAL-MP: Scaling ASTRAL to very large datasets using randomization and parallelization. *Bioinformatics*, 35(20):3961–3969.

[839] Yonezawa, T., Segawa, T., Mori, H., Campos, P., Hongoh, Y., Endo, H., Akiyoshi, A., et al. (2017). Phylogenomics and morphology of extinct paleognaths reveal the origin and evolution of the ratites. *Current Biology*, 27:68–77.

[840] Yu, Y., Barnett, R., and Nakhleh, L. (2013a). Parsimonious inference of hybridization in the presence of incomplete lineage sorting. *Systematic Biology*, 62(5):738–751.

[841] Yu, Y., Degnan, J., and Nakhleh, L. (2012). The probability of a gene tree topology within a phylogenetic network with applications to hybridization detection. *PLOS Genetics*, 8:e1002660.

[842] Yu, Y., Dong, J., Liu, K., and Nakhleh, L. (2014). Maximum likelihood inference of reticulate evolutionary histories. *Proceedings of the National Academy of Sciences of the United States of America*, 111(46):16448–6453.

[843] Yu, Y., and Nakhleh, L. (2015). A maximum pseudo-likelihood approach for phylogenetic networks. *BMC Genomics*, 16:S10.

[844] Yu, Y., Ristic, N., and Nakhleh, L. (2013b). Fast algorithms and heuristics for phylogenomics under ILS and hybridization. *BMC Bioinformatics*, 14(Suppl. 15):S6.

[845] Yu, Y., Than, C., Degnan, J., and Nakhleh, L. (2011). Coalescent histories on phylogenetic networks and detection of hybridization despite incomplete lineage sorting. *Systematic Biology*, 60(2):138–149.

[846] Yule, G. U. (1925). A mathematical theory of evolution, based on the conclusions of Dr. JC Willis, FRS. *Philosophical Transactions of the Royal Society B: Biological Sciences*, 213:21–87.

[847] Zafar, H., Tzen, A., Navin, N., Chen, K., and Nakhleh, L. (2017). SiFit: Inferring tumor trees from single-cell sequencing data under finite-sites models. *Genome Biology*, 18(1):178.

[848] Zanis, M. J., Soltis, D. E., Soltis, P. S., Mathews, S., and Donoghue, M. J. (2002). The root of the angiosperms revisited. *Proceedings of the National Academy of Sciences of the United States of America*, 99(10):6848–6853.

[849] Zhang, C., and Matsen, F. (2019). Variational Bayesian phylogenetic inference. *International Conference on Learning Representations*. https://openreview.net/forum?id=SJVmjjR9FX.

[850] Zhang, C., Rabiee, M., Sayyari, E., and Mirarab, S. (2018a). ASTRAL-III: Polynomial time species tree reconstruction from partially resolved gene trees. *BMC Bioinformatics*, 19(6):153.

[851] Zhang, C., Sayyari, E., and Mirarab, S. (2017). ASTRAL-III: Increased scalability and impacts of contracting low support branches. In Meidanis, J., and Nakhleh, L., editors, *Comparative Genomics*, pages 53–75. Cham, Switzerland: Springer International.

[852] Zhang, C., Scornavacca, C., Molloy, E. K., and Mirarab, S. (2020). ASTRAL-Pro: Quartet-based species-tree inference despite paralogy. *Molecular Biology and Evolution*, 37(11):3292–3307.

[853] Zhang, G. (2015). Genomics: bird sequencing project takes off. *Nature*, 522(7554):34.

[854] Zhang, G., Li, C., Li, Q., Li, B., Larkin, D. M., Lee, C., Storz, J. F., et al. (2014). Comparative genomics reveals insights into avian genome evolution and adaptation. *Science*, 346(6215): 1311–20.

[855] Zhang, L., Chen, F., Zhang, X., Li, Z., Zhao, Y., Lohaus, R., and Chang, X. (2019). The water lily genome and the early evolution of flowering plants. *Nature*, 577:79–84 .

[856] Zhang, Q., Rao, S., and Warnow, T. (2018b). New absolute fast converging phylogeny estimation methods with improved scalability and accuracy. In Parida, L., and Ukkonen, E., editors, *18th International Workshop on Algorithms in Bioinformatics (WABI 2018)*, pages 8:1–8:12. Leibniz International Proceedings in Informatics. Dagsttuhl, Germany: Schloss Dagstuhl Leibniz Zentrum fuer Informatik.

[857] Zhong, B., and Betancur-R., R. (2017). Expanded taxonomic sampling coupled with gene genealogy interrogation provides unambiguous resolution for the evolutionary root of angiosperms. *Genome Biology and Evolution*, 9(11):3154–3161.

[858] Zhou, X., Lutteropp, S., Czech, L., Stamatakis, A., Looz, M. V., and Rokas, A. (2020). Quartet-based computations of internode certainty provide robust measures of phylogenetic incongruence. *Systematic Biology*, 69(2):308–324.

[859] Zhu, J., Liu, X., Ogilvie, H. A., and Nakhleh, L. K. (2019). A divide-and-conquer method for scalable phylogenetic network inference from multilocus data. *Bioinformatics*, 35(14): i370–i378.

[860] Zhu, J., and Nakhleh, L. (2018). Inference of species phylogenies from bi-allelic markers using pseudo-likelihood. *Bioinformatics*, 34:i376–i385.

[861] Zhu, J., Wen, D., Yu, Y., Meudt, H. M., and Nakhleh, L. (2018). Bayesian inference of phylogenetic networks from bi-allelic genetic markers. *PLOS Computational Biology*, 14(1):1–32.

[862] Zhu, J., Yu, Y., and Nakhleh, L. (2016). In the light of deep coalescence: Revisiting trees within networks. *BMC Bioinformatics*, 17(14):415.

[863] Zhu, S., and Degnan, J. H. (2016). Displayed trees do not determine distinguishability under the network multispecies coalescent. *Systematic Biology*, 66(2):283–298.

[864] Zimmermann, T., Mirarab, S., and Warnow, T. (2014). BBCA: Improving the scalability of *BEAST using random binning. *BMC Genomics*, 15(Suppl. 6):S11.

[865] Zwaenepoel, A., and Peer, Y. V. (2019). Inference of ancient whole-genome duplications and the evolution of gene duplication and loss rates. *Molecular Biology and Evolution*, 36(7):1384–1404.

[866] Zwickl, D. (2006). Genetic algorithm approaches for the phylogenetic analysis of large biological sequence datasets under the maximum likelihood criterion. PhD diss., University of Texas at Austin.

[867] Zwickl, D. J., and Hillis, D. M. (2002). Increased taxon sampling greatly reduces phylogenetic error. *Systematic Biology*, 51(4):588–598.

Index